Sea Level Rise

Elisa Fornalé
Editor

Sea Level Rise

Implications for Human Rights, Security, and Peace

Editor
Elisa Fornalé
World Trade Institute, Faculty of Law
University of Bern
Bern, Switzerland

ISBN 978-3-031-89170-0   ISBN 978-3-031-89171-7 (eBook)
https://doi.org/10.1007/978-3-031-89171-7

© The Editor(s) (if applicable) and The Author(s) 2026. This book is an open access publication.

Open Access This book is licensed under the terms of the Creative Commons Attribution 4.0 International License (http://creativecommons.org/licenses/by/4.0/), which permits use, sharing, adaptation, distribution and reproduction in any medium or format, as long as you give appropriate credit to the original author(s) and the source, provide a link to the Creative Commons license and indicate if changes were made.
The images or other third party material in this book are included in the book's Creative Commons license, unless indicated otherwise in a credit line to the material. If material is not included in the book's Creative Commons license and your intended use is not permitted by statutory regulation or exceeds the permitted use, you will need to obtain permission directly from the copyright holder.
The use of general descriptive names, registered names, trademarks, service marks, etc. in this publication does not imply, even in the absence of a specific statement, that such names are exempt from the relevant protective laws and regulations and therefore free for general use.
The publisher, the authors and the editors are safe to assume that the advice and information in this book are believed to be true and accurate at the date of publication. Neither the publisher nor the authors or the editors give a warranty, expressed or implied, with respect to the material contained herein or for any errors or omissions that may have been made. The publisher remains neutral with regard to jurisdictional claims in published maps and institutional affiliations.

Cover credit: Brandi Mueller/Getty images

This Palgrave Macmillan imprint is published by the registered company Springer Nature Switzerland AG
The registered company address is: Gewerbestrasse 11, 6330 Cham, Switzerland

If disposing of this product, please recycle the paper.

Foreword

Almost forty years ago, the United Nations General Assembly, in the Resolution 43/53 of 6 December 1988, took a clear-cut position on the serious impact of anthropogenic greenhouse gases on climate change. The Assembly recognized that "climate change is a common concern of mankind" (para. 1) and referred, *inter alia*, to 'emerging evidence' that 'indicates that the continued rise in atmospheric concentrations of "greenhouse" gases could produce global warming with an eventual rise in sea levels, the effects of which could be disastrous for mankind if timely steps are not taken at all levels' (3rd considerandum). Lastly, it supported the establishment by the World Meteorological Organization and the United Nations Environment Programme of an Intergovernmental Panel on Climate Change (IPCC) to provide 'internationally coordinated scientific assessments of the magnitude, timing and potential environmental and socio-economic impacts of climate change and realistic response strategies' (para. 5).

Since its establishment, the IPCC has analysed and organized the data produced by thousands of scientists around the world, confirming that global warming is caused by anthropogenic greenhouse gases. The IPCC has also identified the multiple effects of anthropogenic greenhouse gases on the seas and oceans, including 'sea level rise, increasing ocean heat content and marine heat waves, ocean deoxygenation, and ocean acidification' (2019 Report, p. 79).

According to the science the sea level rise could reach 60 to 110 cm by 2100 if anthropogenic greenhouse gas emissions continue to increase at the current rate despite the obligations upon States specifically established by the 1992 UN Framework Convention on Climate Change, and the 2015 Paris Agreement. Not only is the impact of the sea level rise considered as 'an existential threat for some Small Islands and some low-lying coasts (*medium confidence*)' (WGII 2022 Report, p. 15) but it is also a threat for the 20% of the world population which lives less than 30 km from a coast, many of them in coastal cities. Coastal areas are both ecosystems and fragile territories where populations and economic activities are concentrated.

Sea level rise as an effect of anthropogenic greenhouse gas emissions raises a number of different serious issues for international law, as it is mainly reflected in the works of the International Law Association (ILA) and the UN International Law Commission (ILC).

The ILA Committee on Sea Level Rise and International Law has been studying the 'possible impacts of sea level rise and the implications under international law of the partial and complete inundation of State territory, or depopulation thereof, in particular of small island and low-lying States' (Resolution 01/2024, 2nd considerandum), with the aim of developing 'proposals for the progressive development of international law in relation to the possible loss of all or of parts of State territory and maritime zones due to sea level rise, including the impacts on statehood, nationality, and human rights' (*Ibidem*). At the 81st Conference of the ILA, held in Athens from 25 to 28 June 2024, it adopted Resolution 01/2024, which contains various conclusions on certain core issues related to sea level rise, such as the stability of baselines, outer limits and delimitation lines of the marine spaces of coastal States despite the modification of their coastlines, the preservation of statehood despite the loss of territory, and the allocation of human rights, duties and responsibilities among States, particularly in situations of cross-border displacement and migration of members of the populations of the affected States.

As for the ILC, the topic of sea level rise has been included in its programme of work more recently, namely in 2019, and the attention of the Study Group on Sea level Rise in Relation to International Law is focusing on various issues of international law, ranging from the law of the sea to treaty law, the immutability and intangibility of boundaries, the permanent sovereignty of States over natural resources, statehood and the protection of persons affected by sea level rise.

It cannot be excluded that useful interpretative elements of international law with respect to sea level rise could result from the advisory opinions of the Inter-American Court of Human Rights and the International Court of Justice. The former has been asked to clarify the scope of State obligations with reference to human rights with respect to climate emergencies. On the other hand, the International Court of Justice is asked, *inter alia*, to identify the obligations of States under international law to ensure the protection of the climate system and other parts of the environment from anthropogenic emissions of greenhouse gases for both States and present and future generations.

In the Advisory Opinion n. 31, adopted by the International Tribunal for the Law of the Sea (ITLOS) on 21 May 2024, sea level rise has been one of the contexts within which the Tribunal defined the specific obligations concerning the protection and preservation of the marine environment.

However, the issue of sea level rise with reference to its effects on existing maritime claims or entitlements has not been addressed by the ITLOS. On this point, it noted that neither the request from the Commission of Small Island States (COSIS) nor the decision by the Commission that approved the request addressed 'the issue of base points, baselines, claims, rights or entitlements to maritime zones established under the Convention, or maritime boundaries, and the corresponding obligations in the context of "physical changes connected to climate change-related sea level rise"'. The Tribunal considered that 'if the Commission had intended to solicit an opinion on the consequences of sea level rise for base points, baselines, claims, rights or entitlements to the maritime zones established under the Convention, or maritime boundaries, and the corresponding obligations, it would have expressly formulated the Request accordingly' (para. 150).

The activism of intergovernmental and non-governmental bodies and international courts and tribunals in relation to the international law to be invoked with regard to sea level rise highlights the ambiguities and contradictions as well as the *lacunae* in international law on the subject. The ambiguities and contradictions can hopefully be resolved through interpretation, while for the *lacunae*, apart from the possible—but with caution—recourse to analogy, the optimal solution would be to elaborate new rules dealing with the issues raised by sea level rise.

In this context, the contributions of scholars may prove very useful in the interpretation of the existing international law and its progressive development. Scholars can propose innovative solutions with less restraints than international courts and tribunals, which are bound to the interpretation and application of existing law, and the ILC, whose projects will be submitted to the UN Member States for their final scrutiny and acceptance. International scholars, both jointly and individually, and through their work, can have an important impact on international law by concurring in its progressive development.

This volume is therefore commendable since it provides a new and interesting contribution to the current debate on international law and sea level rise because it aims at clarifying the relevant legal context, including by examining it critically, and searching for legally founded solutions to the problems raised by sea level rise in international law.

To this end, the volume is organized into three parts, each of which contains several chapters. Part I is devoted to the issue of stability and security on the basis of the assumption that should the risks of sea level rise materialize as currently predicted, the severity of their effects would be such as to impact on international peace and security. I think that this Part gives an added value to the book since it is not often that the topic of sea level rise is addressed from the perspective of the fact that it can pose such threats to international peace and security as to entail an intervention by the UN Security Council.

Part II focuses on the rights of population affected by sea level rise. The topic *per se* is certainly not new, but the various chapters therein deal with it from peculiar and less common perspectives, for instance, that of sea level rise as a possible form of climate-gendered violence or that of the conceptualization of human rights affected by sea level rise in climate litigation. This Part confirms that human rights protection is characterized by many facets which continuously evolve.

Finally, Part III looks at new developments at international level vis-à-vis sea level rise. Here, the attention turns to the most recent State practice in various fields concerned by sea level rise, such as the law of the sea, domestic affairs, and foreign investment. This approach seems to capture new trends in States' actions that might precede new normative developments. The perspective is therefore *de jure condendo*.

In conclusion, readers will find a variety of innovative and intelligent arguments and much to think about in this volume.

Hamburg, Germany

Ida Caracciolo
Judge, International Tribunal
of the Law of the Sea

Ida Caracciolo is judge at the International Tribunal of the Law of the Sea. She is full professor of International Public Law at the University of Campania "Luigi Vanvitelli"; Member of the Italian Group to the Permanent Court of Arbitration; and Alternate arbitrator of the OSCE Court of Arbitration and Conciliation; and member of the Administrative Court of the International Italo-Latin American Organisation (IILA). She is barrister at the bar of Rome. She was *ad hoc* judge at the European Court of Human Rights Legal Expert at the Legal Affairs Service of the Italian Ministry of Foreign Affairs and International Cooperation (1994–2020); Member of the Italian delegation for several bilateral negotiations; Member of the Italian delegation for several organs and committees within international organizations, including the Council of Europe (CAHDI), the European Union (COJUR and COMAR), and the United Nations (the UNGA VI Committee), and for several multilateral negotiations; and Counsel of Italy in cases before the International Court of Justice ("The Legality of the Threat or Use of Nuclear Weapons", 1995–1996; "The Immunity from Legal Process of a Special Rapporteur of the Commission on Human Rights", 1998–1999; "The Legality of Use of Force", 1999–2004), the EU Court of Justice ("Criminal Proceedings against Dante Bigi", 2000–2001), the European Court of Human Rights ("Beyeler v. Italy", 2000–2002), the Tribunal ("Enrica Lexie", 2015; "M/V Norstar", 2016–2018), and the ad hoc Tribunal under Annex VII of UNCLOS ("The Enrica Lexie Incident", 2015–2020). She is the director of a book series and member of the scientific boards of several scientific reviews and book series. She lead and took part to several research projects on International Law and the Law of the Sea funded by the Italian National Research Council. She is the author of numerous publications in Italian, English and French on Public International Law, the Law of the Sea, Human Rights Law and International Humanitarian Law.

Acknowledgements

As editor, I would like to thank all the wonderful authors who have contributed to this joint effort with an incredible degree of commitment and enthusiasm. Their contributions have made this excellent publication possible. I would also like to thank the World Trade Institute at the University of Bern for providing high-level research assistance. In particular, I am grateful to Ms Riccarda Heepen, Ms Martina Pennino and Ms Felicia Meyer for always being a willing source of help. The development of this book project has highly benefited from several academic events we hosted and co-organized at the World Trade Institute in the last years.

One of the key events that encouraged the development of this book has been the Expert Meeting 'Cooperative Duties, Human Rights and Climate Change' that took place on 8 December 2022 at the World Trade Institute, Faculty of Law, University of Bern. I am also extremely grateful for the generous support received from the Swiss National Scientific Foundation (SNSF) (Grant No PP00P1_194808). I also would like to thank the SNSF for co-funding the international conference 'Sea Level Rise and Its International Law Implications for Legal Certainty, Stability and Human Rights' held at the World Trade Institute, Faculty of Law, University of Bern on 9 June 2023 (Grant No IZSEZ0_217578/1). This major event has been co-organized by Prof. Davor Vidas from the Fridtjof Nansen Institute, Norway. I acknowledge all the inspiring contributions and the commitment of the experts that took part in our events. I owe particular gratitude to Ms Sandra Joseph for taking care of organizing these events.

This edited volume is also part of the research conducted within the Horizon Europe project HRJust (*States' Practice of Human Rights Justification: A Study in Civil Society Engagement and Human Rights through the Lens of Gender and Intersectionality*), which aims to address significant and important gaps in human rights regulations and to develop a theory of human rights justifications and a process for Systematic Ongoing Civil Society Engagement as a tool for a gender and intersectional inclusive civil society engagement. HRJust aims to identify gaps in human rights regulations and protection, serving to underpin data for our recommendations to the EU in support of a multinational human rights system and promotion of transnational democratic governance. The HRJust project is coordinated by the University of Gothenburg (SE). This volume includes data collected by the research teams of the World Trade Institute and the Institute of International Relations of Prague involved in Work Package 6 on Climate. We gratefully acknowledge the grant received from the Horizon Europe project HRJust (*States' Practice of Human Rights Justification: A Study in Civil Society Engagement and Human Rights through the Lens of Gender and Intersectionality*), Grant Agreement No. 101094346. And this work has received funding from the Swiss State Secretariat for Education, Research and Innovation (SERI) under the Grant Agreement No. 23.00131.

I deeply appreciate all the support provided by colleagues and friends. Special thanks go to Dr. Federica Cristani for her tremendous support for the development of this book. I would like to express my gratitude to all the staff of Palgrave Press for their expert guidance and our enjoyable working relationship. Finally, many thanks to Mr. Jan Hrubín for assisting with the editing and proofreading of the manuscript.

Contents

1 Sea Level Rise: Implications for Human Rights an Introduction 1
Elisa Fornalé

Part I Sea Level Rise, Stability and Security

2 Sea Level Rise: Scientific Evidence, Socio-Economic Realities, and Adaptation Challenges for Coastal Communities 17
Vilane Gonçalves Sales

3 The International Law Commission's Study Group on Sea Level Rise in Relation to International Law and Its Impact on International Law 61
Massimo Starita

4 The Role of the UN Security Council in Addressing and Providing Responses to Peace and Security Risks Resulting From Sea Level Rise 93
Giuseppe Nesi and Elisa Fornalé

Part II Sea Level Rise and the Rights of Affected Population

5 Sea Level Rise and Human Rights 121
Veronika Bílková

6 Sea Level Rise as a Form of Gendered Climate Violence: International Legal Implications for Migration 145
Sara De Vido

7 Rethinking Sustainable Migration for the Anthropocene 171
Samuel Ballin and Sandra Mantu

8 Human Rights and Justifications in Climate Litigation: A First Attempt at Conceptualization 201
Federica Cristani and Elisa Fornalé

Part III Sea Level Rise: New Developments at International Level

9 Sovereignty, State Cooperation, and Sea Level Rise 233
Tamás Vince Ádány

10 The Advisory Opinion of the International Tribunal for the Law of the Sea 251
Curtis F. J. Doebbler

11 Assessing Innovative Sources for the Loss and Damage Mechanism: The Role and Prospective Regulation of Climate-Friendly Foreign Investment 291
Federica Cristani

Index 333

Notes on Contributors

Tamás Vince Ádány is Professor based at the Pazmany Peter Catholic University in Budapest, Hungary, where he is the Head of the Department of International Law. In 2022, he was a Fulbright Visiting Professor at the University of Notre Dame Law School. His interest is in general international law—the sources of international obligations, responsibility of states, and particularly the position of individuals in international law. He has published extensively on these issues. His current research interest is the potential impact of certain implications of populism on the international legal order. Previously, he worked for the Hungarian government in various professional capacities, and later he formed and ran a small consulting company dealing with international and European legal matters. He holds a law degree, an international law PhD, a habilitation from PPCU, and an MA in international relations from ASERI in Milan, Italy.

Samuel Ballin (they/he) is a PhD candidate at the Centre for Migration Law, Radboud University. Their PhD research examines the economic-ecological resilience and rights of regular migrant workers. Samuel has a Graduate Diploma in Law from the University of Sheffield and an LLM International Migration and Refugee Law from the Vrije Universiteit Amsterdam, and has previously worked as a teacher and a *pro bono* legal adviser. They also coordinate the Netherlands Network of Human Rights Research (NNHRR) Working Group on Human Rights and the Climate Crisis.

Veronika Bílková is the Head of the Department of International Law at the Faculty of Law of Charles University in Prague and Senior Researcher in the Centre for International Law at the Institute of International Relations Prague. She is also the Head of the University Centre for Conflict and Post-Conflict Studies in Prague. She is member and Vice President of the Venice Commission of the Council of Europe (since 2010) and the Management Board of the EU Fundamental Rights Agency (since 2020). She was member of four expert missions on Ukraine established under the OSCE Moscow Mechanism in 2022–2024. Her fields of research include the use of force, international humanitarian law, international human rights law, international criminal law, and the fight against terrorism.

Federica Cristani is the Head of the Centre for International Law at the Institute of International Relations in Prague, where she is also currently involved as WP co-leader in a Horizon Europe HRJust project (States' Practice of Human Rights Justification: a study in civil society engagement and human rights through the lens of gender and intersectionality—GA 101094346). She holds a PhD in international law from the University of Verona (IT). She earlier worked as a post-doctoral researcher at the World Trade Institute of the University of Bern (CH) and at the University of Verona, and has been a Visiting Scholar in various universities and research centres in Hungary, Slovakia, Germany, Denmark, and the United Kingdom. Between 2021 and 2023, she has been a Senior Visiting Scholar at the Arctic Centre of the University of Lapland (FI). She has also been an Adjunct Professor in Bologna (IT) and a Guest Lecturer in Budapest (HU), Bratislava (SK), and Kharkiv (Ukraine). Her main research interests include climate change law, international economic law, and international law of cyberspace.

Sara De Vido is Full Professor of International Law at Ca' Foscari University of Venice, Italy, where she teaches International Law, EU Law, and Human Rights Law. She is an affiliate of the Manchester International Law Centre, United Kingdom, where she co-founded the Women in International Law Network. She is the Delegate of the Rector for Gender Equality and member of the Centre for Human Rights at Ca' Foscari University. She has been working on countering violence against women for years as an expert on the Istanbul Convention and she is legal expert for the EU Network on Preventing Gender-Based Violence and Domestic Violence (2023–2025). Among her most recent works, the book Violence against Women's Health in International Law (Manchester

University Press, Melland Schill Studies in International Law, 2020), and the report "EU Law in Light of the Istanbul Convention: Legal Implications After the Accession" (EELN, 2025). She has recently focused her research on ecocentric and ecofeminist approaches to international law. She co-edited, along with Micaela Frulli (University of Florence), the Commentary on the Istanbul Convention (Elgar Publishing, 2023).

Curtis F. J. Doebbler is Research Professor of Law in the Department of Law at the University of Makeni, Sierra Leone and proprietor of The Law Office of Dr Curtis FJ Doebbler. He holds law degrees from New York Law School (J.D.), Radboud Universiteit (Meestertitle, LL.M.), and the London School of Economics and Political Science (Ph.D.). He has authored twelve books, more than two hundred academic and newspaper articles, and numerous online contributions. He is also a regular commentator on international law on television and radio news programmes. He represents the NGO International-Lawyers.Org at the United Nations and practices international law before domestic and international tribunals, advising and representing both governments and non-governmental actors. He has taught law at universities in the Middle East, Africa, and Europe. His books include The Dictionary of International Law (Rowman & Littlefield Publishers), which was published in March 2018.

Elisa Fornalé is Associate Professor at the World Trade Institute (WTI), the University of Bern. She holds a law degree from the University of Trento, Italy, and a PhD in human rights law from the University of Palermo, Italy. She specializes in international law, human rights, and migration. From 2017 to 2023, she implemented the project 'Framing Environmental Degradation, Human Mobility and Human Development as a Matter of Common Concern' (CLI_M_CO2), which explored the adverse impacts of climate change through a pilot case study of the Small Pacific Island States. From 2021 to 2024, she served as co-Rapporteur of the International Law Association (ILA) Committee on International Law and Sea Level Rise. She is the Work Package Leader for Climate of the HEurope 'Human Rights Justification' project (States' Practice of Human Rights Justification: A Study in Civil Society Engagement and Human Rights through the Lens of Gender and Intersectionality—GA 101094346) and she is leading the SNSF Consolidator Grant ('Resisting Human Erosion'). She is the Gender Coordinator of the Gender Team at the WTI. She initiated the Gender Lecture Series—Know the GAP, and from 2021 to 2024 she implemented the SNSF project 'Gender Equality in the Mirror (GEM)' to advance women's participatory rights.

Sandra Mantu is Professor at the Centre for Migration Law (CMR), Faculty of Law, Radboud University Nijmegen (NL). In 2014, Mantu defended her PhD thesis that dealt with the legal rules and practices of citizenship deprivation in a selection of EU Member States and their link with EU citizenship. She has been involved in several EU-funded projects looking at the legal aspects of EU citizenship and EU migration and mobility frameworks. She is currently a key staff member in the CMR's Jean Monnet Centre of Excellence work programme and the main researcher in EXPULCIT, a project that focuses on the expulsion of EU citizens from host Member States. Her most recent publication is a co-edited volume with Carolus Grutters and Paul Minderhoud titled Migration on the Move—Essays on the Dynamics of Migration (Brill 2017).

Giuseppe Nesi obtained a degree in Law from the University of Catania in 1983 and became a lawyer in 1986. After receiving a Master of Arts in International Affairs from Johns Hopkins University, he obtained a research doctorate (PhD) from Sapienza—University of Rome in 1991. Since 1993 he has taught international law at the Faculty of Law of the University of Trento, where he became full professor in 2001. He has been Dean of the Faculty of Law (2012–2018), and where he has also taught European Union Law, International Protection of Human Rights and Law of International Institutions. Among the founders of the Italian Society of Internation Law and European Union Law (IDI), he is the Board of Editors of the Italian Yearbook of International Law. In the diplomatic and institutional field, he has held various roles, including that of legal expert of the Italian permanent representation at the UN in New York from 2002 to 2010, which culminated in him serving as legal advisor to the President of the United Nations General Assembly Joseph Deiss in its 65th session. Previously, he was a consultant to the parliamentary commission of inquiry into the responsibilities relating to the Cermis massacre in 2000. In 1996 he held the role of legal advisor to the Italian Presidency of the European Union after having been legal advisor to the Presidency of the Conference on Security and Cooperation in Europe (now the OSCE) in 1993. In November 2013, Foreign Minister Emma Bonino appointed him a permanent member of the Interministerial Committee for Human Rights (CIDU). He was elected as a member of the UN International Law Commission on 12 November 2021 with 152 votes. His mandate began in 2023 and will end in 2027.

Vilane Gonçalves Sales is a distinguished Marie Curie fellow at CMCC in Italy, where she contributes climate economics and machine learning expertise to the SEA-LIMITHS project. She holds a PhD in Economics from the University of Birmingham. Formerly a Postdoctoral Researcher on Climate Change Economics at the University of Bern, Dr. Sales has lectured in Environmental Econometrics and served in academic roles across Europe. Her research and analysis advance knowledge on the economic impacts of climate change.

Massimo Starita is Full Professor of International Law at the University of Palermo, where he has taught International Law since 2003. From 2021 to 2024 was the Director of the Master Course in 'Migrations, Rights, Integration'. His research interests include human rights, migrations, climate change, interpretation in international law, and transitional justice. In 2024, he was vice-President of the Italian Society of International Law and he organized the XXVIII Society's national conference. His most recent publications include: 'Sécurité de la navigation et contrôle international', in Journées franco-italiennes—Le contrôle international, Paris (Pédone) 2024; Corso di diritto internazionale (with P. De Sena), Bologna (Il Mulino), 2023; and 'The Impact of Sea Level Rise on Baselines: A Question of Interpretation of the UNCLOS or Evolution of Customary Law?', in Questions of International Law, Zoom-out, 91 (2022), pp. 5–21.

List of Figures

Chapter 2

Fig. 1 Global Mean Sea Level Rise from Satellite Altimetry (1992–2024). Data from all missions (TOPEX/Poseidon, Jason-1, -2, -3, Sentinel-6A) for the global zone, with seasonal signals removed. Extended and adapted from the visualization presented in Guérou et al. (2023) using AVISO + Satellite Altimetry Data, CNES — 22

Chapter 8

Fig. 1 Typologies of States' justification — 216
Fig. 2 Conceptualization of States' justifications — 218

LIST OF TABLES

Chapter 8

Table 1	Building different typologies of justifications	215
Table 2	Structure of the HRJust Climate Claims Dataset	219

CHAPTER 1

Sea Level Rise: Implications for Human Rights an Introduction

Elisa Fornalé

1 Introduction

On 25 September 2024, the President of the General Assembly of the United Nations (UN) held a high-level plenary meeting on 'Addressing the Existential Threats Posed by Sea Level Rise' following the UN General Assembly Resolution of 1 August 2024.[1] This high-level meeting aimed at addressing the adverse impacts of sea level rise by highlighting the need to build a common understanding and promoting cooperative and collaborative actions.

Available data confirm that the sea level will continue to rise (Freestone & Cicek, 2021). Mr Philémon Yunji Yang, in his opening remarks at the United Nations General Assembly, said that 'it is estimated that sea

---

[1] The General Assembly of the United Nations adopted one decision that was submitted on 8 January 2024 (High-level Plenary Meeting on Addressing the Existential Threats Posed by Sea Level Rise: Draft Decision / submitted by the President of the General Assembly, UN Doc A/78/544, adopted on 16 January 2024) and adopted one resolution on 1 August 2024 (UN Doc A/RES/78/319).

E. Fornalé (✉)
World Trade Institute, Faculty of Law, University of Bern, Bern, Switzerland
e-mail: elisa.fornale@unibe.ch

© The Author(s) 2026
E. Fornalé (ed.), *Sea Level Rise*,
https://doi.org/10.1007/978-3-031-89171-7_1

levels will rise by 20 centimetres between 2020 and 2050. Since the start of the twentieth century, sea levels have risen faster than at any other point over the last 3000 years—leaving millions, if not billions, in imminent peril'.[2]

The preparatory work to this meeting focused on some of the most pressing areas of concern, as discussed in particular by the open-ended Study Group of the International Law Commission (ILC) on 'Sea Level Rise in Relation to International Law', to further clarify the legal dimension(s) of this complex case (ILC, 2018; ILC, 2022; ILC, 2024). The UN Secretary-General highlighted that: '[w]e must also address gaps in our international legal framework concerning sea level rise: to ensure continuing access to resources, while protecting existing maritime boundaries; as well as to protect affected persons and—in extreme scenarios—to address the implications related to statehood'.[3]

While infrastructures, financial capacity, and new technologies are crucial, the international legal order could play a major role by identifying new conceptual frameworks 'rather than the imposition, by analogy or precedent, of static forms that were built on the basis of an earlier, no longer valid, situation' (Vidas et al., 2015a, p. 6; Vidas et al., 2015b). Our volume aims to conduct an in-depth analysis of how international law is developing to address emerging challenges and elaborate on how human rights are the central reference points in dealing with mankind's future (Cournil & Colard-Fabregoule, 2012; Human Rights Council, 2018; Fornalé, 2022).

1.1 Sea Level Rise, Human Rights, and Justifications: A New Paradox?

Slow-onset events interfere with the ability of the affected individuals to access and benefit from their human rights (Fornalé, 2023a). This is particularly harmful for people living in what the Special Rapporteur on Contemporary Forms of Racial Discrimination recognizes as 'sacrifice zones' (United Nations Special Rapporteur, 2022a, para. 18–19). This

---

[2] President of the General Assembly, Remarks at the High-Level Plenary Meeting on Addressing the Existential Threats Posed by Sea-level Rise, 25 September 2024, Trusteeship Council Chamber.

[3] United Nations Secretary-General's Remarks to the General Assembly Plenary Meeting on Addressing the Existential Threats Posed by Sea Level Rise, 25 September 2024.

means that the affected populations are increasingly exposed to what some scholars refer to as slow violence—namely 'a violence that occurs gradually and out of sight, a violence of delayed destruction that is dispersed across space and time' (Nixon, 2011, p. 2; Fornalé, 2023b).

In recent years, the protection of persons affected by sea level rise has attracted increasing attention (ILA, 2018; Burson et al., 2023; ILA, 2024). This is confirmed by the rise of climate change litigation involving not only domestic tribunals (Fornalé, 2023a; McAdam, 2013) but also regional and international courts (Mayer & Van Asselt, 2023; Wewerinke-Singh et al., 2021).[4]

To advance a sophisticated understanding of climate change litigation, some significant gaps need to be addressed in order to clarify State practice and arguments used (Rodriguez-Garavito, 2023; Rodriguez-Garavito, 2022; Savaresi & Setzer, 2021; Savaresi, 2020; Peel & Osofsky, 2018). For instance, the existence of these cases could offer the opportunity to explore States' use of justifications in interpreting their protective duties and fulfilling climate commitments (Kumm, 2018). In light of this, there is a degree of flexibility that allows States to balance between domestic prerogatives and individual rights, but this flexibility could be a cause of concern (Besson, 2022; Bílková, 2019). Human rights justifications could represent a positive strategy (such justifications could be invoked to 'found' and 'ground' human rights as opposed to competing policy concerns) but at the same time it could lead to negative outcomes (such as limitations of individual liberties): this is what we could define as the risk of generating a human rights-justification paradox. How can we ensure that legal protection of human rights has priority when balancing between different, even competing interests?[5] Human rights and justifications should not resort to an instrumental approach, or even misinterpretations, in the process of adopting climate change measures.

Accordingly, also the 'distribution of human rights duties and responsibilities among States' that are directly and less directly affected by the

---

[4] Three advisory proceedings have been issued at the international level: the Advisory Opinion of the International Tribunal for the Law of the Sea (requested in December 2022, delivered on 21 May 2024); the Advisory Opinion of the Inter-American Court on Human Rights (requested in January 2023, delivered on 29 May 2025); and the Advisory Opinion of the International Court of Justice (requested in March 2023, delivered on 3 July 2025).

[5] See Chapters 4 and 7 in this Volume.

adverse impacts of climate change needs to be strengthened (ILA, 2024, p. 27). As highlighted by the final report of the International Law Association Committee on International Law and Sea Level Rise, the ability of affected States to meet their human rights duties needs to be better understood to prevent unbalanced obligations (ILA, 2024). Such an imbalance may emerge for instance from the challenges of dealing with climate migration. In line with the 'Sydney Declaration of Principles on the Protection of Persons Displaced in the Context of Sea Level Rise', the duty to cooperate requires States to improve their efforts to strengthen and coordinate measures that guarantee 'human mobility options at domestic, sub-regional, regional and international levels', which should be predictable and durable[6] (ILC, 2018, p. 31; ILA, 2024; ILC, 2024). These are precisely the challenges that our book aims to make visible by enhancing our understanding of this research field and its dynamics.

The contributions to this book examine how to address the threats posed by sea level rise and how to facilitate international cooperation in this regard with an innovative approach to exploring the role of international law.

2 The Aim and Structure of the Book

Our study aims to advance the status of the research on the topic by mapping existing regulatory frameworks, and examine how human rights and climate change rules are interpreted and applied. To strengthen State obligations, it is crucial to identify new venues for influencing the balancing of the different interests at stake and for informing the interpretation of international law. A new perspective could foster the collective capacity to respond and to shape our future in this respect.

The book addresses the following question: How can States collectively respond to present and future slow-onset events, while considering intersectional discrimination, socio-economic disadvantage, and the impact on

---

[6] See also the ILA Resolution adopted in 2024 according to which 'affected and host States are recommended to: i. enhance cooperation by developing human rights-based and gender-responsive bilateral, (sub-)regional, and regional frameworks and agreements on cross-border displacement, migration and planned relocation'. https://www.ila-hq.org/en/documents/ila-resolution-1-committee-on-international-law-and-sea-level-rise-en. The 'Sydney Declaration' is available at https://disasterdisplacement.org/resource/sydney-declaration/.

human mobility, and at the same time endorsing an intergenerational perspective?

The contributing authors aim to critically assess the gaps that persist in addressing the issues raised and encourage revisiting the question of how to respond to these existential threats by exploring underdeveloped topics of inquiry (United Nations Special Rapporteur on Violence Against Women, 2022b). The examination of the gender dimension of these phenomena is one example of this and the role of justifications is another.

The book is structured around three thematic parts: (I) Sea Level Rise, Stability and Security; (II) Sea Level Rise and the Rights of Affected Populations; and (III) Sea Level Rise: New Developments at International Level.

2.1 Sea Level Rise, Stability, and Security

As recalled by the President of the UN General Assembly, the exposure to slow-onset events increases the vulnerability of coastal communities and low-lying countries.[7] The first part of the book includes three chapters that describe the status of research by combining the scientific evidence with the ongoing debates on adaptation strategies

---

[7] 'For those on the frontlines, the impacts of rising seas threaten livelihoods, inflict damage to settlements and critical infrastructure, and can, in its most dramatic manifestations, force the displacement of entire island populations and coastal communities', 25 September 2024, https://news.un.org/en/story/2024/09/1154881. See also: the Advisory Opinion of the Interntional Court of Justice: 'In the Court's view, sea level rise poses challenges in several respects, including of an economic, social, cultural and humanitarian character. It thus finds that the duty to co-operate assumes particular significance in this context, requiring States to take, in co-operation with one another, appropriate measures to address the adverse effects of this serious phenomenon. Such a duty of co-operation is founded on the recognition of the interdependence of States and the ensuing need for solidarity among peoples (see paragraph 308 above). In this regard, co-operation in addressing sea level rise is not a matter of choice for States but a legal obligation' (ICJ, 2025, para. 364); and the Advisory Opinion of the Inter-American Court of Human Rights: 'Indeed, manifestations of climate change such as floods, droughts, heatwaves, sea level rise, and the increase of vector-borne diseases jeopardize the enjoyment of rights such as to life, personal integrity, health, private and family life, property, housing, freedom of movement and residence, water, food, work and social security, culture, and education (infra paras. 392–457). According to scientific advances, these risks together with the magnitude and irreversible nature of the harm that the impacts can cause are, to a great extent, predictable. Moreover, it is predictable that such harm will particularly affect vulnerable individuals and groups' (IACHR, 2025, para. 234).

for sea level rise. Gonçales-Sales's chapter on 'Sea Level Rise: Scientific Evidence, Socio-Economic Realities, and Adaptation Challenges for Coastal Communities' considers the multifaceted challenges for livelihoods of affected communities. In her pragmatic analysis, Gonçales-Sales facilitates the reader's understanding of the threats posed to agriculture, fishing, tourism, and cultural knowledge. She offers a vivid description of how the progressive erosion of coasts as well as inundation is affecting the ability of communities to remain in their homes by increasing their exposure to relocation and displacement. The analysis of the available scientific knowledge in this chapter is valuable for the other two chapters that examine current developments of international law.

The second chapter, 'The International Law Commission's Study Group on Sea Level Rise in Relation to International Law and Its Impact on International Law', by Starita, illustrates the contribution of the International Law Commission (ILC) Study Group on Sea Level Rise in Relation to International Law (hereinafter ILC Study Group) to the studies on this topic. The ILC Study Group was established in 2019 and since then it has contributed to addressing a complex array of issues in relation to international law (e.g. statehood; maritime boundaries delimitation; human rights of affected populations; and international cooperation). The analysis provides an assessment of the four-issue papers adopted by the ILC Study Group together with an introduction to the role played by the International Law Association (ILA) Committee on International Law and Sea Level Rise (hereinafter ILA Committee).

The third chapter, 'The Role of the Security Council in Addressing and Providing Responses to Peace and Security Risks Resulting from Sea level Rise', by Nesi and Fornalé, addresses another crucial dimension of the international debate on sea level rise: its progressive 'securitization' and the role that could be played by the UN Security Council in this regard. In February 2023, the UN Security Council held an open debate to discuss and recognize the increasing risks posed by sea level rise for international peace. The relevance of this topic was later on confirmed by the European Commission, which, in June 2023, adopted a joint communication stating that it would pursue a 'New Outlook on the Climate Security Nexus: Addressing the Impact of Climate Change and Environmental Degradation on Peace, Security and Defence'.[8] In this chapter, the

---

[8] Joint Communication to the European Parliament and the Council, JOIN 2023, 19 final, 28 June 2023.

authors draw attention to the different implications of the adoption of the notion of 'security'. Taking under consideration the 'ever-evolving definition' of a 'threat' to peace and security in international law, they state that competing approaches could arise in trying to identify who is responsible for dealing with security claims (Benton Heath, 2022). As illustrated by Joyeeta and Hilmer (2021, p. 550), this requires a clarification: 'What does the term "security" mean in law and policy?' An expansive interpretation of threats that could include climate change and its adverse impacts could require departing from a 'State-centered approach' to foster a cooperative approach to security. This approach could result in expanding the scope of the mandate of the UN Security Council over time.

2.2 Sea Level Rise and the Rights of Affected Populations

After the initial survey that attempts to evaluate the potential and limitations of the available scientific evidence and its relationship with legal scholarship, the second part of the book introduces four chapters that develop a richer analysis of the building of the legal momentum in the field of human rights, sustainability and human mobility. Drawing on the field of human rights, Bílková's chapter 'Sea Level Rise and Human Rights' offers a significant account of how climate change affects the enjoyment of human rights, and in particular the right to a clean, healthy, and sustainable environment. She explores how slow-onset events could destabilize the capacity of the affected States to comply with human rights instruments and duties. The author demonstrates the relevance of case law at international level to engage with human rights violations and clarify the scope and content of positive obligations.

The next two chapters address the implications of sea level rise for human mobility. De Vido, in her chapter 'Sea Level Rise as a Form of Gendered Climate Violence: International Legal Implications for Migration', examines the gendered dimension of human mobility. By adopting a new vocabulary—see the notions of 'climate violence' and 'chronic emergencies'—the author contributes substantially to the understanding of the gender-climate nexus by adopting an ecofeminist method in international law. She succeeds in this by demonstrating that the reasoning adopted in the well-known Teitiota case lacked a gender perspective and, in her analysis, engages more openly with climate violence by considering the disproportionate impact of sea level rise on women.

Ballin and Mantu's chapter 'Rethinking Sustainable Migration for the Anthropocene' takes a closer look at how the notion of sustainability could contribute to framing our understanding of climate migration. The authors discuss whether and how migration law and the existing approach, particularly as framed by European countries, operate in the related areas. Sea level rise requires alternative venues to ensure sustainability. While there is the tendency to focus on affected countries, this chapter seeks to capture many different perspectives and possibilities that concern European countries as well. The encounter with the Netherlands and its migration law presented here holds important lessons for the future of human mobility.

The chapter on 'Human Rights and Justifications in Climate Litigation: A First Attempt at Conceptualization' by Cristani and Fornalé further explores the role of European countries in dealing with climate change and human rights. It concludes this section by providing new insights into the debate on climate change litigation and introducing an innovative perspective on the debate that builds on the emerging notion of human rights justification. It attempts to answer the following questions: 'What are the different duties—e.g. respecting, protecting, fulfilling, preventing—that are entailed for human rights and justifications? *Vice versa*, how could human rights and justifications be framed to advance States' duties?' This inquiry is motivated by the relevance of State practices and the need to study the complexity of their claims. To this end, the chapter introduces a new dataset (HRJust Climate Claims) that includes 365 cases judged at the domestic, European and international level, together with a zoom in on the decision of the European Court of Human Rights in the *Verein KlimaSeniorinnen Schweiz* case. The chapter presents the methodology and coding of the cases collected. This chapter is not merely descriptive. It provides a preliminary attempt to conceptualize the relationship between human rights and justifications by revealing the typologies adopted by States. It explores an understudied line of inquiry that allows to identify emerging trends and map potential new approaches in addressing the complexity of the theoretical and practical relevance of rights-based climate claims (Rodriguez-Garavito, 2022; Wewerinke-Singh et al., 2021).

2.3 Sea Level Rise—New Developments at International Level

As highlighted in the previous sections, international cooperation plays a central role in facilitating inter-State relations and stability, as well as in maintaining international peace and stability (ILC, 2024; ILA, 2024). Despite the progress made in expanding this principle, a greater precision as to its content and temporal scope could ensure that it would have an impact that is key to addressing global challenges—such as sea level rise (Fisler Damrosch, 2002). Against these challenges, Adany's chapter 'Sovereignty, State Cooperation and Sea Level Rise' explores how sovereignty in its cooperative dimension is changing in the light of the so-called common concerns (Cottier, 2021). As highlighted by the International Court of Justice in its Advisory Opinion' climate change is a common concern. Co-operation is not a matter of choice for States but a pressing need and a legal obligation (ICJ, 2025, para. 308; Shelton, 2009; Brunnée, 2012; Cottier, 2021; Fornalé, 2022). In line with this, Ádány asks about the suitability of existing frameworks for facing new challenges and how the content and significance of cooperative duties in the context of sea level rise can be strengthened.

The second chapter by Doebbler provides a detailed analysis of the Advisory Opinion adopted by the International Tribunal of the Law of the Sea in May 2024. This was an historical moment because it was the first time that an international court dealt with the 'climate crisis' in its advisory role. This chapter helps to contextualize this new development in the climate change debate by looking at the role of UN State Members and non-State actors in the process. Additionally, the chapter reflects on how this significant achievement could play a role in informing new legal developments and rule-making processes in relation to the law of the sea. In particular, Doebbler focuses on the role of the duty of due diligence and demonstrates the relevance of this Advisory Opinion for the other judicial bodies currently dealing with international climate change law.

The final chapter, 'Assessing Innovative Sources for the Loss and Damage Mechanism: The Role and Prospective Regulation of Climate-Friendly Foreign Investment', by Cristani reflects on the potential positive role of foreign investments in addressing climate change. In doing this, the author shows the relevance of the loss and damage mechanism. She provides an historical overview of its creation.[9] The author also provides an innovative analysis of the major progress made towards adopting climate-friendly foreign investment agreements, and she demonstrates the significance of the new generation of these instruments in terms of responding to climate change concerns, specifically those of developing countries.

3   Situating Our Book Project

The manuscript brings together expertise from a variety of fields and several geographical perspectives to tackle these interconnected themes. Our collection seeks to make a fruitful contribution to the global challenge of understanding, analysing, and advancing the common knowledge of the issue.

This book also incorporates the preliminary results of the research conducted within the Horizon Europe project HRJust.[10] The core research question of this project is when and how States use the human rights narrative to justify their regulatory and policy decisions (human rights justifications—HRJs). Accordingly, the project aims to develop a new theory of HRJs, taking climate change, migration, and COVID-19 as the main thematic areas of the research to study and test this theory.

References

Benton Heath, J. (2022). Making Sense of Security. *The American Society of International Law*, 289–339.

Besson, B. (2022). Justifications. In D. Moeckli, S. Shah, S. Sivakumaran & D. Harris (Eds.), *International Human Rights Law* (4th ed., pp. 23–42). Oxford University Press.

---

[9] The parties agreed to 'establish new funding arrangements for assisting developing countries that are particularly vulnerable to the adverse effects of climate change [including sea level rise...] with a focus on addressing loss and damage by providing and assisting in mobilizing new and additional resources' (FCCC/CP/2022/10/Add.1, p. 12).

[10] Horizon Europe project HRJust (States' Practice of Human Rights Justification: A Study in Civil Society Engagement and Human Rights through the Lens of Gender and Intersectionality), Grant Agreement No. 101094346. This work has received funding from the Swiss State Secretariat for Education, Research and Innovation (SERI) under Grant Agreement No. 23.00131. For additional information see: www.hrjust.worldpress.com; www.hrjust-climate-claims.eu; www.hrjust-intersect-observatory.eu.

Bilkova, V. (2019). Populism and Human Rights. In J.E. Nijman & W.G. Werner (Eds.), *Netherlands Yearbook of International Law 2018* (pp. 143–174).

Brunnée, J. (2012) The Global Climate Regime: Whither Common Concern? In H. P. Hestermeyer, D. König, N. Matz-Luck, V. Röben, A. Seibert-Fohr, P.-T. Stoll & S. Vöneky (Eds.), *Coexistence, Cooperation and Solidarity: Liber Amicorum for Rüdiger Wolfrum* (pp. 35–71). Martinus Nijhoff.

Burson, B., Kälin, W., & McAdam, J. (2023). Statehood, Human Rights and Sea Level Rise. *Yearbook of International Disaster Law Online, 4*(1), 265–280. Brill. https://doi.org/10.1163/26662531_00401_013

Cottier, T (ed.) 2021, *The Prospects of Common Concern of Humankind in International Law*, Cambridge University Press. Cambridge.

Cournil, C., & Colard-Fabregoule, C. (2012). *Changements environnementaux globaux et droits de l'homme*. Bruylant.

Fisler Damrosch, L. (2002). Obligations of Cooperation in the International Protection of Human Rights. In J. Delbrück & U. E. Heinz (Eds.), *International Law of Cooperation and State Sovereignty* (pp. 15–43). Duncker & Humblot.

Fornalé, E. (2022). Collective Action, Common Concern and Climate-Induced Migration. In S. Behrman & A. Kent (Eds.), *Climate Refugees: Global, Local and Critical Approaches* (pp. 107–127). Earth System Governance Series, Cambridge University Press.

Fornalé, E. (2023a). Vulnerability, Intertemporality, and Climate Litigation. *Nordic Journal of Human Rights, 41*(4), 357–377. https://doi.org/10.1080/18918131.2023.2225973

Fornalé, E. (2023b). Slow Violence, Gender Equality and Climate Agency in Times of 'Polycrisis'. *Revista General de Derecho Europeo, 61*. https://www.iustel.com/v2/revistas/detalle_revista.asp?id_noticia=426572&d=1

Freestone, D., & Cicek, D. (2021). *Legal Dimensions of Sea Level Rise: Pacific Perspectives*. World Bank. https://openknowledge.worldbank.org/handle/10986/35881

Horn, L. (2015). Climate Change and the Future Role of the Concept of the Common Concern of Humankind, *AJEL, 2*, 24–56.

Human Rights Council (HRC). (2018, March 22). *The Slow Onset Effects of Climate Change and Human Rights Protection for Cross-Borders Migrants* (A/HRC/37/CRP.4). https://www.ohchr.org/sites/default/files/Documents/Issues/ClimateChange/SlowOnset/A_HRC_37_CRP_4.pdf

IACtHR. (2025, May 29). *Advisory Opinion AO-32/25 Requested by the Republic of Chile and the Republic of Colombia*. Climate Emergency and Human Rights. https://jurisprudencia.corteidh.or.cr/en/vid/1084981967

ICJ. (2025, July 23). *Advisory Opinion*. Obligations of States in Respect of Climate Change. https://www.icj-cij.org/case/187

International Law Commission (ILC). (2018). *Annex II, Sea Level Rise in Relation to International Law, Bogdan Aurescu, Yacouba Cissé, Patrícia Galvão Teles, Nilüfer Oral, Juan José Ruda Santolaria* (UN Doc A/73/10). https://legal.un.org/ilc/reports/2018/english/annex_B.pdf

ILC. (2022, April 19). *Sea Level Rise in Relation to International Law: Second Issues Paper, by Patrícia Galvão Teles and Juan José Ruda Santolaria, Co-Chairs of the Study Group on Sea Level Rise in Relation to International Law* (UN Doc A/CN.4/752). http://legal.un.org/docs/?symbol=A/CN.4/752

ILC. (2024, February 19). *Sea Level Rise in Relation to International Law: Additional paper to the Second Issues Paper (2022) by Patrícia Galvão Teles and Juan José Ruda Santolaria* (75th session of the ILC (2024) (UN Doc A/CN.4/774). https://documents.un.org/doc/undoc/gen/n24/044/43/pdf/n2404443.pdf

International Law Association (ILA). (2018). *Sydney Conference (2018). International Law and Sea Level Rise*. https://disasterlaw.ifrc.org/sites/default/files/media/disaster_law/2021-03/DraftReport_SeaLevelRise2018.pdf

International Law Association (ILA). (2024, May). *International Law and Sea Level Rise Committee: Final Committee Report Athens*. https://www.ila-hq.org/en_GB/committees/international-law-and-sea-level-rise

Joyeeta, G., & Hilmer, B. (2021). Climate Change and Security. In R. Geiss Robin & N. Melzer (Eds.), *The Oxford Handbook of the International Law of Global Security* (pp. 548–565). Oxford University Press.

Kumm, M. (2018). The Turn to Justification: On the Structure and Domain of Human Rights Practice. In A. Etinson (Ed.), *Human Rights: Moral or Political?* (pp. 238–261). Oxford University Press.

Mayer, B., & Van Asselt, H. (2023). The Rise of International Climate Litigation. *Review of European, Comparative & International Environmental Law, 32*(2), 175–184. https://doi.org/10.1111/reel.12515

McAdam, J. (2013). The Emerging New Zealand Jurisprudence on Climate Change, Disasters and Displacement'. *Migration Studies, 3*, 131.

Nixon, R. (2011). *Slow Violence and the Environmentalism of the Poor*. Harvard University Press. https://psycnet.apa.org/doi/10.4159/harvard.9780674061194

Peel, J., & Osofsky, H. M. (2018). A Rights Turn in Climate Change Litigation? *Transnational Environmental Law, 7*(1), 37–67. https://doi.org/10.1017/S2047102517000292

Rodriguez-Garavito, C. (2023). Climatizing Human Rights: Economic and Social Rights for the Anthropocene. In K. Young & M. Langdorf (Eds.), *Oxford Handbook of Economic and Social Rights*. Oxford University Press. https://doi.org/10.1093/oxfordhb/9780197550021.013.68

Rodriguez-Garavito, C. (2019). Empowered Participatory Jurisprudence, Experimentation, Deliberation and Norms in Socioeconomic Rights Adjudication. In

K. G. Young (Ed.), *The Future of Economic and Social Rights* (pp. 233–258). Cambridge University Press.

Rodríguez-Garavito, C. (2022). *Litigating the Climate Emergency: How Human Rights, Courts, and Legal Mobilization Can Bolster Climate Action*. Cambridge University Press.

Savaresi, A. (2020). The Use of Human Rights Arguments in Climate Change Litigation and Its Limitations. In D. Ismangil, K. Van der Schaaf & L. van Troost (Eds.). *Climate Change, Justice and Human Rights, Changing Perspectives on Human Rights* (pp. 49–55). Strategic Studies.

Savaresi, A., & Setzer, J. (2021). Rights-Based Litigation in the Climate Emergency: Mapping the Landscape and New Knowledge Frontiers. *Journal of Human Rights and the Environment*. https://ssrn.com/abstract=3928385

Shelton, D. (2009). Common Concern of Humanity. *Iustum Aequum Salutare, 1*, 33–40.

Soltau, F. (2016). Common Concern of Humankind. In K. R. Gray, R. Tarasofsky & C. Carlarne (Eds.), *The Oxford Handbook of International Climate Change Law* (pp. 3–92). Oxford University Press.

United Nations Special Rapporteur on Violence Against Women. (2022b, July 11). *Report on Violence Against Women and Girls in the Context of the Climate Crisis: Including Environmental Degradation and Related Disaster Risk Mitigation and Response* (UN Doc A/77/136). https://documents.un.org/doc/undoc/gen/n22/418/07/pdf/n2241807.pdf

United Nations Special Rapporteur on Contemporary Forms of Racism, Racial Discrimination, Xenophobia and Related Intolerance. (2022a, October 25). *Report on Contemporary Forms of Racism, Racial Discrimination, Xenophobia and Related Intolerance, E. Tendayi Achiume, on Ecological Crisis, Climate Change and Racial Justice* (UN Doc A/77/549).

Vidas, D., Fauchald, O. K., Jensen, O., & Tvedt, M. W. (2015a). International Law for the Anthropocene? Shifting Perspective in Regulation of the Oceans, Environment and Genetic Resources. *Antropocene, 9*, 1–13. https://doi.org/10.1016/j.ancene.2015.06.003

Vidas, D., Zalasiewicz, J., & Williams, M. (2015b). What Is the Anthropocene - and Why Is It Relevant for International Law? *Yearbook of International Environmental Law, 25*(1), 3–23. https://doi.org/10.1093/yiel/yvv062

Wewerinke-Singh, M., Aguon, J., & Hunter, J. (2021). Bringing Climate Change Before the International Court of Justice: Prospects for Contentious Cases and Advisory Opinions. In I. Alogna, C. Bakker, & J.-P. Gauci (Eds.), *Climate Change Litigation: Global Perspectives* (pp. 393–414). https://doi.org/10.1163/9789004447615_018.

Open Access This chapter is licensed under the terms of the Creative Commons Attribution 4.0 International License (http://creativecommons.org/licenses/by/4.0/), which permits use, sharing, adaptation, distribution and reproduction in any medium or format, as long as you give appropriate credit to the original author(s) and the source, provide a link to the Creative Commons license and indicate if changes were made.

The images or other third party material in this chapter are included in the chapter's Creative Commons license, unless indicated otherwise in a credit line to the material. If material is not included in the chapter's Creative Commons license and your intended use is not permitted by statutory regulation or exceeds the permitted use, you will need to obtain permission directly from the copyright holder.

PART I

Sea Level Rise, Stability and Security

CHAPTER 2

Sea Level Rise: Scientific Evidence, Socio-Economic Realities, and Adaptation Challenges for Coastal Communities

Vilane Gonçalves Sales

1 INTRODUCTION

Climate change has driven some of the most significant transformations in the landscape of the planet. One of the most striking consequences is the relentless rise in global sea levels, which reshapes coastlines and threatens communities worldwide. The SLR is a well documented and extensively researched global phenomenon. Various studies have identified factors contributing to this increase, including the thermal expansion of ocean waters, melting of glaciers and ice sheets, and changes in ocean mass (Maitland et al., 2024). Human influence has unequivocally warmed the atmosphere, ocean, and land, leading to a global mean sea level (GMSL) increase of 0.20 m between 1901 and 2018. The average rate of sea level rise has accelerated over time, from 1.3 mm per year between 1901 and 1971 to 1.9 mm per year between 1971 and 2006, and further to 3.7 mm per year between 2006 and 2018 (Hamlington et al., 2024). While the

---

V. G. Sales (✉)
Euro-Mediterranean Center on Climate Change (CMCC), Venice, Italy
e-mail: vilane.goncalvessales@cmcc.it

© The Author(s) 2026
E. Fornalé (ed.), *Sea Level Rise*,
https://doi.org/10.1007/978-3-031-89171-7_2

current trend in global mean sea level rise is positive, indicating a rise of around 3.2 mm/year, the increase is not uniform across the globe, with variations observed in different regions (Box et al., 2018; Strassburg et al., 2015). Wright et al. (2018) report that by mid-century, relative sea levels in some coastal cities could rise by up to 1.2 m compared to 2000 levels. Vousdoukas et al. (2018) provided some probabilistic projections of extreme sea levels (ESLs), forecasting a global average 100-year ESL increase of 34–76 cm under a moderate emission scenario and 58–172 cm under a business-as-usual scenario by 2100.

As data on global sea level rise reveals alarming trends, the importance of comprehensive research becomes increasingly evident. Sea level rise has interconnected impacts across environmental, social, economic, and cultural dimensions, creating cascading challenges. Environmentally, it threatens coastal ecosystems like wetlands, estuaries, and coral reefs, which are crucial for biodiversity, carbon sequestration, and storm protection, while also causing habitat loss, increased salinity, and erosion, which disrupt ecological balance and biodiversity (Mills et al., 2019; Natesan & Parthasarathy, 2010; Riera-Spiegelhalder et al., 2023). Economically, it poses significant risks, including infrastructure damage, rising adaptation costs, and reduced investments in affected areas, with ripple effects on industries, property values, and local economies (Bosello et al., 2011; Nováčková & Tol, 2017). Socially, sea level rise leads to the displacement of coastal communities, loss of livelihoods, and increased vulnerability to extreme weather events, straining social infrastructure and services, and threatening cultural heritage sites and identities (Barnett et al., 2020; Chen et al., 2024). The complexity and interconnectedness of these impacts underscore the need for integrated research and strategies that address the immediate and long-term challenges posed by sea level rise across multiple sectors (Kopp et al., 2017).

This chapter posits that effectively addressing sea level rise requires a holistic, interdisciplinary approach that integrates scientific evidence, socio-economic realities, and adaptive strategies. By examining the intricate relationships between environmental changes, social vulnerabilities, and economic impacts, researchers and policymakers can develop more robust and equitable solutions to this global challenge. The chapter begins with a presentation of the current scientific evidence of sea level rise, including measurement techniques and future projections worldwide. It then explores the socio-economic realities faced by affected communities. The discussion then turns to the challenges and strategies for adaptation, considering both infrastructure and policy frameworks. Finally, the chapter concludes with a synthesis of global perspectives and local realities, emphasizing the need for context-specific solutions and international cooperation in the face of this global phenomenon.

2 Scientific Evidence of Sea Level Rise

The Earth's climate system is currently experiencing unprecedented changes, with sea level rise emerging as one of the most tangible and immediate threats to coastal communities worldwide. While the planet has undergone natural climate variations throughout its history, the current rate and scale of sea level rise are unparalleled, largely due to human activities. Understanding this critical difference requires a comprehensive examination of both current data and historical climate records. Paleoclimate research, the study of past climates throughout Earth's history, provides an essential context for interpreting contemporary climate changes. By analysing natural climate recorders such as tree rings, ice cores, sediment layers, and fossils, scientists can reconstruct past climate conditions and sea level fluctuations (Kopp et al., 2017; Lambeck et al., 2014). These historical reconstructions demonstrate that while sea levels have indeed varied naturally over geological timescales, the current rate of change far exceeds anything observed in the recent geological past.

For instance, studies of the Early Pliocene epoch (5.3 to 3.6 million years ago) have revealed sea level fluctuations of up to 20–40 m, indicating substantial variations in ice volume over extended periods (Hansen et al., 2013). During the Last Glacial Maximum, approximately 20,000 years ago, sea levels were about 120 m lower than today due to vast amounts of water trapped in ice sheets (Yin et al., 2021). As the climate warmed naturally over thousands of years, these ice sheets melted, causing sea levels to rise gradually. However, these changes occurred over millennia, allowing ecosystems and species to adapt gradually. In contrast, the sea level rise observed over the past century, and particularly in recent decades, is occurring at a dramatically accelerated pace. The current rate of sea level rise, driven primarily by human-induced global warming, has been happening over decades, giving natural systems and human societies little time to adapt (Church & Gregory, 2019).

2.1 Causes of Sea Level Rise

Sea level rise is driven by a complex interplay of global and regional factors, with the primary contributors being thermal expansion of ocean water, loss of land ice, and changes in land water storage (Cazenave & Cozannet, 2014; Hamlington et al., 2020). As global temperatures rise due to human-induced climate change, the oceans absorb much of this heat, leading to thermal expansion—a process where seawater expands as it warms, contributing significantly to the observed rise in sea levels. In addition to thermal expansion, the accelerated melting of land-based ice sheets in Greenland and Antarctica is adding substantial amounts of freshwater to the oceans, further elevating sea levels (Smith et al., 2020; Gadi, 2023). Although a complete melting of these ice sheets would lead to dramatic sea level increases of up to 7 m for Greenland and 3–5 m for Antarctica, even smaller losses are already having a significant impact (Hamlington et al., 2020). Regional variations in sea level are further influenced by non-climatic factors such as the Earth's response to both past and present ice melt, which involves gravitational forces, the Earth's deformation, and rotational changes (Cazenave & Cozannet, 2014). Additionally, vertical land motions (VLMs), including uplift and subsidence caused by glacial isostatic adjustment, tectonic activity, sediment loading, and human activities, contribute to local variations in sea level (Cazenave & Remy, 2011).

2.2 Measurement Techniques and Challenges

Understanding the causes of sea level rise is essential, but equally important is the ability to accurately measure these changes to assess their impact and predict future trends. To do this, scientists rely on advanced measurement techniques that provide precise data on global mean sea level rise. These techniques, including satellite altimetry and tide gauge records, are crucial for quantifying the rapid changes currently being observed and for addressing the challenges associated with monitoring sea level rise on both global and regional scales.

Satellite altimetry is a sophisticated technique used to measure sea level changes from space, offering unparalleled precision in tracking global mean sea level (GMSL) variations. By emitting radar pulses from satellites and measuring the distance to the ocean surface, scientists can determine the sea level with high accuracy, while accounting for factors such as atmospheric delays (Guérou et al., 2023). Recent analyses utilizing this method have provided crucial insights into GMSL changes over a 29-year period, namely from 1993 to 2021. The data reveals that GMSL has risen at an average rate of 3.00 mm per year, indicating a consistent and alarming upward trend. More importantly, an acceleration in this rise estimated at 0.12 mm per year has been observed, suggesting that the pace of sea level increase is itself accelerating, which could lead to more rapid changes in the future (Guérou et al., 2023).

Figure 1 illustrates the global mean sea level rise from 1992 to 2024, combining data from multiple satellite missions (TOPEX/Poseidon, Jason-1, -2, -3, and Sentinel-6A). This representation is an extension of the visualization presented by Guérou et al. (2023) utilizing AVISO + Satellite Altimetry Data from CNES, and uses data for the global zone with seasonal signals removed to emphasize the long-term trend.

While satellite altimetry offers a comprehensive view of global mean sea level changes from space, tide gauges remain the primary source of coastal sealevel observations, though their global distribution is uneven (Marcos et al., 2019). These long-standing instruments, located along coastlines worldwide, not only validate satellite measurements but also offer high-resolution insights into regional and local sea level trends, capturing the effects of tides, storms, and other short-term fluctuations. However, the use of tide gauge data for estimating global sea level rise (SLR) comes with certain challenges. The global network of tide gauges is unevenly distributed, with a higher concentration in the Northern Hemisphere, thus potentially introducing biases in global sea level estimates due to regional variations in SLR (Natarov et al., 2017). Additionally, tide gauges measure relative sea level, which can be affected by land movements such as subsidence or uplift, leading to offsets in reference levels and complicating the derivation of accurate global trends. Satellite altimetry has proven valuable in detecting and correcting these offsets (Ray et al., 2023). Moreover, tide gauge data are autocorrelated, meaning that successive measurements are not independent, which can lead to underestimated uncertainties when using conventional statistical methods (Wang, 2023). Despite these limitations, tide gauges have provided valuable long-term records of sea level changes. A statistical analysis reveals that globally, sea levels rose at an average rate of 0.39–1.03 mm/year (Beenstock et al., 2015).

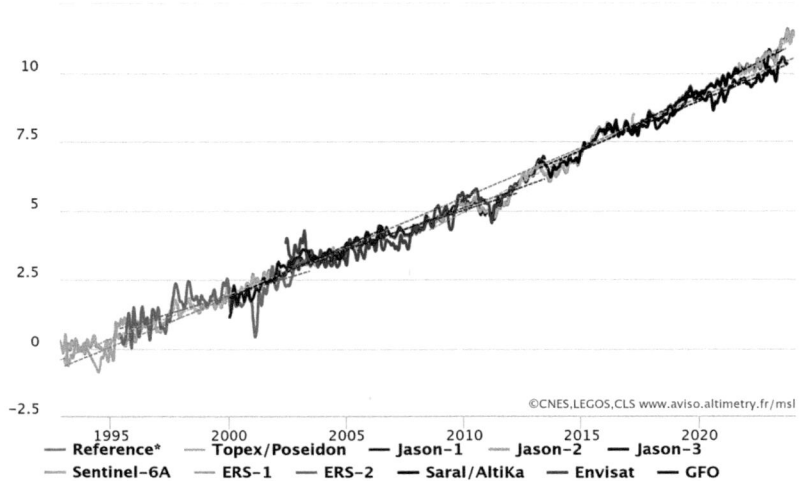

Fig. 1 Global Mean Sea Level Rise from Satellite Altimetry (1992–2024). Data from all missions (TOPEX/Poseidon, Jason-1, -2, -3, Sentinel-6A) for the global zone, with seasonal signals removed. Extended and adapted from the visualization presented in Guérou et al. (2023) using AVISO + Satellite Altimetry Data, CNES

2.3 Future Projections and Uncertainties

Building upon the understanding of current sea level rise measurements and their inherent challenges, the focus now shifts to the critical task of projecting future sea level changes. While advanced techniques like satellite altimetry and tide gauge records provide crucial data on the observed sea level rise, projecting these trends into the future introduces additional layers of complexity and uncertainty.

The Intergovernmental Panel on Climate Change (IPCC) plays a pivotal role in synthesizing and communicating the current state of knowledge on sea level rise projections, which are fundamental for informing policy decisions and adaptation strategies worldwide (Anzidei et al., 2020; Hinkel et al., 2015; Ma et al., 2015; Siegert et al., 2020). However, the complexity of the Earth's climate system and the multitude of factors influencing sea level dynamics introduce significant uncertainties into these projections (Willis & Church, 2012; Le Cozannet et al., 2017; Bakker et al., 2017; Kopp et al., 2023). These uncertainties extend beyond measurement challenges to include complex ice sheet dynamics and potential climate tipping points (Golledge, 2020; Wahl et al., 2017).

The Synthesis Report of the IPCC's Sixth Assessment Report (AR6) provides the most recent sea level rise projections, incorporating the latest advancements in scientific understanding and modelling techniques (IPCC, 2023). According to this report, the global mean sea level rise relative to the 1995–2014 baseline is projected to range from 0.28–0.55 m by 2100 under a very low emission scenario (SSP1-1.9) to 0.63–1.01 m under a very high emission scenario (SSP5-8.5). These projections underscore the substantial influence of greenhouse gas emissions on both the rate and magnitude of sea level rise. Importantly, the AR6 Synthesis Report emphasizes the long-term nature of sea level rise, projecting continued increases for centuries to millennia due to ongoing deep ocean warming and ice sheet melting (IPCC, 2023). Even if global warming is limited to 1.5 °C, sea levels are expected to rise by approximately 2–3 m over the next 2000 years, with a potential rise of 2–6 m if the warming reaches 2 °C. The report also acknowledges the possibility of sea level rise exceeding these likely ranges due to deep uncertainties in ice sheet processes, particularly under high emission scenarios. In the most extreme case, a sea level rise approaching 2 m by 2100 and potentially exceeding 15 m by 2300 cannot be ruled out under the very high emissions scenario (SSP5-8.5), though such projections are associated with low confidence (IPCC, 2023).

3 Socio-Economic Realities

As rising seas reshape coastlines, they simultaneously transform the socio-economic fabric of coastal communities, presenting complex problems that span economic, social, and cultural dimensions. This section explores the multifaceted socio-economic realities of sea level rise, examining its economic impacts on coastal regions and the social implications for affected communities. Through an analysis of diverse case studies, it is illustrated how these impacts manifest in different contexts, from urban coastal areas to small island developing states, while the adaptive strategies and policy interventions are highlighted.

3.1 Economic Impacts on Coastal Regions

Coastal regions, often referred to as the *key engines of the global economy*, are uniquely vulnerable to the socio-economic impacts of sea level rise (SLR) (Durand et al., 2022). These areas, which host a significant concentration of the world's population and economic activity—including 11 of the 15 largest megacities—face increasing risks as sea levels continue to rise (United Nations, Department of Economic and Social Affairs, 2022). The economic ramifications of SLR are multifaceted and complex, driven by the exposure of assets, the costs associated with adaptation measures, and the challenges posed by vertical land motions and regional variations in sea level (Cazenave & Moreira, 2022).

Although the literature often focuses on the physical mechanisms and projections of SLR, the socio-economic consequences will be significant, particularly in densely populated coastal zones. The uncertainty inherent in SLR projections further complicates decision-making processes, as it can lead to both under-investment and over-investment in protective measures, with substantial economic implications (Cazenave & Moreira, 2022). The economic impacts of sea level rise on coastal regions can be broadly categorized into direct and indirect costs. Direct costs primarily involve immediate physical damage to infrastructure and property, while indirect costs encompass the broader, often long-term effects on economic sectors such as tourism and fisheries. This distinction is crucial for understanding the full scope of economic challenges posed by rising seas and for developing comprehensive adaptation strategies.

3.1.1 Direct Costs

Direct costs in this case refer to the immediate and tangible financial impacts directly attributable to sea level rise (SLR). These costs include expenses incurred from physical damage to infrastructure, loss of land, and investments required to implement protective measures aimed at mitigating the effects of SLR (Bosello et al., 2011). One of the most significant direct costs is the damage to coastal infrastructure, including the destruction or impairment of buildings, roads, bridges, and other essential structures due to flooding, storm surges, and coastal erosion. For instance, Neumann et al. (2015a; b) estimated that the joint effects of storm surge and SLR could impact over \$150 billion in property and 600,000 people in California by 2100.

A substantial portion of these direct costs comes from the construction and ongoing maintenance of protective infrastructure, such as levees, seawalls, and barriers, which are crucial for defending coastal communities. The financial investment required for these measures is considerable, with Depsky et al. (2023) emphasizing that without such investments, the economic impact of sea level rise would be even more devastating. Jevrejeva et al. (2018) projected that if global temperatures rise by 2 °C by 2100 without additional adaptation measures, annual flood costs could escalate to USD 11.7 trillion, or 2% of global GDP—an alarming increase from the USD 10.2 trillion projected with just a 1.5 °C rise. Similarly, Depsky et al. (2023) highlighted that global net present value (NPV) costs associated with SLR could range from USD 600 billion to 3.4 trillion under optimal adaptation scenarios, largely due to the permanent loss of valuable coastal land and immobile capital. Without proactive adaptation, the average annual global costs of climate-driven SLR could reach USD 3.1 to 6.9 trillion by 2100 under a 4 °C warming scenario, reflecting the immense financial burden posed by these losses.

However, while these protective measures are indispensable for safeguarding coastal regions, they can also lead to unintended consequences. Hummel et al. (2021) demonstrated that implementing these protections, particularly in areas like San Francisco Bay, can inadvertently exacerbate flooding and increase economic damages in other regions. For instance, protecting one segment of the shoreline in San Francisco Bay was shown to potentially lead to an additional USD 723 million in damages from a single flood event. This dual nature of protective infrastructure underscores the complexity of addressing the direct costs of sea level rise and highlights the need for comprehensive, system-wide approaches to coastal protection.

3.1.2 Indirect Costs

Indirect costs encompass the broader economic impacts that arise as a consequence of SLR but are not directly tied to immediate physical damage. These costs include long-term effects on economic activity, productivity, and public finances that often result from disruptions to infrastructure, loss of economic opportunities, and shifts in population dynamics (Bosello et al., 2011). For instance, SLR can lead to reduced GDP growth due to the loss of productive land, displacement of populations, and decreased investment in vulnerable areas (Hummel et al., 2021). Additionally, the costs of relocating populations and mobile capital away from at-risk coastal areas contribute significantly to overall indirect costs, including expenses related to housing, infrastructure, and loss of livelihoods, as well as non-market relocation costs such as the emotional distress of displacement and the loss of social networks (Depsky et al., 2023). Pycroft et al. (2016) further highlight the significant indirect costs associated with migration due to SLR, as forced relocation can disrupt economic stability and affect capital stock, thereby reducing future economic productivity and growth.[1]

Moreover, localized shoreline protection efforts, such as those examined by Hummel et al. (2021) in San Francisco Bay, can inadvertently exacerbate flooding in unprotected areas, leading to substantial economic damages. The construction of levees and seawalls, while protecting specific segments of the shoreline, can disrupt natural water flow and tidal patterns, increasing water levels and flooding in other regions. The economic damage in these unprotected areas can be significant, with estimates reaching up to $723 million for a single flood event. Hummel et al. (2021) emphasize that these indirect costs are not uniformly distributed, with socially vulnerable populations often bearing a disproportionate burden. These communities, which may lack the resources for adequate adaptation measures, are more susceptible to the risks posed by sea level rise and related flooding.

Furthermore, adaptation costs can strain public finances, leading to increased public deficits and reducing resources available for other essential public services. Parrado et al. (2020) discussed how these financial pressures could decrease economic growth by limiting resources for private investment and capital accumulation. Additionally, the loss of ecosystem services provided by coastal wetlands, such as flood protection, carbon sequestration, and biodiversity support, incurs further indirect costs by increasing the vulnerability of coastal areas to damage and reducing their resilience to climate change impacts (Depsky et al., 2023).

The indirect costs of sea level rise extend beyond general economic impacts to affect specific sectors crucial to many coastal economies. Two sectors particularly vulnerable to these indirect costs are tourism and fisheries. In the tourism sector, SLR poses significant challenges, especially in regions where beach-based tourism is a primary economic driver. As SLR leads to coastal erosion, beach retreat, and increased flood risks, the

---

[1] Non-market costs refer to the economic impacts that are not easily quantified in traditional financial or market terms (van Kooten, 2013).

tourism infrastructure—such as hotels, resorts, and recreational areas—becomes increasingly vulnerable. For example, in the Bahamas, a 1-m rise in sea level could place 83% of the tourism infrastructure at risk, leading to substantial economic losses in revenue and employment (Pathak et al., 2021a; b). Similarly, in the Mexican Caribbean, SLR scenarios project economic losses ranging from $330 million to $2.3 billion USD, depending on the extent of sea level rise (Ruiz-Ramírez et al., 2019). These losses are exacerbated by the potential reduction in beach width, which is crucial for maintaining tourist appeal. For instance, the beach tourism revenue in Egypt's Red Sea resorts is expected to decline by up to 30% by 2100 under the worst-case SLR scenarios, resulting in daily revenue losses of up to 897,000 USD (Sharaan et al., 2020). As the SLR continues, the indirect costs associated with the decline in tourism—such as reduced local income, diminished tax revenues, and increased costs for beach nourishment and infrastructure relocation—will significantly impact the economic stability of coastal communities reliant on tourism.

Similarly, the fisheries sector faces substantial indirect costs due to sea level rise, primarily through the disruption of coastal and marine ecosystems that support fish populations and related economic activities. SLR-induced changes, such as increased salinity, altered coastal habitats, and ocean acidification, can lead to shifts in species distribution and productivity, affecting the availability of commercially valuable fish stocks (Weatherdon et al., 2016). These ecological changes translate into economic consequences, as reduced fish catches can diminish the supply of local seafood, which in turn impacts related sectors, such as tourism (Danylchuk et al., 2023). In coastal communities where fishing and tourism are intertwined, the indirect costs are compounded. For example, the loss of local seafood provision could decrease tourism-related expenditure by $8 million per year and reduce consumer surplus, thereby lowering visitor numbers and overall tourism revenue (Pascoe et al., 2023). Moreover, the degradation of coastal wetlands, which serve as critical nursery grounds for many fish species, further exacerbates the decline in fisheries, leading to long-term economic challenges for coastal communities that depend on both fisheries and tourism for their livelihoods (Newton et al., 2020). These sector-specific impacts compound the broader indirect costs of SLR, highlighting the complex and far-reaching economic consequences of rising sea levels for coastal communities.

3.2 Social Implications for Coastal Communities

The social implications of sea level rise (SLR) extend far beyond economic impacts, profoundly affecting human lives, community structures, and social equity, particularly in vulnerable coastal communities. These communities, often characterized by low-income populations, minority groups, and elderly residents, face disproportionate risks from rising seas, which exacerbates existing social vulnerabilities (Best et al., 2023; Martinich et al., 2013).

3.2.1 Displacement, Migration, and Mental Health

One of the most immediate consequences of SLR is the displacement and migration of coastal populations, with vulnerable communities often bearing the brunt of these impacts. In Bangladesh, for instance, communities believe they can cope with an inundation of up to 30.5 cm, but a 61 cm rise would force them to abandon their homes entirely (Ali & Syfullah, 2017). This process of managed retreat is not merely a physical relocation but a complex social phenomenon with significant psychological and emotional impacts, particularly for vulnerable groups.

Studies in Ghana have shown that planned relocation negatively impacts well-being, increases anxiety, and lowers perceptions of safety among relocated individuals (Abu et al., 2024). These mental health consequences are often compounded by the loss of community connections and identity, which leads to the researchers emphasizing the need for holistic relocation strategies that incorporate psychological support, community-building initiatives, and cultural preservation efforts. Ensuring that the relocation is conducted in a culturally sensitive manner can mitigate some of the negative impacts and help maintain the social fabric of displaced communities (Piggott-McKellar et al., 2020). Displacement from ancestral lands and communities not only leads to physical relocation but also results in a loss of cultural heritage and social networks, which are vital for community resilience and mental health. The forced migration or displacement due to SLR can exacerbate mental health issues, including anxiety, depression, and post-traumatic stress disorder (PTSD), particularly among older adults and those with pre-existing health conditions (Simms et al., 2021).

3.2.2 Health, Food Security, and Environmental Justice

SLR poses significant health and safety risks, particularly for vulnerable populations throughout the globe. Immediate physical dangers are manifested through increased flooding, which not only threatens lives but also contributes to long-term health issues. Flooding exacerbates the contamination of water supplies, leading to a higher prevalence of water-borne diseases such as cholera and dysentery, which are particularly devastating in regions with limited access to healthcare and clean water (Bloetscher et al., 2016). Moreover, saltwater intrusion into freshwater resources can compromise drinking water quality, further aggravating health risks for coastal communities.

The impacts of SLR also extend to food security, as the salinization of agricultural lands reduces crop yields, leading to malnutrition and food scarcity in affected regions. This is especially problematic in low-lying coastal areas of developing countries, where agriculture is a primary source of livelihood and food supply (Riera-Spiegelhalder et al., 2023). Furthermore, the physical infrastructure that supports healthcare delivery is itself at risk from SLR. Flooding and storm surges can damage or destroy healthcare facilities, limiting access to essential services during and after extreme weather events. In areas like Southeast Asia and the Pacific Islands, where healthcare infrastructure is already under strain, the destruction of facilities due to SLR-related events can have catastrophic consequences for public health (Hens et al., 2018).

The health impacts of SLR are disproportionately severe for socially vulnerable populations, which raises critical environmental justice concerns. For instance, in the Gulf region of the United States, socially vulnerable coastal communities, including Black and Hispanic populations, as well as areas with higher concentrations of renters and older adults, face heightened risks of isolation, displacement, and abandonment as a result of SLR (Best et al., 2023; Martinich et al., 2013). These populations are less likely to receive the necessary investments in protective infrastructure, leading to a higher likelihood of severe health outcomes.

3.2.3 Cultural Heritage and Indigenous Knowledge

SLR threatens to erase centuries of cultural heritage in coastal areas, with particularly devastating impacts on indigenous communities whose traditional ways of life are closely tied to specific coastal ecosystems. The loss of these lands and resources not only affects cultural identity but also risks the loss of valuable environmental stewardship practices and adaptive strategies that have been honed over generations (Savo et al., 2017).

The erosion of indigenous knowledge systems further compounds the socio-economic challenges faced by these communities, reducing their ability to adapt to environmental changes. To preserve cultural heritage and enhance resilience, it is crucial to integrate traditional knowledge into broader SLR adaptation efforts and actively involve these communities in decision-making processes (Aswani et al., 2019).

The threat posed by SLR to cultural heritage is profound, with coastal sites around the world facing increasing risks of flooding, erosion, and eventual submersion. For instance, UNESCO World Heritage Sites in the Mediterranean, such as those identified by Reimann et al. (2018), are already experiencing significant threats, with projections indicating that flood risks could increase by 50% and erosion risks by 13% by 2100. Similarly, coastal fortifications in the Canary Islands, as highlighted by Sánchez et al. (2020), are vulnerable to rising sea levels, which necessitates the incorporation of climate change adaptation measures into cultural heritage management to preserve these historical structures. In the case of China, Li et al. (2022) and Chen et al. (2024) have shown that a significant number of coastal archaeological sites are at risk, with estimates suggesting that thousands of sites could be endangered under various sea level rise scenarios. The situation is mirrored in other parts of the world, such as India, as Gade (2022) underscores the urgent need to protect both built and unbuilt cultural heritage from the impacts of SLR in that country. The implications extend beyond physical damage to these sites, as cultural heritage embodies the identity, history, and practices of communities, and its loss can lead to irreversible socio-cultural damage.

3.2.4 *Community Engagement and Governance*

Addressing the complex social implications of SLR requires fair, inclusive, and culturally sensitive adaptation strategies that prioritize the needs of vulnerable coastal communities. Research on Botany Bay, Australia, has shown that different groups within the same community can have varying levels of concern and engagement regarding SLR adaptation, which necessitates multifaceted approaches to community engagement and decision-making (Kreller, 2021).

The lack of faith in government intervention observed in communities like King Salmon, California, highlights the critical need for sustained engagement and transparent communication to build trust between vulnerable communities and governing bodies (Richmond & Kunkel, 2024). This trust is especially important for marginalized groups that may have historically been excluded from decision-making processes. By fostering long-term relationships and ensuring that community voices are heard, governments can develop more effective and equitable adaptation strategies.

3.2.5 Adaptation Limits and Global Context

While local adaptation strategies are essential, it is crucial to recognize their limits in the face of unchecked sea level rise. Vulnerable coastal communities, especially those with limited resources, may reach thresholds beyond which adaptation is no longer viable. Magnan et al. (2022) emphasize that even the most ambitious local adaptation efforts will fail if global greenhouse gas emissions are not simultaneously mitigated. This underscores the interconnectedness of local adaptation with global climate policies and highlights the need for a comprehensive approach that would include both mitigation and adaptation. Developing countries, particularly in regions like Southeast Asia and Africa, face heightened risks due to their high population exposure and limited adaptive capacity (Neumann et al., 2015a; b; Barbier, 2014). A global strategy that would combine short-term emergency responses with long-term investments in resilience-building is essential to protecting these vulnerable populations and ensuring equitable adaptation outcomes. The social implications of SLR for coastal communities, particularly vulnerable ones, are complex and far-reaching, requiring multifaceted, equitable, and inclusive approaches to adaptation. These strategies must consider not only the physical impacts of rising seas but also the psychological, cultural, and social dimensions of this global challenge.

3.2.6 Case Studies

To fully understand the social implications of SLR, it is essential to examine specific case studies from around the world, namely from places where coastal communities are facing the direct and varied impacts of this global phenomenon. By exploring the social implications, the adaptive strategies employed, the vulnerabilities exposed, and the resilience demonstrated in these regions, it is possible to gain a deeper insight into the complex realities of SLR and the critical need for context-specific interventions that address both immediate threats and long-term sustainability.

Bangladesh

Coastal communities in Bangladesh are acutely vulnerable to SLR, facing challenges like salinity increase, land erosion, and permanent inundation (Ali & Syfullah, 2017). Studies highlight the severe impacts of it on livelihoods, particularly in polder-enclosed areas where local adaptation techniques may only be effective up to a 30.5 cm inundation, beyond which forced relocation becomes inevitable. Cyclones also exacerbate the vulnerability, with storm surges expected to expand the vulnerable zones significantly by 2050, which will require substantial investments in adaptation (Dasgupta et al., 2013). Community resilience programmes, like the Vulnerability to Resilience (V2R) in coastal Bangladesh, have successfully engaged locals in hazard-resilient activities, improving their access to infrastructure and reducing vulnerability (Ahmed et al., 2016). Local knowledge is also playing a crucial role in adaptation, as seen in the Bhola and Satkhira districts, where community-led initiatives are effectively integrating traditional knowledge with broader national adaptation strategies (Sultana & Luetz, 2022).

Pacific Islands

The Solomon Islands have been profoundly impacted by sea level rise, with several vegetated reef islands disappearing as a result of the combined effects of rising seas and increased wave exposure. This environmental change has necessitated the relocation of entire communities, prompting the implementation of proactive adaptation strategies that leverage traditional knowledge to enhance resilience (Albert et al., 2016; Leon et al., 2015). In Fiji, the relocation of coastal communities like Vunidogoloa due to sea level rise and severe flooding marks one of the earliest instances of climate-induced resettlement in the South Pacific. This relocation was not just a logistical challenge but also a deeply cultural and emotional one, as it involved moving away from customary lands that are integral to the community's identity and way of life. The relocation of Vunidogoloa highlighted the importance of inclusive engagement and the need to address the land-people bond, governance issues, and funding challenges in such resettlements (Charan et al., 2017; McNamara & Des Combes, 2015). Additionally, in Vanua Votua, the compounded effects of frequent flooding and coastal mining have severely disrupted local livelihoods, undermining the resilience of communities and highlighting the need for comprehensive strategies that address both environmental and socio-economic vulnerabilities (Varea et al., 2022).

Senegal

Coastal communities along the West African coast, including those in Senegal, are increasingly vulnerable to sea level rise, facing issues like erosion, flooding, and land loss. Adaptation efforts are urgently needed there, as these communities are already experiencing significant socio-economic impacts due to these changes (Daca et al., 2024). In Senegal, the use of satellite remote sensing has been instrumental in assessing the vulnerability of coastal regions to flooding, helping to inform better management practices (Mendoza et al., 2023).

Uruguay

In Uruguay, innovative community-based and ecosystem-based adaptation measures have been implemented to increase resilience to extreme events and sea level rise. These efforts include capacity building, introducing innovations in coastal management, and strengthening socio-institutional structures (Carro et al., 2018; Leal Filho et al., 2018).

4 Adaptation Challenges and Strategies

Adapting to the multifaceted impacts of sea level rise and coastal flooding presents numerous challenges that require a comprehensive, integrated approach. Coastal regions worldwide face unique vulnerabilities, which necessitate adaptation strategies tailored to local characteristics and conditions (Barnett et al., 2013). The complexity of these challenges is underscored by the need for a combination of structural and non-structural measures, as highlighted in Dedekorkut-Howes et al. (2020), who point out significant gaps in the comparative costs and benefits of various strategies. Infrastructure systems are particularly at risk, with many regions, especially those in developing countries, facing significant challenges in upgrading and maintaining their robust infrastructure due to limited financial resources (Mozumder et al., 2023).

Azevedo de Almeida and Mostafavi (2016) identify several critical adaptation measures, including protection, accommodation, and retreat, but also emphasize the significant implementation challenges these measures face. Additionally, economic constraints, as highlighted in Theokritoff et al. (2023), pose a significant barrier to effective adaptation, particularly in regions where resources are scarce. Furthermore, van den Hurk et al. (2022) discuss the importance of adaptive pathways and effective science-policy interfaces in places where uncertainty and socio-economic consequences add layers of complexity to adaptation efforts. The literature consistently indicates that while there are significant advances in understanding and planning for sea level rise, as evidenced by Toimil et al. (2020), there remains a need for increased observations, better projections, and more comprehensive risk frameworks to address the dynamic and interlinked challenges of coastal adaptation.

Moreover, social and cultural barriers, governance issues, the lack of comprehensive climate data and developed and developing countries' different capacities to respond to sea level rise further complicate global adaptation efforts (Barnett et al., 2013). Developed countries, with their advanced technological capabilities and robust financial resources, are generally better equipped to implement high-cost, sophisticated adaptation measures, such as engineered flood defences and advanced monitoring systems (Nicholls et al., 2014). In contrast, developing countries often face severe resource constraints, which limit their ability to invest in such infrastructure and force them to rely more heavily on low-cost, community-based, and ecosystem-based adaptation strategies (Leal Filho et al., 2018). Furthermore, the socio-economic impacts of sea level rise are likely to be more severe in developing countries, where a higher proportion of the population is directly dependent on climate-sensitive sectors such as agriculture and fisheries (Hens et al., 2018). This disparity highlights the need for international cooperation and support mechanisms to bridge the gap in adaptation capacities.

Taking these facts into consideration underscores the importance of tailoring adaptation strategies not only to the unique geographic and climatic conditions of each region but also to their specific socio-economic contexts (Dedekorkut-Howes et al., 2021). A multifaceted approach incorporating infrastructure development, ecosystem-based adaptation, community education, policy improvements, and, where necessary, relocation, is essential in this case (Bongarts Lebbe et al., 2021). Infrastructure development, including the construction of sea walls, levees, and drainage systems, remains a vital strategy for protecting vulnerable areas, particularly in regions with the financial resources to invest in such measures (Azevedo de Almeida & Mostafavi, 2016). Simultaneously, ecosystem-based adaptation, such as the restoration and preservation of mangroves and wetlands, provides natural barriers against sea level rise and storm surges while offering additional benefits like biodiversity conservation and carbon sequestration (Narayan et al., 2016).

While empowering communities through education and awareness (Kreller, 2021) and improving policy and governance structures (Lawrence et al., 2020) are crucial components of effective adaptation strategies, a detailed analysis of these aspects is beyond the scope of this chapter. This complex array of strategies highlights the need for a comprehensive and context-sensitive approach to adapting to sea level rise, as different regions and communities must navigate the diverse challenges of implementing these measures effectively (Magnan et al., 2022). This section will focus primarily on infrastructure and ecosystem-based adaptation, socio-economic adaptation, and managed retreat strategies, while acknowledging the broader context of community engagement and governance in adaptation efforts.

4.1 Infrastructure and Ecosystem-Based Adaptation

Infrastructure and ecosystem-based adaptations are crucial strategies in addressing sea level rise. Hard engineering solutions or *grey infrastructure*, such as seawalls, dikes, and storm surge barriers, provide robust and immediate protection against coastal flooding and erosion. However, these structures can be expensive and often have significant environmental impacts (Schoonees et al., 2019; Gittman et al., 2016). The substantial costs associated with infrastructure improvements and other adaptation measures often exceed local budgets, forcing difficult trade-offs between immediate needs and long-term adaptation planning. Despite these challenges, hard engineering remains critical, particularly where immediate protection is necessary (Azevedo de Almeida & Mostafavi, 2016). In contrast, soft engineering approaches, often utilizing green infrastructure, include techniques such as beach nourishment, dune restoration, and conservation of wetlands and mangroves. These strategies focus on enhancing natural resilience by working with natural processes, offering additional benefits like biodiversity conservation and carbon sequestration (Narayan et al., 2016). While generally having a lower environmental impact than hard infrastructure, their effectiveness can vary depending on local conditions and may require more time to establish (Cohn et al., 2021). These approaches are part of the broader category of nature-based solutions (NbS) for coastal adaptation.

Nature-based solutions (NbS) offer a sustainable approach by leveraging natural processes. Ecosystem-based adaptation (EbA) strategies, such as mangrove and wetland conservation, are a specific type of NbS focused on climate adaptation. These approaches can be more cost-effective than traditional structures, and they provide ongoing ecosystem services (Narayan et al., 2016). Mangroves and wetlands are particularly effective in reducing wave energy, stabilizing shorelines, and mitigating floods (Koh et al., 2018; Menéndez et al., 2020; Narayan et al., 2016). Coral reefs also play a crucial role, serving as natural breakwaters and supporting marine ecosystems critical for coastal livelihoods (Koh et al., 2018; Narayan et al., 2016). However, the effectiveness of these natural defences is highly context dependent. Their success relies on local environmental conditions and human factors, which often limits their applicability in urbanized areas with restricted space for large-scale habitat restoration (Saleh & Weinstein, 2016; Singhvi et al., 2022; Kuwae & Crooks, 2021).

Increasingly, the integration of soft and hard engineering into hybrid approaches is gaining popularity. These approaches leverage the strengths of both methods, providing robust protection while minimizing environmental impacts. For example, combining seawalls with mangrove restoration can enhance coastal resilience and provide ecological benefits (Kindeberg et al., 2022). The adaptive gradients framework developed by Hamin et al. (2018) advocates for such hybrid solutions that incorporate natural, nature-based, structural, and non-structural approaches (Menéndez et al., 2020).

Implementation of these integrated strategies requires to take under consideration local conditions and to develop collaborative planning. As Powell et al. (2018) note, the success of such strategies depends on effectively combining green and grey infrastructure while considering ecological and community benefits. However, *green infrastructure* is still not being utilized as extensively as traditional *grey infrastructure*, which indicates a gap in adopting more sustainable approaches (Menéndez et al., 2020). A balanced, integrated approach that combines the strengths of both hard and nature-based solutions that are tailored to the specific conditions and needs of each coastal area is essential. This approach can address the multifaceted challenges posed by sea level rise, particularly in diverse and complex coastal environments (Kuwae & Crooks, 2021).

4.2 Socio-Economic Adaptation

Socio-economic adaptation plays a key role in enhancing the resilience of coastal communities facing sea level rise. Socio-economic adaptations, including economic diversification, climate-resilient urban planning, and community engagement, are essential components of a comprehensive approach to coastal resilience. Economic diversification is a key strategy to reduce dependence on vulnerable sectors and build resilience against environmental changes. For instance, the development of the Indonesian seaweed industry has strengthened market resilience and provided alternative livelihoods (Tabrani et al., 2022; Langford, 2023). Similarly, the Blue Economy in West Africa, supported by earth observation services, offers promising avenues for sustainable fisheries, aquaculture, and marine tourism (Foli et al., 2022). A successful diversification, however, requires a supportive policy environment and strategic investments. Public-private partnerships (PPPs) can be instrumental in attracting foreign direct investment and promoting economic diversification, particularly in Small Island Developing States (SIDS), where economic vulnerabilities are often pronounced (Hinkel et al., 2018). These partnerships must be carefully designed to ensure inclusivity and contribute to urban regeneration without exacerbating socio-economic inequalities.

In addition, climate-resilient urban planning strategies are pivotal in adapting to sea level rise impacts, especially in densely populated coastal cities. Integrating decentralized water management into urban planning has proven effective in enhancing flood resilience and ensuring sustainable water resources. Coastal cities in the Netherlands and Taiwan have successfully implemented decentralized rainwater management systems, reducing flood risks and improving freshwater availability (Schuetze & Chelleri, 2013). Moreover, incorporating cultural heritage into urban planning can bolster economic resilience by preserving unique cultural assets, which in turn attract tourism and provide economic benefits, which is a strategy particularly relevant for SIDS (Allam & Jones, 2019; Hinkel et al., 2018).

The success of these urban planning strategies depends heavily on the involvement of local communities and stakeholders in the decision-making process. Bongarts Lebbe et al. (2021) emphasize that participatory engagement and multi-scalar governance are essential for designing and implementing effective coastal adaptation strategies. Community engagement is a cornerstone of successful adaptation strategies for coastal resilience. Engaging communities in the planning and implementation of adaptation measures ensures that local knowledge and needs are incorporated into the strategies, leading to sustainable outcomes (Lawlor, 2021). Participatory approaches, such as the use of Participatory GIS and stakeholder engagement, have been shown to build resilience by involving diverse community members in decision-making processes (Yusuf et al., 2018). This approach enhances the legitimacy and acceptance of adaptation measures. Education and awareness are also vital, as higher levels of knowledge empower communities to engage in decision-making processes and adopt innovative solutions to climate-related challenges (Bennett et al., 2016).

Finally, socio-economic factors significantly influence the adaptive capacity of coastal communities. Economic resources play a crucial role, as communities with ample financial means are better equipped to invest in infrastructure, technology, and other adaptation measures (Hinkel et al., 2018). However, many coastal communities, particularly in developing countries, face financial constraints that limit their ability to implement necessary adaptations. By addressing the socio-economic factors that influence adaptive capacity and involving communities in the adaptation process, coastal regions can better prepare for the impacts of sea level rise and ensure long-term sustainability.

4.3 Managed Retreat Strategies

Managed retreat, once considered a last resort, is increasingly recognized as a proactive and strategic approach to climate change adaptation in coastal areas (Haasnoot et al., 2021; Mach & Siders, 2021). This strategy involves the planned relocation of people and assets away from areas at high risk of sea level rise and associated hazards. However, implementing managed retreat is complex, as it involves multifaceted challenges that span social, economic, and cultural dimensions (Haasnoot et al., 2019).

A key consideration in managed retreat is the acquisition of properties in high-risk areas. This process often involves transferring development rights to safer locations, requiring careful planning and consideration of local government attitudes and potential challenges such as funding and community resistance (Robb et al., 2020). Successful implementation of managed retreat programmes depends heavily on community and stakeholder engagement. An early involvement of these groups in the planning process helps address concerns, build trust, and ensure that the retreat strategies align with community needs and values (Hashida & Dundas, 2024; Lauer et al., 2021). In addition, the social justice implications of managed retreat are significant and require careful attention. There are concerns that the decision-making processes in these programmes may disproportionately impact low-income or minority communities, potentially perpetuating existing social inequities (Siders, 2019).

Coastal cultural heritage preservation presents another challenge in managed retreat strategies. Balancing the need for retreat with the desire to maintain one's cultural identity requires innovative approaches, such as digitization to document and protect cultural assets, and engaging local communities in preservation efforts (Skublewska-Paszkowska et al., 2022; Stanco et al., 2011). This is particularly important for indigenous communities, whose traditional ways of life are often intrinsically linked to specific coastal ecosystems (Savo et al., 2016).

Financial considerations play a crucial role in managed retreat. Ensuring equitable outcomes requires providing adequate support for affected populations, including financial assistance and access to social services (Felipe Pérez & Tomaselli, 2021). Innovative approaches to this include using transferable development rights and exploring varied funding mechanisms to facilitate retreat (Mach et al., 2019; Robb et al., 2020). Implementing managed retreat also requires addressing the *uncertainty contagion*—the complex interplay of scientific, governance-related, financial, political, social, and cultural uncertainties that can lead to inaction or resistance (Hanna et al., 2020). However, in some cases, particularly where the risks are too great, planned relocation and resettlement of communities from high-risk areas to safer locations may be a necessity (Haasnoot et al., 2021).

Managed retreat, while challenging, offers coastal communities a unique opportunity to reimagine their future in the face of rising seas. This strategy, when approached with sensitivity and foresight, can lead to transformative adaptation by preserving communities, cultures, and ways of life, rather than merely relocating people and buildings (Mach & Siders, 2021). The path forward is complex, requiring careful navigation of property rights, genuine community engagement, and innovative financial solutions to ensure that the most vulnerable are not left behind (Siders, 2019; Felipe Pérez and Tomaselli, 2021). With thoughtful planning and a commitment to social justice, managed retreat can evolve from a last resort into a powerful tool for building resilient, sustainable coastal communities that adapt proactively to climate change.

4.4 Challenges in Implementation

Implementing sea level rise adaptation strategies presents a complex set of challenges that span financial, political, scientific, and social domains. These barriers often intersect and compound each other, and effectively overcoming them thus requires a multifaceted approach.

A critical issue highlighted in the literature is the technological limits to adaptation, which can constrain the effectiveness of measures designed to protect coastal communities. For instance, Hinkel et al. (2018) discuss how technological solutions are not always sufficient to address the complex and varied impacts of sea level rise across different regions, especially where infrastructure must adapt to rapidly changing conditions. This limitation is compounded by economic and financial barriers which restrict the ability of societies to invest in the necessary adaptive infrastructure, and this is the case particularly in less affluent regions where the cost of adaptation may be prohibitive.

The challenge of social conflicts also plays a significant role in hindering adaptation efforts. Conflicting interests within communities can impede the implementation of adaptation strategies, as different groups may prioritize different responses based on their own unique vulnerabilities and values. This issue is particularly acute in areas where adaptation measures, such as managed retreat or the construction of protective barriers, have uneven social impacts that can potentially exacerbate existing inequalities (Hinkel et al., 2018). Moreover, the differences in coastal and social characteristics across regions necessitate tailored approaches to adaptation, further complicating the implementation process as one-size-fits-all solutions are often inadequate (Mustonen et al., 2022).

The need for hybrid approaches and participatory policies in designing adaptive strategies is emphasized by Bongarts Lebbe et al. (2021), who note that integrating both natural and built infrastructure is essential for addressing the dynamic and long-term nature of sea level rise. However, the lack of concrete, implemented examples of effective coastal adaptation solutions presents a significant barrier. Most adaptation efforts are small-scale and uncoordinated, which limits their overall impact. Additionally, the lack of attention to coastal adaptation in the literature, particularly regarding the complexities of urban and coastal areas, further hampers the development of effective strategies (Dedekorkut-Howes et al., 2020).

Multicriteria decision analysis (MCDA) is a valuable tool in planning adaptive pathways, but its application is limited by the complex and uncertain nature of sea level rise projections (Skidmore & Cohon, 2023). As Haasnoot et al. (2019) indicate, adaptation options diminish as sea levels rise unless a proactive approach is adopted. However, the generic pathways proposed do not always consider local governance or socio-economic conditions, which are crucial for a successful implementation. This gap suggests that while MCDA can support flexible adaptation planning, it must be complemented by local knowledge and governance frameworks to be truly effective.

Short-term risk reduction strategies often dominate current coastal adaptation planning processes, as noted by Keeler et al. (2018). This focus can delay or inhibit the adoption of more proactive and discontinuous long-term strategies, such as relocation (Haasnoot et al., 2019). The emphasis on immediate risk reduction may lead to unsustainable investments and a lack of preparedness for more drastic future measures. Moreover, a failure to anticipate real estate market responses to these risk reduction measures could result in significant social and financial costs when more comprehensive adaptation strategies become necessary (Woodruff et al., 2020).

Lastly, the regional adaptation to sea level rise is hindered by the lack of boundary-spanning approaches that would engage multiple organizations and community stakeholders. John and Yusuf (2019) emphasize the importance of such approaches in overcoming barriers to effective climate adaptation. However, the absence of integrated efforts across different sectors and scales often leads to fragmented and ineffective adaptation responses, which underscores the need for better coordination and collaboration across various levels of governance (Biesbroek et al., 2013; Nalau et al., 2015). These challenges demonstrate the multifaceted nature of implementing adaptive strategies for sea level rise, and thus the problem requires a coordinated effort across technological, economic, social, and governance-related dimensions to ensure effective and equitable outcomes (Hinkel et al., 2018; Magnan et al., 2022).

5 Conclusions

This chapter has provided a comprehensive overview of sea level rise, integrating scientific evidence with socio-economic realities and adaptation strategies. By examining the current state of knowledge across these domains, it highlights the complex and interconnected nature of the challenges posed by sea level rise.

5.1 Synthesis of Key Findings

Scientific evidence clearly indicates an accelerating rise in sea levels that is predominantly due to human-induced climate change (Hamlington et al., 2024; Strassburg et al., 2015; Box et al., 2018). Between 1901 and 2018, global mean sea levels rose by 0.20 m, with the average rate of increase accelerating from 1.3 mm per year between 1901 and 1971 to 3.7 mm per year between 2006 and 2018 (Maitland et al., 2024; Wright et al., 2018). Satellite altimetry data for the period from 1993 to 2021 corroborates this trend, showing an average rise of 3.00 mm per year, with an acceleration of 0.12 mm per year (Guérou et al., 2023). Future projections, as outlined in the IPCC's Sixth Assessment Report (AR6), suggest a global mean sea level rise of 0.28–0.55 m by 2100 under a very low emission scenario, and 0.63–1.01 m under a very high emission scenario (IPCC, 2023). These projections emphasize the long-term nature of sea level rise, indicating increases that will continue for centuries, even if global warming is limited to 1.5 °C (Church & Gregory, 2019; Golledge, 2020).

The socio-economic impacts of sea level rise are multifaceted, and significantly affect coastal regions and communities (Durand et al., 2022; Cazenave & Moreira, 2022). Economically, the direct costs include damage to infrastructure, property loss, and the expenses associated with protective measures (Neumann et al., 2015a; b; Depsky et al., 2023), while the indirect costs involve long-term effects on economic activity, productivity, public finances, and disruptions to sectors, such as tourism and fisheries (Bosello et al., 2011; Newton et al., 2020; Pascoe et al., 2023). Socially, sea level rise leads to displacement, migration, and mental health challenges (Abu et al., 2024; Piggott-McKellar et al., 2020), with health risks escalating due to flooding, water contamination, and compromised healthcare infrastructure (Bloetscher et al., 2016). Food security is also threatened by saltwater intrusion into agricultural lands (Riera-Spiegelhalder et al., 2023). These impacts raise serious environmental justice concerns, as vulnerable communities disproportionately experience these challenges (Best et al., 2023; Martinich et al., 2013). Cultural heritage, especially for indigenous communities, is at risk due to the loss of ancestral lands and traditional knowledge (Aswani et al., 2019;

Savo et al., 2017). Addressing these issues requires a range of adaptation strategies, including infrastructure- and ecosystem-based approaches (Barnett et al., 2013; Narayan et al., 2016), socio-economic adaptations like economic diversification and climate-resilient urban planning (Tabrani et al., 2022; Schuetze & Chelleri, 2013), and managed retreat from high-risk areas (Haasnoot et al., 2021; Mach & Siders, 2021), all of which must be informed by a scientific understanding and tailored to specific local contexts (Dedekorkut-Howes et al., 2020; Magnan et al., 2022).

5.2 Future Directions and Considerations

Emerging trends in sea level rise research and policy demonstrate a growing shift towards more integrated, technologically advanced, and adaptive approaches. Cities are increasingly incorporating sophisticated climate projections into their urban planning processes, as evidenced by Caprario et al. (2022), who present a framework for integrating sea level rise scenarios into coastal city resilience planning. This trend is further supported by interdisciplinary approaches like urban design in combination with stakeholder engagement, which creates a holistic adaptation strategy (Meguro et al., 2024). Additionally, advanced technologies such as machine learning are being utilized to develop 'geographically differentiated climate change mitigation strategies', particularly in urban areas (Milojevic-Dupont & Creutzig, 2021). The integration of urban growth prediction models with future flood risk scenarios, as highlighted by Kim and Newman (2020), underscores the importance of dynamic scenario planning in creating more nuanced and adaptable long-term plans. Nicholls et al. (2021) further emphasize the need for regular updates to coastal risk and adaptation assessments so that the assessments would incorporate new sea level rise information reflecting the evolving nature of these challenges and the necessity for flexible, future-oriented policy approaches.

As the understanding of sea level rise continues to evolve, several key areas for further research and data collection have emerged, which could significantly enhance global responses to this challenge. Advances in satellite technology and machine learning are revolutionizing the measurement and projection of sea level rise. Hamlington et al. (2023) underscore the importance of satellite monitoring in providing crucial data indicators that support the development of dynamic adaptation policy pathways for coastal communities. However, Nieves et al. (2021)

note that while machine learning models are effective in predicting near-future regional sea level changes, especially in areas influenced by internal climate variability, there is still a need for more sophisticated data-driven techniques that would improve long-term projections. Another critical area for future research is improving extreme sea level (ESL) estimates. Wahl et al. (2017) point out that uncertainties in ESL estimates often exceed those in global sea level rise projections, which highlights the need for more accurate methodologies that would provide a complete picture of coastal risks and inform comprehensive adaptation planning.

Moreover, while global sea level rise projections have improved, there remains a pressing need for more accurate regional and local projections. Over the last decade, it has been demonstrated that these projections can vary significantly due to 'factors such as non-climatic background uplift or subsidence, oceanographic effects, and the geoid' and lithosphere's response to shrinking land ice (Kopp et al., 2014). Future research should focus on refining these projections by considering the complex interplay of various factors affecting regional sea levels (Vecchio et al., 2024). Additionally, the complex nature of sea level rise necessitates interdisciplinary research approaches. By integrating insights from climate science, oceanography, geology, urban planning, and social sciences, future studies can provide a more comprehensive understanding of sea level rise impacts and potential adaptation strategies. This holistic approach could lead to more effective and sustainable solutions for coastal communities.

The interdisciplinary nature of sea level rise research and policy is crucial for addressing the multifaceted challenges it presents. One significant approach involves modelling coastlines as coupled human-natural systems, integrating environmental and resource economics with other disciplines to better understand and predict the impacts of sea level rise. This method supports coastal climate adaptation policies that consider both market and non-market values, as well as spatial dynamics. Advances in artificial intelligence (AI) and machine learning, as demonstrated by Nieves et al. (2021), enhance the accuracy of sea level rise predictions by processing large datasets from various sources, thereby improving the precision of future projections. The development of leading indicators of resilience for critical infrastructure systems, focusing on structural characteristics that embody system resilience, provides insights for adaptive decision-making across multiple sub-systems and stakeholders. Additionally, citizen science and indigenous knowledge play crucial roles in understanding local sea level rise impacts and informing adaptation strategies, as these approaches empower local communities to contribute valuable on-the-ground data that can inform climate resilience efforts (Tengö et al., 2021).

References

Abu, M., Heath, S. C., Adger, W. N., Codjoe, S. N. A., Butler, C., & Quinn, T. (2024). Social Consequences of Planned Relocation in Response to Sea Level Rise: Impacts on Anxiety, Well-Being, and Perceived Safety. *Scientific Reports*, *14*(1), 3461. https://doi.org/10.1038/s41598-024-53277-9

Ahmed, B., Kelman, I., Fehr, H. K., & Saha, M. (2016). Community Resilience to Cyclone Disasters in Coastal Bangladesh. *Sustainability*, *8*(8), 805. https://doi.org/10.3390/su8080805

Albert, S., Leon, J. X., Grinham, A. R., Church, J. A., Gibbes, B. R., & Woodroffe, C. D. (2016). Interactions Between Sea Level Rise and Wave Exposure on Reef Island Dynamics in the Solomon Islands. *Environmental Research Letters*, *11*(5), Article 054011. https://doi.org/10.1088/1748-9326/11/5/054011

Ali, M. S., & Syfullah, K. (2017). Effect of Sea Level Rise Induced Permanent Inundation on the Livelihood of Polder Enclosed Beel Communities in Bangladesh: People's Perception. *Journal of Water and Climate Change*, *8*(2), 219–234. https://doi.org/10.2166/wcc.2016.236

Allam, Z., & Jones, D. (2019). Climate change and Economic Resilience through Urban and Cultural Heritage: The Case of Emerging Small Island Developing States Economies. *Economies*, *7*(2), 62. https://doi.org/10.3390/economies7020062

Anzidei, M., Doumaz, F., Vecchio, A., Serpelloni, E., Pizzimenti, L., Civico, R., Greco, M., Martino, G., & Enei, F. (2020). Sea Level Rise Scenario for 2100 AD in the Heritage Site of Pyrgi (Santa Severa, Italy). *Journal of Marine Science and Engineering*, *8*(2), 64. https://doi.org/10.3390/jmse8020064

Aswani, S., Howard, J. A. E., Gasalla, M. A., Jennings, S., Malherbe, W., Martins, I. M., Salim, S. S., Van Putten, I. E., Swathilekshmi, P. S., Narayanakumar, R., & Watmough, G. R. (2019). An Integrated Framework for Assessing Coastal Community Vulnerability across Cultures, Oceans and Scales. *Climate and Development*, *11*(4), 365–382. https://doi.org/10.1080/17565529.2018.1442795

AVISO+. (n.d.). *Mean Sea Level Products*. Retrieved July 12, 2024 from https://www.aviso.altimetry.fr/en/data/products/ocean-indicators-products/mean-sea-level/data-acces.html

Azevedo de Almeida, B., & Mostafavi, A. (2016). Resilience of Infrastructure Systems to Sea Level Rise in Coastal Areas: Impacts, Adaptation Measures, and

Implementation Challenges. *Sustainability*, *8*(11), 1115. https://doi.org/10.3390/su8111115

Bakker, A. M., Louchard, D., & Keller, K. (2017). Sources and Implications of Deep Uncertainties Surrounding Sea Level Projections. *Climatic Change*, *140*, 339–347. https://doi.org/10.1007/s10584-016-1864-1

Barbier, E. B. (2014). A Global Strategy for Protecting Vulnerable Coastal Populations. *Science*, *345*(6202), 1250–1251. https://doi.org/10.1126/science.1254629

Barnett, J., Waters, E., Pendergast, S., & Puleston, A. (2013). Barriers to Adaptation to Sea Level Rise. *National Climate Change Adaptation Research Facility*, *85*. https://nccarf.edu.au/wp-content/uploads/2019/03/Barnett_2013_Barriers_to_adaptation_to_sea_level_rise.pdf

Barnett, R. L., Charman, D. J., Johns, C., Ward, S. L., Bevan, A., Bradley, S. L., Camidge, K., Fyfe, R. M., Gehrels W. R., Gehrels, M. J., Hatton, J., Khan, N. S., Marshall, P., Maezumi, S. Y., Mills, S., Mulville, J., Pérez, M., Roberts, H. M., Scourse, J. D., Shepherd, F., & Stevens, T. (2020). Nonlinear Landscape and Cultural Response to Sea Level Rise. *Science Advances*, *6*(45). https://doi.org/10.1126/sciadv.abb6376

Beenstock, M., Felsenstein, D., Frank, E., & Reingewertz, Y. (2015). Tide Gauge Location and the Measurement of Global Sea Level Rise. *Environmental and Ecological Statistics*, *22*, 179–206. https://doi.org/10.1007/s10651-014-0293-4

Bennett, N. J., Blythe, J., Tyler, S., & Ban, N. C. (2016). Communities and Change in the Anthropocene: Understanding Social-Ecological Vulnerability and Planning Adaptations to Multiple Interacting Exposures. *Regional Environmental Change*, *16*, 907–926. https://doi.org/10.1007/s10113-015-0839-5

Best, K., He, Q., Reilly, A. C., Niemeier, D. A., Anderson, M., & Logan, T. (2023). Demographics and Risk of Isolation Due to Sea Level Rise in the United States. *Nature Communications*, *14*(1), 7904. https://doi.org/10.1038/s41467-023-43835-6

Biesbroek, G. R., Klostermann, J. E., Termeer, C. J., & Kabat, P. (2013). On the Nature of Barriers to Climate Change Adaptation. *Regional Environmental Change*, *13*(5), 1119–1129. https://doi.org/10.1007/s10113-013-0421-y

Bloetscher, F., Polsky, C., Bolter, K., Mitsova, D., Garces, K. P., King, R., Carballo, I. C., & Hamilton, K. (2016). Assessing Potential Impacts of Sea Level Rise on Public Health and Vulnerable Populations in Southeast Florida and Providing a Framework to Improve Outcomes. *Sustainability*, *8*(4), 315. https://doi.org/10.3390/su8040315

Bongarts Lebbe, T., Rey-Valette, H., Chaumillon, É., Camus, G., Almar, R., Cazenave, A., Claudet, J., Rocle, N., Meur-Férec, C., Viard, F., Mercier, D., Dupuy, C., Ménard, F., Rossel, B. A., Mullineaux, L., Sicre, M.-A., Zivian,

A., Gaill, F., & Euzen, A. (2021). Designing Coastal Adaptation Strategies to Tackle Sea Level Rise. *Frontiers in Marine Science, 8*. https://doi.org/10.3389/fmars.2021.740602

Bosello, F., Roson, R., Nicholls, R. J., Richards, J., & Tol, R. S. J. (2011). Economic Impacts of Climate Change in Europe: Sea Level Rise. *Climatic Change, 112*(1), 63–81. https://doi.org/10.1007/s10584-011-0340-1

Box, J. E., Thomson, L. I., Burgess, D. O., Wouters, B., O'Neel, S., Mernild, S. H., & Colgan, W. T. (2018). Global Sea-Level Contribution from Arctic Land Ice: 1971–2017. *Environmental Research Letters, 13*(12), Article 125012. https://doi.org/10.1088/1748-9326/aaf2ed

Caprario, J., Tasca, F. A., Santana, P. L., Azevedo, L. T. S., & Finotti, A. R. (2022). Framework for Incorporating Climate Projections in the Integrated Planning and Management of Urban Infrastructure. *Urban Climate, 41*(3). https://doi.org/10.1016/j.uclim.2021.101060

Carro, I., Seijo, L., Nagy, G. J., Lagos, X., & Gutiérrez, O. (2018). Building Capacity on Ecosystem-Based Adaptation Strategy to Cope with Extreme Events and Sea Level Rise on the Uruguayan Coast. *International Journal of Climate Change Strategies and Management, 10*(4), 504–522. https://doi.org/10.1108/IJCCSM-07-2017-0149

Cazenave, A., & Cozannet, G. L. (2014). Sea Level Rise and Its Coastal Impacts. *Earth's Future, 2*(2), 15–34. https://doi.org/10.1002/2013EF000188

Cazenave, A., & Remy, F. (2011). Sea Level and Climate: Measurements and Causes of Changes. *Wiley Interdisciplinary Reviews: Climate Change, 2*(5), 647–662. https://doi.org/10.1002/wcc.139

Cazenave A., Moreira L. (2022) Contemporary sea-level changes from global to local scales: a review. *Proceedings of the Royal Society A: Mathematical Physical and Engineering Sciences 478*(2261) https://doi.org/10.1098/rspa.2022.0049

Charan, D., Kaur, M., & Singh, P. (2017). Customary Land and Climate Change Induced Relocation: A Case Study of Vunidogoloa Village, Vanua Levu, Fiji. *Climate Change Adaptation in Pacific Countries*, 19–33. https://doi.org/10.1007/978-3-319-50094-2_2

Chen, Z., Gao, Q., Li, X., Yang, X., & Wang, Z. (2024). Beyond Inundation: A Comprehensive Assessment of Sea Level Rise Impact on Coastal Cultural Heritage in China. *Heritage Science, 12*(1), 121. https://doi.org/10.1186/s40494-024-01233-1

Church, J. A., & Gregory, J. M. (2019). Sea Level Change. In J. K. Cochran, H. J. Bokuniewicz, & P. L. Yager (Eds.), *Encyclopedia of Ocean Sciences* (3rd ed., pp. 493–499). Academic Press. https://doi.org/10.1016/B978-0-12-409548-9.10820-6

Cohn, J. L., Copp Franz, S., Mandel, R. H., Nack, C. C., Brainard, A. S., Eallonardo, A., & Magar, V. (2021). "Strategies to work towards long-term sustainability and resiliency of nature-based solutions in coastal environments: a review and case studies", *Integrated Environmental Assessment and Management* 18(1):123–134. https://doi.org/10.1002/ieam.4484

Le Cozannet, G., Manceau, J. C., & Rohmer, J. (2017). Bounding Probabilistic Sea Level Projections within the Framework of the Possibility Theory. *Environmental Research Letters*, 12(1), Article 014012. https://doi.org/10.1088/1748-9326/aa5528

Dada, O. A., Almar, R., & Morand, P. (2024). Coastal Vulnerability Assessment of the West African Coast to Flooding and Erosion. *Scientific Reports*, 14(1), 890. https://doi.org/10.1038/s41598-023-48612-5

Danylchuk, A. J., Griffin, L. P., Ahrens, R., Allen, M. S., Boucek, R. E., Brownscombe, J. W., Casselberry, G. A., Danylchuk, S. C., Filous, A., Goldberg, T. L., Perez, A. U., Rehage, J. S., Santos, R. O., Shenker, J., Wilson, J. K., Adams, A. J., & Cooke, S. J. (2023). Cascading Effects of Climate Change on Recreational Marine Flats Fishes and Fisheries. *Environmental Biology of Fishes*, 106(2), 381–416. https://doi.org/10.1007/s10641-022-01333-6

Dasgupta, S., Huq, M., Khan, Z. H., Ahmed, M. M. Z., Mukherjee, N., Khan, M. F., & Pandey, K. (2013). Cyclones in a Changing Climate: The Case of Bangladesh. *Climate and Development*, 6(2), 96–110. https://doi.org/10.1080/17565529.2013.868335

Dedekorkut-Howes, A., Torabi, E., & Howes, M. (2020). When the Tide Gets High: A Review of Adaptive Responses to Sea Level Rise and Coastal Flooding. *Journal of Environmental Planning and Management*, 63(12), 2102–2143. https://doi.org/10.1080/09640568.2019.1708709

Dedekorkut-Howes, A., Torabi, E., & Howes, M. (2021). Planning for a Different Kind of Sea Change: Lessons from Australia for Sea Level Rise and Coastal Flooding. *Climate Policy*, 21(2), 152–170. https://doi.org/10.1080/14693062.2020.1819766

Depsky, N., Bolliger, I., Allen, D., Choi, J. H., Delgado, M., Greenstone, M., Hamidi, A., Houser, T., Kopp, R. E., & Hsiang, S. (2023). DSCIM-Coastal v1. 1: An Open-Source Modeling Platform for Global Impacts of Sea Level Rise. *Geoscientific Model Development*, 16(14), 4331–4366. https://doi.org/10.5194/gmd-16-4331-2023

Durand, G., van den Broeke, M. R., Le Cozannet, G., Edwards, T. L., Holland, P. R., Jourdain, N. C., Marzeion, B., Mottram, R., Nicholls, R. J., Pattyn, F., Paul, F., Slangen, A. B. A., Winkelmann, R., Burgard, C., Van Calcar, C. J., Barré, J.-B., Bataille, A., & Chapuis, A. (2022). Sea Level Rise: From Global Perspectives to Local Services. *Frontiers in Marine Science*, 8. https://doi.org/10.3389/fmars.2021.709595

Felipe Pérez, B., & Tomaselli, A. (2021). Indigenous Peoples and Climate-Induced Relocation in Latin America and the Caribbean: Managed Retreat as a Tool or a Threat? *Journal of Environmental Studies and Sciences, 11*(3), 352–364. https://doi.org/10.1007/s13412-021-00693-2

Foli, B. A. K., Williams, I. K., Boakye, A. A., Azumah, D. M. Y., Agyekum, K. A., & Wiafe, G. (2022). Earth Observation Services in Support of West Africa's Blue Economy: Coastal Resilience and Climate Impacts. *Remote Sensing in Earth Systems Sciences, 5*, 59–70. https://doi.org/10.1007/s41976-021-00058-x

Gade, K. (2022). Sea Level Rise and Its Impacts on Indian Coastal Cultural Heritage: Climate Change Threats to Unique Built and Unbuilt Cultural Heritage. *International Journal of Architectural Heritage, 5*(1), 1–9. https://doi.org/10.37628/ijah.v5i1.886

Gadi, R., Rignot, E., & Menemenlis, D. (2023). "Modeling ice melt rates from seawater intrusions in the grounding zone of petermann gletscher, greenland", *Geophysical Research Letters 50*(24), https://doi.org/10.1029/2023gl105869

Gittman, R. K., Scyphers, S. B., Smith, C. S., Neylan, I. P., & Grabowski, J. H. (2016). 'Ecological consequences of shoreline hardening: a meta-analysis', *BioScience 66*(9):763–773. https://doi.org/10.1093/biosci/biw091

Golledge, N. R. (2020). Long-Term Projections of Sea-Level Rise from Ice Sheets. *Wiley Interdisciplinary Reviews: Climate Change, 11*(2). https://doi.org/10.1002/wcc.634

Guérou, A., Meyssignac, B., Prandi, P., Ablain, M., Ribes, A., & Bignalet-Cazalet, F. (2023). Current Observed Global Mean Sea Level Rise and Acceleration Estimated From Satellite Altimetry and the Associated Measurement Uncertainty. *Ocean Science, 19*(2), 431–451. https://doi.org/10.5194/os-19-431-2023

Haasnoot, M., Brown, S., Scussolini, P., Jimenez, J. A., Vafeidis, A. T., & Nicholls, R. J. (2019). Generic Adaptation Pathways for Coastal Archetypes under Uncertain Sea-Level Rise. *Environmental Research Communications, 1*(7). https://doi.org/10.1088/2515-7620/ab1871

Haasnoot, M., Lawrence, J., & Magnan, A. K. (2021). Pathways to Coastal Retreat. *Science, 372*(6548), 1287–1290. https://doi.org/10.1126/science.abi6594

Hamlington, B. D., Gardner, A. S., Ivins, E., Lenaerts, J. T., Reager, J. T., Trossman, D. S., Zaron, E. D., Adhikari, S., Arendt, A., Aschwanden, A., Beckley, B. D., Bekaert, D. P. S., Blewitt, G., Caron, L., Chambers, D. P., Chandanpurkar, H. A., Christianson, K., Csatho, B., Cullather, R. I., & Willis, M. J. (2020). Understanding of Contemporary Regional Sea-Level Change and the Implications for the Future. *Reviews of Geophysics, 58*(3). https://doi.org/10.1029/2019RG000672

Hamlington, B. D., Tripathi, A., Rounce, D. R., Weathers, M., Adams, K. H., Blackwood, C., Carter, J., Collini, R. C., Engemann, L., Haasnoot, M., &

Kopp, R. E. (2023). Satellite Monitoring for Coastal Dynamic Adaptation Policy Pathways. *Climate Risk Management*, 42. https://doi.org/10.1016/j.crm.2023.100555

Hanna, C., White, I., & Glavovic, B. (2020). The Uncertainty Contagion: Revealing the Interrelated, Cascading Uncertainties of Managed Retreat. *Sustainability*, 12(2), 736. https://doi.org/10.3390/su12020736

Hansen, J., Sato, M., Russell, G., & Kharecha, P. (2013). Climate Sensitivity, Sea Level and Atmospheric Carbon Dioxide. *Philosophical Transactions of the Royal Society A: Mathematical, Physical and Engineering Sciences*, 371(2001). https://doi.org/10.1098/rsta.2012.0294

Hashida, Y., & Dundas, S. J. (2024). Barriers to Coastal Managed Retreat: Evidence from New Jersey's Blue Acres Program. *Marine Resource Economics*, 39(3), 179–205. https://doi.org/10.1086/729868

Hamlington, B. D., Bellas-Manley, A., Willis, J. K., Fournier, S., Vinogradova, N., Nerem, R. S., Piecuch, C. G., et al., 2024. "The rate of global sea level rise doubled during the past three decades", *Communications Earth and Environment* 5(1), https://doi.org/10.1038/s43247-024-01761-5

Hamin, E. M., Abunnasr, Y., Roman Dilthey, M., Judge, P. K., Kenney, M. A., Kirshen, P., Sheahan, T. C., et al. (2018). "Pathways to coastal resiliency: the adaptive gradients framework", *Sustainability* 10(8):2629. https://doi.org/10.3390/su10082629

Hens, L., Thinh, N. A., Hanh, T. H., Cuong, N. S., Lan, T. D., Van Thanh, N., & Le, D. T. (2018). Sea-Level Rise and Resilience in Vietnam and the Asia-Pacific: A Synthesis. *Vietnam Journal of Earth Sciences*, 40(2), 126–152. https://doi.org/10.15625/0866-7187/40/2/11107

Hinkel, J., Aerts, J. C., Brown, S., Jiménez, J. A., Lincke, D., Nicholls, R. J., Scussolini, P., Sanchez-Arcilla, A., Vafeidis, A., & Addo, K. A. (2018). The Ability of Societies to Adapt to Twenty-First-Century Sea-Level Rise. *Nature Climate Change*, 8(7), 570–578. https://doi.org/10.1038/s41558-018-0176-z

Hinkel, J., Jaeger, C., Nicholls, R. J., Lowe, J., Renn, O., & Shi, P. (2015). Sea-Level Rise Scenarios and Coastal Risk Management. *Nature Climate Change*, 5(3), 188–190. https://doi.org/10.1038/nclimate2505

Hummel, M. A., Griffin, R., Arkema, K. K., & Guerry, A. D., (2021). "Economic evaluation of sea-level rise adaptation strongly influenced by hydrodynamic feedbacks", *Proceedings of the National Academy of Sciences* 118(29), https://doi.org/10.1073/pnas.2025961118

Van den Hurk, B., Bisaro, A., Haasnoot, M., Nicholls, R. J., Rehdanz, K., & Stuparu, D. (2022). Living With Sea-Level Rise in North-West Europe: Science-Policy Challenges Across Scales. *Climate Risk Management*, 35. https://doi.org/10.1016/j.crm.2022.100403

Intergovernmental Panel on Climate Change (IPCC). (2023). *Climate Change 2023: Synthesis Report, Summary for Policymakers.* Contribution of Working Groups I, II and III to the Sixth Assessment Report (pp. 1–34). https://doi.org/10.59327/ipcc/ar6-9789291691647.001

Jevrejeva, S., Jackson, L. P., Grinsted, A., Lincke, D., & Marzeion, B. (2018). Flood Damage Costs Under the Sea Level Rise With Warming of 1.5 C and 2 C. *Environmental Research Letters, 13*(7), Article 074014. https://doi.org/10.1088/1748-9326/aacc76

John, B. S. III., & Yusuf, J. E. (2019). Perspectives of the Expert and Experienced on Challenges to Regional Adaptation for Sea Level Rise: Implications for Multisectoral Readiness and Boundary Spanning. *Coastal Management, 47*(2), 151–168. https://doi.org/10.1080/08920753.2019.1564951

Keeler, A. G., McNamara, D. E., & Irish, J. L. (2018). Responding to Sea Level Rise: Does Short-Term Risk Reduction Inhibit Successful Long-Term Adaptation? *Earth's Future, 6*(4), 618–621. https://doi.org/10.1002/2018EF000828

Kim, Y., & Newman, G. (2020). Advancing Scenario Planning Through Integrating Urban Growth Prediction with Future Flood Risk Models. *Computers, Environment and Urban Systems, 82.* https://doi.org/10.1016/j.compenvurbsys.2020.101498

Kindeberg, T., Almström, B., Skoog, M., Olsson, P. A., Hollander, J. (2022). "Toward a multifunctional nature-based coastal defense: a review of the interaction between beach nourishment and ecological restoration", *Nordic Journal of Botany*(1), 2023. https://doi.org/10.1111/njb.03751

Koh, H. L., Teh, S. Y., Kh'Ng, X. Y., & Raja Barizan, R. S. (2018). Mangrove Forests: Protection Against and Resilience to Coastal Disturbances. *Journal of Tropical Forest Science, 30*(5), 446–460. https://doi.org/10.26525/jtfs2018.30.5.446460

Van Kooten, G. C. (2013). How Economists Measure Wellbeing: Social Cost-Benefit Analysis. In *Climate Change, Climate Science and Economics* (pp. 49–65). Springer. https://doi.org/10.1007/978-94-007-4988-7_6

Kopp, R. E., Garner, G. G., Hermans, T. H., Jha, S., Kumar, P., Slangen, A. B., Turilli, M., Edwards, T. L., Gregory, J. M., Koubbe, G., Levermann, A., Merzky, A., Nowicki, S., Palmer, M. D., & Smith, C. (2023). The Framework for Assessing Changes to Sea-Level (FACTS) v1.0-rc: A Platform for Characterizing Parametric and Structural Uncertainty in Future Global, Relative, and Extreme Sea-Level Change. *EGUsphere*, 1–34. https://doi.org/10.5194/egusphere-2023-14

Kopp, R. E., Horton, R. M., Little, C. M., Mitrovica, J. X., Oppenheimer, M., Rasmussen, D. J., Strauss, B. H., & Tebaldi, C. (2014). Probabilistic 21st and 22nd Century Sea-Level Projections at a Global Network of Tide-Gauge Sites. *Earth's Future, 2*(8), 383–406. https://doi.org/10.1002/2014EF000239

Kopp, R. E., Strauss, B. H., Bader, D. A., Kulp, S., Deconto, R. M., Oppenheimer, M., Horton, R. M., Hay, C. C., & Pollard, D. (2017). Evolving Understanding of Antarctic Ice-Sheet Physics and Ambiguity in Probabilistic Sea-Level Projections. *Earth's Future, 5*(12), 1217–1233. https://doi.org/10.1002/2017ef000663

Kreller, A. M. (2021). Transforming Fair Decision-Making about Sea-Level Rise in Cities: The Values and Beliefs of Residents in Botany Bay, Australia. *Environmental Values, 30*(1), 7–42. https://doi.org/10.3197/096327120X15752810323959

Kuwae, T., & Crooks, S. (2021). Linking Climate Change Mitigation and Adaptation through Coastal Green–Gray Infrastructure: A Perspective. *Coastal Engineering Journal, 63*(3), 188–199. https://doi.org/10.1080/21664250.2021.1935581

Langford, A., (2023). "Globalisation and livelihood transformations in the indonesian seaweed industry", https://doi.org/10.4324/9781003183860

Lauer, H., Delos Reyes, M., & Birkmann, J. (2021). Managed Retreat as Adaptation Option: Investigating Different Resettlement Approaches and Their Impacts—Lessons From Metro Manila. *Sustainability, 13*(2), 829. https://doi.org/10.3390/su13020829

Lambeck, K., Rouby, H., Purcell, A., Sun, Y., & Sambridge, M. (2014). Sea Level and Global Ice Volumes from the Last Glacial Maximum to the Holocene. *Proceedings of the National Academy of Sciences, 111*(43), 15296–15303. https://doi.org/10.1073/pnas.1411762111

Lawlor, P. (2021). The Role of the Community and Voluntary Sector in Identifying Vulnerabilities to Climate Change in Coastal Areas and Implementing Climate Adaptation Responses. *Administration, 69*(4), 83–108. https://doi.org/10.2478/admin-2021-0029

Lawrence, J., Boston, J., Bell, R., Olufson, S., Kool, R., Hardcastle, M., & Stroombergen, A. (2020). Implementing Pre-Emptive Managed Retreat: Constraints and Novel Insights. *Current Climate Change Reports, 6*, 66–80. https://doi.org/10.1007/s40641-020-00161-z

Leal Filho, W., Modesto, F., Nagy, G. J., Saroar, M., YannickToamukum, N., & Ha'apio, M. (2018). Fostering Coastal Resilience to Climate Change Vulnerability in Bangladesh, Brazil, Cameroon and Uruguay: A Cross-Country Comparison. *Mitigation and Adaptation Strategies for Global Change, 23*, 579–602. https://doi.org/10.1007/s11027-017-9750-3

Leon, J. X., Hardcastle, J., James, R., Albert, S., Kereseka, J., & Woodroffe, C. D. (2015). Supporting Local and Traditional Knowledge with Science for Adaptation to Climate Change: Lessons Learned from Participatory Three-Dimensional Modeling in BoeBoe, Solomon Islands. *Coastal Management, 43*(4), 424–438. https://doi.org/10.1080/08920753.2015.1046808

Li, Y., Jia, X., Liu, Z., Zhao, L., Sheng, P., & Storozum, M. J. (2022). The Potential Impact of Rising Sea Levels on China's Coastal Cultural Heritage: A GIS Risk Assessment. *Antiquity*, *96*(386), 406–421. https://doi.org/10.15184/aqy.2022.1

Ma, Z., Chen, N., Yang, J., & Chen, N. (2015). Coastal Sea Level Projections with Improved Accounting for Vertical Land Motion. *Scientific Reports*, *5*(1). https://doi.org/10.1038/srep16085

Mach, K. J., Kraan, C. M., Hino, M., Siders, A. R., Johnston, E. M., & Field, C. B. (2019). Managed Retreat Through Voluntary Buyouts of Flood-Prone Properties. *Science Advances*, *5*(10). https://doi.org/10.1126/sciadv.aax8995

Mach, K. J., & Siders, A. R. (2021). Reframing Strategic, Managed Retreat for Transformative Climate Adaptation. *Science, 372*(6548), 1294–1299. https://doi.org/10.1126/science.abh1894

Magnan, A. K., Oppenheimer, M., Garschagen, M., Buchanan, M. K., Duvat, V. K., Forbes, D. L., Lambert, E., Petzold, J., Renaud, F. G., Sebesvari, Z., Van der Wal, R. S. W., Hinkel, J., & Pörtner, H.-O. (2022). Sea Level Rise Risks and Societal Adaptation Benefits in Low-Lying Coastal Areas. *Scientific Reports*, *12*(1), Article 10677. https://doi.org/10.1038/s41598-022-14303-w

Maitland, D. O., Taylor, M. A., Raj, R. P., Stephenson, T. S., Bonaduce, A., Nisancioglu, K. H., & Richter, K. (2024). Determining Sea-Level Rise in the Caribbean: A Shift from Temperature to Mass Control. *Scientific Reports*, *14*(1). https://doi.org/10.1038/s41598-024-60201-8

Marcos, M., Wöppelmann, G., Matthews, A., Ponte, R. M., Birol, F., Ardhuin, F., Coco, G., Gómez, A. S., Ballu, V., Testut, L., Chambers, D., & Stopa, J. E. (2019). Coastal Sea Level and Related Fields from Existing Observing Systems. *Surveys in Geophysics*, *40*, 1293–1317. https://doi.org/10.1007/s10712-019-09513-3

Martinich, J., Neumann, J., Ludwig, L., & Jantarasami, L. (2013). Risks of Sea Level Rise to Disadvantaged Communities in the United States. *Mitigation and Adaptation Strategies for Global Change*, *18*, 169–185. https://doi.org/10.1007/s11027-011-9356-0

McNamara, K. E., & Des Combes, H. J. (2015). Planning for Community Relocations Due to Climate Change in Fiji. *International Journal of Disaster Risk Science*, *6*, 315–319. https://doi.org/10.1007/s13753-015-0065-2

Meguro, W., Briones, J., Failano, G., & Fletcher, C. H. (2024). A Science and Community-Driven Approach to Illustrating Urban Adaptation to Coastal Flooding to Inform Management Plans. *Sustainability*, *16*(7). https://doi.org/10.3390/su16072849

Mendoza, E. T., Salameh, E., Sakho, I., Turki, I., Almar, R., Ojeda, E., Deloffre, J., Frappant, F., & Laignel, B. (2023). Coastal Flood Vulnerability Assessment: A Satellite Remote Sensing and Modeling Approach. *Remote Sensing Applications: Society and Environment, 29.* https://doi.org/10.1016/j.rsase.2023.100923

Menéndez, P., Losada, I. J., Torres-Ortega, S., Narayan, S., & Beck, M. W. (2020). The Global Flood Protection Benefits of Mangroves. *Scientific Reports, 10*(1), 1–11. https://doi.org/10.1038/s41598-020-61136-6

Mills, L., Janeiro, J., & Martins, F. (2019). The Impact of Sea Level Rise in the Guadiana Estuary. In J. M. F. Rodrigues, P. J. S. Cardoso, J. Monteiro, R. Lam, V. V. Krzhizhanovskaya, M. H. Lees, J. J. Dongarra, & P. M. A. Sloot (Eds.), *Computational Science: ICCS 2019* (Vol. 11539, pp. 287–300). Springer. https://doi.org/10.1007/978-3-030-22747-0_23

Milojevic-Dupont, N., & Creutzig, F. (2021). Machine Learning for Geographically Differentiated Climate Change Mitigation in Urban Areas. *Sustainable Cities and Society, 64.* https://doi.org/10.1016/j.scs.2020.102526

Mozumder, M. M. H., Schneider, P., Islam, M. M., Deb, D., Hasan, M., Monzer, M. A., & Nur, A. A. U. (2023). Climate Change Adaptation Strategies for Small-Scale Hilsa Fishers in the Coastal Area of Bangladesh: Social, Economic, and Ecological Perspectives. *Frontiers in Marine Science, 10.* https://doi.org/10.3389/fmars.2023.1151875

Mustonen, T., Harper, S., Peci, G., Castan Broto, V., Lansbury, N., Okem, A., Ayanlade, S., Ayanlade, A., & Dawson, J. (2022). The Role of Indigenous Knowledge and Local Knowledge in Understanding and Adapting to Climate Change. *IPCC Climate Change,* 2713–2807. https://www.researchgate.net/publication/362432216

Nalau, J., Preston, B. L., & Maloney, M. C. (2015). Is Adaptation a Local Responsibility? *Environmental Science & Policy, 48,* 89–98. https://doi.org/10.1016/j.envsci.2014.12.011

Narayan, S., Beck, M. W., Reguero, B. G., Losada, I. J., Van Wesenbeeck, B., Pontee, N., Sanchirico, J. N., Ingram, J. C., Lange, G.-M., & Burks-Copes, K. A. (2016). The Effectiveness, Costs and Coastal Protection Benefits of Natural and Nature-Based Defences. *PloS One, 11*(5). https://doi.org/10.1371/journal.pone.0154735

Natarov, S. I., Merrifield, M. A., Becker, J. M., & Thompson, P. R. (2017). Regional Influences on Reconstructed Global Mean Sea Level. *Geophysical Research Letters, 44*(7), 3274–3282. https://doi.org/10.1002/2016GL071523

Natesan, U., & Parthasarathy, A. (2010). The Potential Impacts of Sea Level Rise along the Coastal Zone of Kanyakumari District in Tamilnadu, India. *Journal of Coastal Conservation, 14*(3), 207–214. https://doi.org/10.1007/s11852-010-0103-6

Neumann, J. E., Emanuel, K., Ravela, S., Ludwig, L., Kirshen, P., Bosma, K., & Martinich, J. (2015b). Joint Effects of Storm Surge and Sea-Level Rise on US Coasts: New Economic Estimates of Impacts, Adaptation, and Benefits of Mitigation Policy. *Climatic Change*, *129*, 337–349. https://doi.org/10.1007/s10584-014-1304-z

Neumann, B., Vafeidis, A. T., Zimmermann, J., & Nicholls, R. J. (2015a). Future Coastal Population Growth and Exposure to Sea-Level Rise and Coastal Flooding: A Global Assessment. *PloS One*, *10*(3). https://doi.org/10.1371/journal.pone.0118571

Newton, A., Icely, J., Cristina, S., Perillo, G. M., Turner, R. E., Ashan, D., Cragg, S., Luo, Y., Tu, C., Li, Y., Zhang, H., Ramesh, R., Forbes, D. L., Solidoro, C., Béjaoui, B., Gao, S., Pastres, R., Kelsey, H., Taillie, D., & Kuenzer, C. (2020). Anthropogenic, Direct Pressures on Coastal Wetlands. *Frontiers in Ecology and Evolution*, *8*, 144. https://doi.org/10.3389/fevo.2020.00144

Nicholls, R., Hanson, S., Lowe, J., Slangen, A., Wahl, T., Hinkel, J., & Long, A. (2021). Integrating New Sea-Level Scenarios into Coastal Risk and Adaptation Assessments: An Ongoing Process. *WIREs Climate Change*, *12*(6). https://doi.org/10.1002/wcc.706

Nicholls, R. J., Hanson, S. E., Lowe, J. A., Warrick, R. A., Lu, X., & Long, A. J. (2014). Sea-Level Scenarios for Evaluating Coastal Impacts. *Wiley Interdisciplinary Reviews: Climate Change*, *5*(1), 129–150. https://doi.org/10.1002/wcc.253

Nieves, V., Radin, C., & Camps-Valls, G. (2021). Predicting Regional Coastal Sea Level Changes With Machine Learning. *Scientific Reports*, *11*(1), 7650. https://doi.org/10.1038/s41598-021-87460-z

Nováčková, M., & Tol, R. S. J. (2017). Effects of Sea Level Rise on Economy of the United States. *Journal of Environmental Economics and Policy*, *7*(1), 85–115. https://doi.org/10.1080/21606544.2017.1363667

Parrado, R., Bosello, F., Delpiazzo, E., Hinkel, J., Lincke, D., & Brown, S. (2020). Fiscal Effects and the Potential Implications on Economic Growth of Sea-Level Rise Impacts and Coastal Zone Protection. *Climatic Change*, *160*(2), 283–302. https://link.springer.com/article/10.1007/s10584-020-02664-y

Pascoe, S., Paredes, S., & Coglan, L. (2023). The Indirect Economic Contribution of Fisheries to Coastal Communities through Tourism. *Fishes*, *8*(3), 138. https://doi.org/10.3390/fishes8030138

Pathak, A., van Beynen, P. E., Akiwumi, F. A., & Lindeman, K. C. (2021a). Climate Adaptation Within the Tourism Sector of a Small Island Developing State: A Case Study From the Coastal Accommodations Subsector in the Bahamas. *Business Strategy & Development*, *4*(3), 313–325. https://doi.org/10.1002/bsd2.160

Pathak, A., van Beynen, P. E., Akiwumi, F. A., & Lindeman, K. C. (2021b). Impacts of Climate Change on the Tourism Sector of a Small Island Developing State: A Case Study for the Bahamas. *Environmental Development*, *37*. https://doi.org/10.1016/j.envdev.2020.100556

Piggott-McKellar, A. E., Pearson, J., McNamara, K. E., & Nunn, P. D. (2020). A Livelihood Analysis of Resettlement Outcomes: Lessons for Climate-Induced Relocations. *Ambio*, *49*(9), 1474–1489. https://doi.org/10.1007/s13280-019-01289-5

Powell, E. J., Tyrrell, M. C., Milliken, A., Tirpak, J. M., & Staudinger, M. D. (2018). A review of coastal management approaches to support the integration of ecological and human community planning for climate change. *Journal of Coastal Conservation* *23*(1), 1–18. https://doi.org/10.1007/s11852-018-0632-y

Pycroft, J., Abrell, J., & Ciscar, J. C. (2016). The Global Impacts of Extreme Sea-Level Rise: A Comprehensive Economic Assessment. *Environmental and Resource Economics*, *64*, 225–253. https://doi.org/10.1007/s10640-014-9866-9

Ray, R. D., Widlansky, M. J., Genz, A. S., & Thompson, P. R. (2023). Offsets in Tide-Gauge Reference Levels Detected by Satellite Altimetry: Ten Case Studies. *Journal of Geodesy*, *97*(12), 110. https://doi.org/10.1007/s00190-023-01800-7

Reimann, L., Vafeidis, A. T., Brown, S., Hinkel, J., & Tol, R. S. (2018). Mediterranean UNESCO World Heritage at Risk from Coastal Flooding and Erosion Due to Sea-Level Rise. *Nature Communications*, *9*(1), 4161. https://doi.org/10.1038/s41467-018-06645-9

Richmond, L., & Kunkel, K. (2024). Living in the 'Blue Zone' of a Sea-Level Rise Inundation Map: Community Perceptions of Coastal Flooding in King Salmon, California. *Climate Risk Management*, *44*. https://doi.org/10.1016/j.crm.2024.100596

Riera-Spiegelhalder, M., Arampatzis, S., Enseñado, E. M., Campos-Rodrigues, L., Dekker-Arlain, J. D., Vervoort, K., & Papadopoulou, O. (2023). Socio-Economic Assessment of Ecosystem-Based and Other Adaptation Strategies in Coastal Areas: A Systematic Review. *Journal of Marine Science and Engineering*, *11*(2), 319. https://doi.org/10.3390/jmse11020319

Robb, A., Stocker, L., Payne, M., & Middle, G. J. (2020). Enabling Managed Retreat From Coastal Hazard Areas Through Property Acquisition and Transferable Development Rights: Insights From Western Australia. *Urban Policy and Research*, *38*(3), 230–248. https://doi.org/10.1080/08111146.2020.1768842

Ruiz-Ramírez, J. D., Euán-Ávila, J. I., & Rivera-Monroy, V. H. (2019). Vulnerability of Coastal Resort Cities to Mean Sea Level Rise in the Mexican

Caribbean. *Coastal Management, 47*(1), 23–43. https://doi.org/10.1080/08920753.2019.1525260

Sánchez, E. G., Sánchez, H. G., & Ribalaygua, C. (2020). Cultural Heritage and Sea Level Rise Threat: Risk Assessment of Coastal Fortifications in the Canary Islands. *Journal of Cultural Heritage, 44*, 211–217. https://doi.org/10.1016/j.culher.2020.02.005

Savo, V., Lepofsky, D., Benner, J. P., Kohfeld, K. E., Bailey, J., & Lertzman, K. (2016). Observations of Climate Change among Subsistence-Oriented Communities Around the World. *Nature Climate Change, 6*(5), 462–473. https://doi.org/10.1038/nclimate2958

Savo, V., Morton, C., & Lepofsky, D. (2017). Impacts of Climate Change for Coastal Fishers and Implications for Fisheries. *Fish and Fisheries, 18*(5), 877–889. https://doi.org/10.1111/faf.12212

Saleh, F., Weinstein, M. P., (2016). "The role of nature-based infrastructure (nbi) in coastal resiliency planning: a literature review", *Journal of Environmental Management 183*:1088–1098. https://doi.org/10.1016/j.jenvman.2016.09.077

Schoonees, T., Gijón Mancheño, A., Scheres, B., Bouma, T. J., Silva, R., Schlurmann, T., & Schüttrumpf, H. (2019). "Hard structures for coastal protection, towards greener designs", *Estuaries and Coasts 42*(7):1709–1729. https://doi.org/10.1007/s12237-019-00551-z

Schuetze, T., & Chelleri, L. (2013). Integrating Decentralized Rainwater Management in Urban Planning and Design: Flood Resilient and Sustainable Water Management Using the Example of Coastal Cities in the Netherlands and Taiwan. *Water, 5*(2), 593–616. https://doi.org/10.3390/w5020593

Sharaan, M., Somphong, C., & Udo, K. (2020). Impact of SLR on Beach-Tourism Resort Revenue at Sahl Hasheesh and Makadi Bay, Red Sea, Egypt; A Hedonic Pricing Approach. *Journal of Marine Science and Engineering, 8*(6). https://doi.org/10.3390/jmse8060432

Singhvi, A., Luijendijk, A., van Oudenhoven, A. P.E. (2022). "The grey – green spectrum: a review of coastal protection interventions", *Journal of Environmental Management, 311*:114824. https://doi.org/10.1016/j.jenvman.2022.114824

Siders, A. R. (2019). Social Justice Implications of US Managed Retreat Buyout Programs. *Climatic Change, 152*(2), 239–257. https://doi.org/10.1007/s10584-018-2272-5

Siegert, M. J., Alley, R. B., Rignot, E., Englander, J., & Corell, R. W. (2020). Twenty-First Century Sea-Level Rise Could Exceed IPCC Projections for Strong-Warming Futures. *One Earth, 3*(6), 691–703. https://doi.org/10.1016/j.oneear.2020.11.002

Simms, J. R., Waller, H. L., Brunet, C., & Jenkins, P. (2021). The Long Goodbye on a Disappearing, Ancestral Island: A Just Retreat from Isle de

Jean Charles. *Journal of Environmental Studies and Sciences*, 11(3), 316–328. https://doi.org/10.1007/s13412-021-00682-5

Skidmore, T. A., & Cohon, J. L. (2023). A Multicriteria Decision Analysis Framework for Developing and Evaluating Coastal Retreat Policy. *Integrated Environmental Assessment and Management*, 19(1), 83–98. https://doi.org/10.1002/ieam.4662

Skublewska-Paszkowska, M., Milosz, M., Powroznik, P., & Lukasik, E. (2022). 3D Technologies for Intangible Cultural Heritage Preservation—Literature Review for Selected Databases. *Heritage Science*, 10(1), 3. https://doi.org/10.1186/s40494-021-00633-x

Smith, B., Fricker, H. A., Gardner, A., Medley, B., Nilsson, J., Paolo, F. S., Holschuh, N., Adusumilli, S., Brunt, K., Csatho, B., Harbeck, K., Markus, T., Neumann, T., Siegfried, M. R., & Zwally, H. J. (2020). Pervasive Ice Sheet Mass Loss Reflects Competing Ocean and Atmosphere Processes. *Science*, 368(6496), 1239–1242. https://doi.org/10.1126/science.aaz5845

Stanco, F., Battiato, S., & Gallo, G. (2011). *Digital Imaging for Cultural Heritage Preservation. Analysis, Restoration, and Reconstruction of Ancient Artworks*. CRC Press. https://doi.org/10.1201/b11049

Strassburg, M. W., Nababan, B., Lumban Gaol, J., Kim, K.-Y., Manurung, P., Hamlington, B. D., Vignudelli, S., & Leben, R. R. (2015). Sea Level Trends in Southeast Asian Seas. *Climate of the Past*, 11(5), 743–750. https://doi.org/10.5194/cp-11-743-2015

Sultana, N., & Luetz, J. M. (2022). Adopting the Local Knowledge of Coastal Communities for Climate Change Adaptation: A Case Study from Bangladesh. *Frontiers in Climate*, 4. https://doi.org/10.3389/fclim.2022.823296

Tabrani, T., Angkasa, W. I., Rahmatika, D. N., Firmansyah, F., & Rahajo, T. B. (2022). Diversification Strategies of Indonesian Seaweed Commodities to Strengthen Markets, Competitiveness, and Economic Resilience of Coastal Communities. *Return: Study of Management, Economics and Business*, 1(3), 127–133. https://doi.org/10.57096/return.v1i3.21

Tengö, M., Austin, B. J., Danielsen, F., & Fernández-Llamazares, Á. (2021). Creating Synergies between Citizen Science and Indigenous and Local Knowledge. *BioScience*, 71(5), 503–518. https://doi.org/10.1093/biosci/biab023

Theokritoff, E., van Maanen, N., Andrijevic, M., Thomas, A., Lissner, T., & Schleussner, C. F. (2023). Adaptation Constraints in Scenarios of Socio-Economic Development. *Scientific Reports*, 13(1), Article 19604. https://doi.org/10.1038/s41598-023-46931-1

Toimil, A., Losada, I. J., Nicholls, R. J., Dalrymple, R. A., & Stive, M. J. (2020). Addressing the Challenges of Climate Change Risks and Adaptation in Coastal

Areas: A Review. *Coastal Engineering*, *156*. https://doi.org/10.1016/j.coastaleng.2019.103611

United Nations, Department of Economic and Social Affairs. (2022). *World Population Projected to Reach 9.8 Billion in 2050, and 11.2 Billion in 2100*. https://www.un.org/en/desa/world-population-projected-reach-98-billion-2050-and-112-billion-2100

Varea, R., Varea, R., Kant, R., & Farrelly, T. (2022). Qi no tu i baba ni qwali (Living Down by the River): Impacts of Flooding and Mining on Ecosystems and Livelihoods. *Frontiers in Marine Science*, *9*. https://doi.org/10.3389/fmars.2022.954062

Vecchio, A., Anzidei, M., & Serpelloni, E. (2024). Sea Level Rise Projections Up to 2150 in the Northern Mediterranean Coasts. *Environmental Research Letters*, *19*(1), Article 014050. https://doi.org/10.1088/1748-9326/ad127e

Vousdoukas, M. I., Verlaan, M., Mentaschi, L., Voukouvalas, E., Jevrejeva, S., Jackson, L. P., & Feyen, L. (2018). Global Probabilistic Projections of Extreme Sea Levels Show Intensification of Coastal Flood Hazard. *Nature Communications*, *9*(1). https://doi.org/10.1038/s41467-018-04692-w

Wahl, T., Haigh, I. D., Nicholls, R. J., Arns, A., Dangendorf, S., Hinkel, J., & Slangen, A. B. (2017). Understanding Extreme Sea Levels for Broad-Scale Coastal Impact and Adaptation Analysis. *Nature Communications*, *8*(1). https://doi.org/10.1038/ncomms16075

Wang, G. (2023). The 95 per Cent Confidence Interval for the Mean Sea-Level Change Rate Derived from Tide Gauge Data. *Geophysical Journal International*, *235*(2), 1420–1433. https://doi.org/10.1093/gji/ggad311

Weatherdon, L. V., Magnan, A., Rogers, A. D., Sumaila, U. R., & Cheung, W. W. L. (2016). Observed and Projected Impacts of Climate Change on Marine Fisheries, Aquaculture, Coastal Tourism, and Human Health: An Update. *Frontiers in Marine Science*, *3*. https://doi.org/10.3389/fmars.2016.00048

Willis, J., & Church, J. A. (2012). Regional Sea-Level Projection. *Science*, *336*(6081), 550–551. https://doi.org/10.1126/science.1220366

Woodruff, S. C., Mullin, M., & Roy, M. (2020) Is Coastal Adaptation a Public Good? The Financing Implications of Good Characteristics in Coastal Adaptation. *Journal of Environmental Planning and Management*, *63*(12), 2082–2101. https://doi.org/10.1080/09640568.2019.1703656

Wright, L. D., Nichols, C. R., & Syvitski, J. P. M. (2018). Sea Level Rise: Recent Trends and Future Projections. In *Coastal and Marine Resource Management* (pp. 47–57). Springer. https://doi.org/10.1007/978-3-319-75453-6_3

Yin, Q., Zhen-bin, W., Goosse, H., & Hodell, D. (2021). Insolation Triggered Abrupt Weakening of Atlantic Circulation at the End of Interglacials. *Science*, *373*(6558), 1035–1040. https://doi.org/10.1126/science.abg1737

Yusuf, J. E., Rawat, P., Considine, C., Covi, M., St John, B., Nicula, J. G., & Anuar, K. A. (2018). Participatory GIS as a Tool for Stakeholder Engagement in Building Resilience to Sea Level Rise: A Demonstration Project. *Marine Technology Society Journal, 52*(2), 45–55. https://doi.org/10.4031/MTSJ.52.2.12

Open Access This chapter is licensed under the terms of the Creative Commons Attribution 4.0 International License (http://creativecommons.org/licenses/by/4.0/), which permits use, sharing, adaptation, distribution and reproduction in any medium or format, as long as you give appropriate credit to the original author(s) and the source, provide a link to the Creative Commons license and indicate if changes were made.

The images or other third party material in this chapter are included in the chapter's Creative Commons license, unless indicated otherwise in a credit line to the material. If material is not included in the chapter's Creative Commons license and your intended use is not permitted by statutory regulation or exceeds the permitted use, you will need to obtain permission directly from the copyright holder.

CHAPTER 3

The International Law Commission's Study Group on Sea Level Rise in Relation to International Law and Its Impact on International Law

Massimo Starita

1 Introduction

In recent years, sea levels have risen rapidly, accentuating already existing phenomena of international significance (such as human migration), posing new problems (such as the progressive submergence of the territories of States with low-lying coastlines), and creating conditions for future problems (such as the ultimate complete loss of territory of some particularly vulnerable States, in particular, some Pacific Island States).

Faced with such challenges, the international law of the sea has not changed with the same speed. The International Law Commission (ILC) has taken on the need to promote an acceleration of this change. In the following pages, the working method followed so far by the ILC and, more specifically, by its Study Group on Sea level Rise in Relation to

M. Starita (✉)
University of Palermo, Palermo, Italy
e-mail: massimo.starita@unipa.it

© The Author(s) 2026
E. Fornalé (ed.), *Sea Level Rise*,
https://doi.org/10.1007/978-3-031-89171-7_3

International Law will be critically analysed. More precisely, the chapter will identify and discuss two questions: (1) whether suggesting 'practical solutions' to States for dealing with the impacts of sea level rise (SLR) on maritime zones (i.e. one of the two main tasks of the Study Group) falls within the ILC's mandate, and (2) whether the practical solutions suggested by the Study Group have already had an impact on State practice and *opinio juris*.

In analysing these two questions, the chapter will focus on the only sub-topic among those considered by the Study Group which has reached a sufficient level of maturity to make considerations that have at least a minimum level of reliability about it. This topic concerns the extension of maritime zones and boundaries and the impact of SLR on them. Both the Study Group and the United Nations General Assembly (UNGA) Sixth Committee held an intense debate on this subject, in particular, on the Pacific Island States' claim to retain their rights to maritime areas as they now stand despite SLR.

This chapter is structured as follows. After Sects. 2, 3, and 4 briefly outline the general context of the Study Group's work, that is, the changing role of the ILC, and describe the relationship between the Study Group's work and this evolutionary framework, Sect. 5 deals with the question of whether the ILC is competent, under the UN Charter (United Nations, 1945) and the ILC Statute (UNGA, 1947), to fulfil the role it has assigned itself, a question that will be answered positively. Sections 6 and 7 then explore the impact that the work of the Study Group has had. In this respect, the chapter's main argument is that the impact of the work of the Study Group cannot be attributed to the main 'practical solution' indicated by the Study Group, which consists in the idea that the Pacific Island States' interest in maintaining their sovereignty and jurisdiction over maritime zones could be assured by adopting a teleological interpretation of the articles of the 1982 UN Convention on the Law of the Sea (UNCLOS) (United Nations, 1982)[1] relating to a technical issue, i.e. how to draw the baselines of maritime zones. The Study Group's main contribution lies, instead, in promoting a process of creating a general rule, though this process has not yet been consolidated. Finally, the chapter ends with some concluding remarks (Sect. 8).

---

[1] United Nations Convention on the Law of the Sea (UNCLOS), 10 December, 1982, https://www.un.org/depts/los/convention_agreements/texts/unclos/unclos_e.pdf.

2 The Changing Role of the International Law Commission

The role of the ILC has changed in recent years. This evolution has been pushed by several factors. Among them, two key elements deserve to be mentioned here. On the one hand, the task of codifying customary law, i.e. its main statutory task,[2] has largely been fulfilled (Galvao Teles, 2016, p. 215; Murase, 2021, p. 217). From this point of view, it would be hard not to share the view that 'the Commission has done an excellent job in many respects' and that 'much of its output is considered to be the cornerstone of the contemporary international legal order' (Boisson de Chazournes, 2021, p. 135). On the other hand, the prominence of customary law has diminished due to the consolidation in the international legal order of areas essentially regulated by universal and/or regional treaties. In other words, it would be difficult to present contemporary international law as still characterized by an essentially customary matrix (De Sena & Starita, 2023).

In this new situation, the ILC has been reorganizing its objectives, its tasks, and its methods of work, even in the absence of any formal changes to its mandate. This was achieved through strategic choices concerning the subjects of the Commission's work. *Firstly*, the Commission has taken on topics with an overt character of progressive development. The examples are many. The work on the Statute of an International Criminal Court, finalized in 1994, with its impact on the Rome Statute concluded four years later, has recently been considered 'a seminal moment that turned the ILC in this direction' (Voulgaris, 2022, p. 771). *Secondly*, for some time now the ILC has been working not only on the 'codification and progressive development of customary international law', but also on the elaboration of texts intended to function as guidelines addressed to all those called upon to apply international law and, in particular, domestic judges.[3] Think, for example, of the *Draft Conclusions on Subsequent Agreements and Subsequent Practice in Relation to the Interpretation of*

---

[2] Under Article 1 of the Statute, reflecting Article 13 of the UN Charter, 'the International Law Commission shall have for its object the promotion of the progressive development of international law and its codification' (UNGA, 1947, ILC Statute).

[3] It must be underscored the diversity of the final outputs of the works of the Commission: guidelines, guides to practice, draft declarations, draft principles, draft articles, draft conclusions, resolutions, model rules, and so on.

Treaties and the *Draft Conclusions on Identification of Customary International Law*, both adopted in 2018 (ILC, 2018b, Chapters IV and V; ILC, 2018c). Both of those draft conclusions concern methodological issues and recommend methodological approaches to judges. However, the practice of issuing guidelines started before, as shown by the Guide to Practice on Reservations to Treaties adopted in 2011.[4] *Thirdly*, in still other cases, the ILC limited itself to research studies, such as in the case of the issue of fragmentation of international law, which it entrusted to a Study Group, and the output of its work was the adoption of the 'Conclusions of the Study-Group' on this issue (ILC, 2011, pp. 22–38). This attitude of the ILC has aroused considerable interest among scholars. On the one hand, the Commission's tendency to finalize new forms of texts, such as guidelines and studies, has been criticized as pushing beyond the original mandate of the Commission. In this perspective, it has been noted that 'simply clarifying the contents of international law rules is not what is expected of the Commission' (Murase, 2021, p. 221). In a similar vein, it has been observed that 'there is a danger that conclusions, including the reports of study groups, or even guidelines and principles, could lead to the classification of the products of the Commission's work as doctrine of the most qualified publicists of different nations, within the meaning of Paragraph 1(d) of Article 38 of the Statute of the International Court of Justice' (Kamto, 2021, p. 205).

On the other hand, in the last few years, other scholars have welcomed this attitude of the Commission. As for the guidelines, it has been observed that there is a need today for a 'restatement' of international law, stemming also from the increased role of judges in the interpretation and assessment of international law. As in the practice in some common law States (in particular the United States of America), in the international realm as well, restating the law seeks to respond to needs of legal certainty. Both inside and outside the Commission, restating international law has been seen as a useful task for the latter (ILC, 2006, Add.1, para. 504; Peters, 2021, pp. 1391–1392).[5]

---

[4] For the guidelines and commentaries constituting the Guide to Practice on Reservation to Treaties, see ILC (2011), Part three.

[5] Under this approach, the restatement function of the ILC is considered as a strategy to contain the dangers of foreign relations law gaining the upper hand over international law. Furthermore, the feasibility of this strategy would stem from similarities between

According to another, more openly 'progressive' approach, the ILC today could not fail to take on the need to identify the new 'challenges' of the international community and assist States in responding adequately to the new complexities of the world situation, including by means of new working methods. In this perspective, the ILC 'should be as inclusive and plural as possible... also because not all States are as vocal as others in commenting [on] the ILC work' (Galvão Teles, 2016, p. 216). According to yet another view, a considerable part of the work of the ILC is an exercise of 'legislation' that is based more on policy considerations than on pre-existing practice and *opinio juris*. Nonetheless, even these final products would fall within the mandate of the ILC, 'when the ILC is conscious of the question it sets out to answer and the requisite methodology that it employs to do so' (Voulgaris, 2022, pp. 780–788).

Important though this issue may be, engaging in a deep discussion of these different positions is beyond the scope of this short chapter. Just one quick observation will help the following discussion of the work of the Study Group on 'Sea level Rise in Relation to International Law'. To put it simply, we argue that the whole issue should be assessed not so much in terms of the Commission's competence, but in terms of the impact of its action. There are two considerations to be made in this regard.

On the one hand, the position that challenges the competence of the Commission to work on texts other than draft codification articles or draft conventions is difficult to hold from a strictly legal point of view. Indeed, not only is the term 'progressive development' characterized by a considerable breadth, with the consequence that the Commission's new orientation can be brought back to the concept of 'evolutionary interpretation' of treaties. Furthermore, there is also a substantial support by States for the Commission's new trend. Beyond some criticism from individual States, the UN General Assembly has generally endorsed the Commission's tendency to reshape its role through the constant endorsement—in resolutions that have often been adopted without a vote—of the ILC's work program.

---

international law and common law, as both are 'constituted by case law and thinly codified'. However, in common law systems, in particular in the United States, the practice of restatements of the law is closely connected with the rule of precedent and it could be asked whether international law is really akin to common law *in its entirety or only in certain areas* (De Sena & Starita, 2023, pp. 121-174).

On the other hand, the fact that this new orientation of the ILC represents an evolutionary interpretation of the original mandate without going beyond it, should not distract attention from the necessity of an impact assessment of the ILC's work. In other words, it becomes important to understand whether the array of new products of the ILC, such as guidelines, restatements, really guides the interpretation and application of international law and what authority is granted to them by national judges, international courts, treaty expert bodies, and governments. This is the angle from which the work of the ILC on 'Sea Level Rise in Relation to International Law' will be approached here. Before explaining our position, it seems useful to explain why the work of the Study Group can be placed in this overall context and what nuances it presents in this context.

3 The Study Group on 'Sea Level Rise in Relation to International Law' as a Further Development of the New Trends in the ILC's Practice

The work of the ILC on the topic of 'Sea Level Rise in Relation to International Law' fits into the overarching evolutionary context described in the previous Section and is, at the same time, a further development of this evolutionary process.

Firstly, the inclusion of this topic in the long-term program of work was much more based on progressive development than on codification objectives and represents a perfect example of the new policy of the ILC, namely that of addressing some current challenges in the contemporary world somewhat in advance of State practice and *opinio juris*. Not only did the ILC expressly consider State practice as merely 'emerging' and not established, but it also considered it with reference to only some of the many issues raised by sea level rise. Consequently, the ILC let a Study Group, set up internally, ascertain *whether* and in which areas (of the three selected: law of the sea, migration, statehood) the topic may be ripe for codification or progressive development.[6]

---

[6] According to the 2018 recommendation of the Working-Group on the long-term program of work (2018 Syllabus), 'there is an emerging State practice with regard to issues related to the law of the sea (such as maintaining baselines, construction of artificial islands, and coastal fortifications) and the protection of persons affected by sea-level rise (such as the relocation of local communities within the country or to other countries, and

Second, the concept of progressive development is used in a sense that does not correspond to what is textually stated in Article 15 of the Statute of the International Law Commission, under which 'the expression 'progressive development of international law' is used for convenience as meaning the preparation of draft conventions on subjects which have not yet been regulated by international law or in regard to which the law has not yet been sufficiently developed in the practice of States' (UNGA, 1947, ILC Statute; UNGA, 1950; UNGA, 1955; UNGA, 1981). While sea level rise is clearly a subject in regard to which the law has not yet been sufficiently developed in the practice of States, the perspective of the preparation of a draft convention has been poorly developed from the outset. The main objectives of the Study Group, instead, are defined as twofold: studying possible legal effects of sea level rise and proposing 'practical solutions' to States.

As for the first aim, according to the 2018 Syllabus, identifying possible legal effects or implications of the phenomenon of sea level rise means performing 'a mapping exercise of the legal questions raised by sea level rise and its interrelated issues' in three areas referred to as key issues (i.e. law of the sea; statehood; and protection of persons affected) (ILC, 2018a, para. 18; ILC, 2020, para. 45). While this first objective would make the work of the Study Group akin to a doctrinal study, the second aim stated in the 2018 Syllabus has a concreteness-oriented nature, since it consists of promoting 'practical solutions in order to respond effectively to the issues prompted by sea level rise'. This second objective should be developed in relation to the above-mentioned three key issues (law of the sea, migration, statehood).[7]

---

the creation of humanitarian visa categories)...The topic is feasible because the work of the study group will be able to identify areas ripe for possible codification and progressive development of international law and where there are gaps' (ILC, 2018a). In 2019, the Commission recommended the inclusion of the topic in the long-term program of work of the Commission (ILC, 2018c, para. 369). The General Assembly took note of this recommendation in Para. 9 of Resolution 73/265 (UNGA, 2018).

[7] 'These questions should be examined through an indepth analysis of existing international law, including treaty and customary international law, in accordance with the mandate of the Commission, which is the progressive development of international law and its codification. This effort could contribute to the endeavours of the international community to ascertain the degree to which current international law is able to respond to these issues and where there is a need for States to develop practicable solutions in order to respond effectively to the issues prompted by sea level rise' (ILC, 2018a, para. 5).

It is this idea of pointing out 'practical solutions' to States that represents a further development in the broader process of rethinking the tasks of the ILC, with its recent practice being essentially characterized, as noted above, by the drafting of guidelines or restatements whose recipients are mainly judges or quasi-judicial bodies. Although the meaning of the term 'practical solutions' is not specified in the Syllabus, it seems difficult to consider them equivalent to restatements or guidelines, i.e. to texts mainly addressed to judges. Developing practicable answers in order to respond effectively to a certain issue (such as sea level rise) seems to be a job for governments and diplomats. Furthermore, it could be reasonably argued that a practical solution does not necessarily correspond to existing legal rules. Such a perspective could assume either the existence of a gap in the legal system (a position that will be developed later in this chapter) or the existence of a rule that no longer corresponds to the needs of international society and should therefore be changed. In both cases, the weight of ethical and political assessments is considerable. In the first scenario, a gap could be filled through agreements between States in a spirit of mutual concession taking into account one or more general principles, as well as on the basis of considerations of equity. But the evolution of existing law also starts from considerations of the same nature before being channelled into the procedures for legal change recognized by the legal system.

It is not surprising, therefore, that in the subsequent work of the Study Group, the practical solutions evoked in the 2018 Syllabus have been interpreted as a tool for social progress and as a policy-oriented exercise. Since the first report of the Co-Chairs, it has been clear that the Study Group attributed much weight to the policy consideration of responding to the needs of local populations particularly affected by (or particularly vulnerable to) the SLR and the inundation of territories. However, it is also interesting to highlight that this political stance of the Study Group has been justified by legal principles.

4   Blending Policy Considerations and Legal Principles

In the area of the international law of the sea, the papers delivered by the Study Group mix overt political preferences for the concerns of Member States particularly affected by SLR with legal principles—in particular those protecting legal certainty and stability—and interpretative choices of

UNCLOS based on the said legal principles (Anggadi, 2022). According to the First Issues Paper prepared by the Co-Chairs of the Study Group in 2020, the position that a normal baseline is ambulatory[8] 'does not respond to the concerns of the Member States impacted by sea level rise or the need to preserve the legal stability, security, certainty and predictability' (ILC, 2020, para. 82).

On the contrary, the First Issues Paper maintains that 'an approach responding adequately to these concerns is one based on the preservation of baselines and outer limits of the maritime zones measured therefrom, as well as of the entitlements of the coastal State' (ILC, 2020, para. 104, sub. (e)). This consideration, along with the observation that UNCLOS does not prohibit *expressis verbis* such preservation, leads the Co-Chairs to state that 'nothing prevents Member States from depositing notifications, in accordance with the Convention, regarding the baselines and outer limits of maritime zones measured from the baselines and, after the negative effects of sea level rise occur, [from] stop[ping] updating these notifications in order to preserve their entitlements' (ILC, 2020, para. 104, sub. (f)). Such a statement, in the Study Group's perspective, is even more important when one takes into account that the First Issues Paper maintains that no customary norm, either regional or general, has yet been formed (ILC, 2020, para. 104, sub. (i)). In other words, the basic idea of the Study Group is shifting from codification and progressive development of customary international law to interpretation of UNCLOS.

This approach had already been chosen by the International Law Association (ILA). Based on the suggestions of its Committee on 'International Law and Sea Level Rise', the ILA Conference stated in Resolution 5/2018 of 24 August 2018 that: 'on the grounds of legal certainty and stability, provided that the baselines and the outer limits of maritime zones of a coastal or an archipelagic State have been properly determined

---

[8] The position that normal baseline is ambulatory was defended by the ILA Committee on 'baselines under the international law of the sea', in its final report adopted at the Sofia Conference in 2012. In this report, the ILA Committee concluded that 'the existing law of the normal baselines applies in situations of significant coastal changes caused by both territorial gain and territorial loss. Coastal States may protect and preserve territory through physical reinforcement, but not through the legal fiction of a charted line that is unrepresentative of the actual low-water line' (ILA, 2012). As we will shall see in a moment, this proposition was substantially contradicted a few years later by a subsequent ILA Committee.

in accordance with the 1982 Law of the Sea Convention, these baselines and limits should not be recalculated should sea level change affect the geographical reality of the coastline' (ILA, 2018, p. 29). The underlying idea is that Articles 5 (concerning the 'normal baselines', i.e. the low-water line) and 7 (on the 'straight baselines') of UNCLOS, being silent on the specific problem of the ambulatory or fixed nature of baselines, in principle, accept both interpretations. That being said, the ILA Committee recommended an understanding of the Convention aimed at favouring the preservation of baselines and entitlements to maritime zones. Such an evolutionary interpretation would reduce 'legal uncertainties regarding maritime boundaries and the limits of maritime zones at a time when many coastal States are facing the challenges of SLRimpacts' (ILA, 2018, pp. 13/21). The ILA work had a significant impact on State practice. On 6 August 2021, the eighteen States members of the Pacific Islands Forum adopted a declaration on preserving maritime zones in the face of climate change-related sea level rise where they proclaimed 'that our maritime zones, as established and notified to the Secretary-General of the United Nations in accordance with the Convention, and the rights and entitlements that flow from them, shall continue to apply, without reduction, notwithstanding any physical changes connected to climate change-related sea level rise' (Pacific Islands Forum, 2021).

In an Additional Paper adopted in 2023, the Co-Chairs of the ILC Study Group confirmed this approach in the light of a number of submissions and statements presented by Member States to the ILC and/or in the Sixth Committee of the General Assembly. The 2023 Additional Paper underlines three main points: (1) many States referred to the issue of legal stability and stated that it is connected with the solution of fixed baselines (or fixed outer limits of maritime zones measured from them) (ILC, 2023, paras. 83 ff.); (2) a consistent number of States highlighted the need to interpret the United Nations Convention on the Law of the Sea in such a manner as to respond to the effects of sea level rise, mostly in the sense that the Convention does not forbid the freezing of baselines (ILC, 2023, paras. 89 ff.); and (3) only very few States focused on the formation of a new customary rule (ILC, 2023, paras. 95 ff.). In other words, the Co-Chairs find in the debate in the Sixth Committee a clear confirmation of the approach taken three years earlier and restate the 'practical solution' of fixing baselines, a solution driven, as mentioned earlier, both by the need to meet the political and economic interests of the States most vulnerable to the sea level rise and by legal principles of stability and certainty.

To sum up, the main practical solution proposed by the Co-Chairs is that States should accept the 2021 Pacific Island States' Declaration on maintaining their exclusive rights to their maritime areas despite sea level rise through an evolutionary interpretation of UNCLOS rules on the drawing of baselines (and outer limits), according to which coastal States are left free to 'freeze' baselines (and outer limits).

5 Assessing the ILC's Competence to Propose 'Practical Solutions' to States

The Study Group's approach, which marginalizes customary law and emphasizes the interpretation of UNCLOS, raises two questions. The *first* one concerns the Commission's mandate. One could criticize the Study Group's work by arguing that the Commission is a technical body, and not a political one. As we could see, this criticism is not new, since it has already been made—more generally—against the Commission's reorientation of its mandate in its recent work (see Sect. 2). It could be argued, however, that the Study Group wished to play a role that is even more on the borderline with the ILC's mandate. The 'practical solutions' formula employed in the Syllabus, in fact, not only rejects any ambition to codify customary law, but also, one might say, places itself at the very boundaries of the notion of progressive development, as it seems to recall other concepts of reference, from equity to mutual concessions to political expediency.

On closer inspection, this first problem could be solved quite easily. Indeed, at least two considerations clearly point in the direction of the ILC's competence to render practical solutions to States. Firstly, it is backed by States. It is important to remark that in UNGA Resolution 73/265, which was adopted without a vote on 22 December 2018, the UN General Assembly entrusted the topic 'Sea Level Rise in Relation to International Law' to the ILC, taking note of the report of the ILC on the work of its seventieth session, including the approach that the Commission would follow, which was centred on the search for 'practical solutions'.[9] It can therefore be said that the Commission is, in a sense, backed by the UN General Assembly as a whole, in this 'repositioning' operation within the flexible limits of its mandate. In other words, the overall attitude followed by States in the General Assembly suggests that the Commission is competent both to propose practical solutions, and to decide on the working method best suited to this purpose.

A second element that reasonably leads to tracing the drawing up of practical solutions to States within the ILC's mandate is derived from the nature of the solutions that were concretely specified by the Study Group in its work. Indeed, it can be argued that the orientation adopted by the Study Group to marginalize customary law and, at the same time, focus on interpretation of UNCLOS, can in some ways be included in the concept of progressive development. The term 'progressive development', used in Article 13 of the UN Charter and in the Statute of the ILC, can be considered as sufficiently *generic*[10] to include not only the promotion of a new convention, but also an evolutionary interpretation of conventional law. The evolutionary character of the interpretation proposed by the Study Group stems from an analysis of the general practice of States.[11] Neither of the two alternative approaches that would lessen the consequences of sea level rise on the maritime zones of low-lying States—either maintaining baselines, and consequently rendering the internal water landward of baselines larger; or maintaining the outer limits of maritime zones, and consequently rendering the maritime zones

---

[9] See UNGA (2018, para. 9), in which the General Assembly 'takes note of Paragraphs 368 to 370 of the report of the International Law Commission and notes, in particular, the inclusion of the topics 'Universal criminal jurisdiction' and 'Sea level rise in relation to international law' in the long-term programme of work of the Commission, and in this regard calls upon the Commission to take into consideration the comments, concerns and observations expressed by Governments during the debate in the Sixth Committee'. The syllabus of the topic is annexed to the report of the International Law Commission.

[10] The reasoning of the ICJ in *Navigational and Related Rights* (ICJ, 2009, p. 242) can be used here *mutatis mutandis*: "where the parties have used generic terms in a treaty, the parties necessarily having been aware that the meaning of the terms was likely to evolve overtime, and where the treaty has been entered into for a very long period or is 'of continuing duration', the parties must be presumed, as a general rule, to have intended those terms to have an evolving meaning". For an introduction to the question see Bjorge (2014).

[11] The question of the ambulation of baselines in the light of sea level rise caused by climate change has been largely discussed. A selected bibliography can be found in ILC (2023).

measured from the baseline larger—correspond to a well-established practice of the UNCLOS States parties. Consequently, one could agree with the observation that "accepting either of these approaches would require changes in the interpretation of the LOSC rules" (Busch, 2021, p. 318). As emphasized in the ILA 2018 Report, any proposal aimed at maintaining existing entitlements to maritime zones 'might involve a change in the current interpretation of the rules of UNCLOS as applied to such situations' (ILA, 2018, p. 13 Starita, 2022. In this perspective, the evolutionary nature of the Study Group's interpretative proposal can be seen in the circumstance that it would push towards the adoption of a new interpretative practice different from the one prevailing to date. Moreover, even if one were to adopt a different position according to which no clear line of interpretation would emerge from the State practice on the matter, the question of the ILC's competence would not be affected. Indeed, assuming that this were the case, one could in any case speak of an evolutionary interpretation. The Study Group's proposal would, in fact, push an ambiguous practice in a specific interpretative direction.

6 Assessing the Impact of the Work of the Study Group

The *second* question raised by the ILC Study Group's approach has to do with the impact of the Study Group's 'practical solutions' on the evolution of international law and the ILC's ability to effectively play the new role it is carving out for itself. Can it be said that an evolutionary process is underway to reshape the different interests of States in the face of the problems posed by the sea level rise? And if so, how? Through the affirmation of an evolutionary interpretation of existing conventional norms, as suggested by the Study Group, or the emergence of a new customary norm? It seems trivial to note that the more the current developments in the practice of States correspond to the practical solutions suggested by the Study Group, the more the 'leading' role of the ILC in proposing a new law to meet the challenges of the contemporary world will be consolidated. Regarding this second question, the following considerations can be made.

It seems important, first of all, to recall that the perspective of favouring fixed baselines through an interpretation of UNCLOS had already been advanced by the ILA in its—undoubtedly pioneering—work on international law and sea level rise (see Sect. 4). It can be said that there is a sort of continuum between the work of the ILA and that of the Study Group, from a chronological point of view, and a certain overlap between them, from the point of view of content. This should be born in mind in an impact-analysis of this topic. It is hard to deny that the work of ILA had already had an impact on the practice of the States particularly vulnerable to the phenomenon of the sea level rise caused by climate change, in particular by encouraging the adoption of the Pacific Islands Forum 'proclamation' of the continued application of maritime zones notwithstanding sea level rise. But it remained to be seen what the attitude of third States was in relation to the Pacific Island States' claim. It is in this respect that the impact of the Study Group's work in the development of international State practice and opinio juris has been considerable. This impact cannot be seen in a generalized adherence of States to the 'interpretative' approach, but rather in promoting the process of the formation of an ad hoc customary international norm. Indeed, the debate in the UNGA Sixth Committee proves that the approach of reducing the question of the impact of sea level rise on territories and marine zones to a simple interpretation of a couple of articles of UNCLOS has failed.[12] On the one hand, some States that are not parties to UNCLOS, such as the United States, have explicitly pledged not to challenge the baselines and outer limits as defined by the Pacific Island States under the Declaration, without linking their position to the Study Group's proposed interpretation of UNCLOS as a legal basis.[13] On the

---

[12] Even before the publication of the record of the debate in the Sixth Committee, this author had already criticized the Study Group's approach. The criticism of such a perspective, which seeks to deny the centrality of customary law in the international law of the sea, is based on three different considerations. Firstly, in the interpretation proposed by the ILA and the Study Group, the conventional rules cannot be imposed on States that are not parties to UNCLOS. Secondly, the game at stake does not consist of small fluctuations (which could be resolved by resorting to the criterion of fixed lines instead of moving lines), but of appreciable expansions of the areas subject to the sovereignty or the jurisdiction of the coastal State. Thirdly, UNCLOS does not establish a 'closed system' in which the function of interpreting and applying its rules to disputes between States is exercised by only one court or tribunal. Therefore, an 'evolutionary understanding of UNCLOS as to the question of the ambulatory/fixed character of baselines (or outer limits of maritime zones), could not happen suddenly, but only by way of a progressive process with the participation of different international tribunals, conciliation commissions, and, above all, States' (Starita, 2022).

[13] See in particular the statement made by the representative of the United States which stated that 'his Government had undertaken not to challenge lawfully established baselines

other hand, not all States parties to UNCLOS have endorsed the interpretative perspective. In a joint statement, for example, the Nordic countries (Denmark, Finland, Iceland, Norway, and Sweden), although favourable towards the basic claim of the atoll States, observed that the concept of legal stability "must be approached with caution, with full respect for the United Nations Convention on the Law of the Sea and taking into consideration all possible implications, including those concerning existing rights and obligations under international law". Furthermore, the Nordic countries added that "there is no explicit provision in the Convention requiring States parties to update their published baselines and outer limits of maritime zones. However, they had also observed that there was a difference between legally freezing baselines and not updating published baselines".[14]

However, despite the problems posed by the "interpretative approach", the work of the Study Group did influence the international community by promoting the process of the formation of a new customary rule (though this process has not yet come to an end and will not necessarily come to an end). If there has been an impact of this work, it is, *firstly*, because the ILC's endorsement of the 2021 Declaration of the Pacific Island States and the Proclamation therein (along with the ILA's interpretative thesis) has ensured for the 2021 Proclamation an "institutional" space for discussion; *secondly*, because it gave it a greater political-legal weight than it would probably have had 'by its own force'; *thirdly*, because it has allowed a large number of endorsements of the Pacific Island States' proclamation to emerge, and has also provided usable arguments at least among the States Parties to UNCLOS; and, *fourthly*, because it has enhanced the probing value of the lack of reaction by some States to the 2021 Proclamation as a form of acquiescence to it.

First, the reports of the Study Group have had an 'institutionalizing' effect on the process of acceptance by third States of the legitimacy of the Pacific Island States' proclamation. In other words, despite their interlocutory nature, the reports of the Study Group on the topic of sea level rise and international law, and the proposals for 'practical

and maritime zone limits that were not updated despite sea-level rise caused by climate change' (UNGA, 2023b, p. 12).

[14] See the statement of the representative of Denmark, speaking on behalf of the Nordic countries (UNGA, 2023a, pp. 12–13).

solutions' therein, provided a basis for the discussion of States in the Sixth Committee of the UN General Assembly. As a consequence, the Sixth Committee (as well as the direct dialogue with the ILC through comments on the Study Group's reports) has been the privileged place for States to express public statements as to whether the Pacific Island States' claim is permitted or prohibited by international law. While it is true that the Sixth Committee has always been the main public space where governments comment on the Commission's work, it must be observed that usually these comments concern the correspondence of the ILC's proposals to customary law or progressive lines of development. In this case, however, since the Study Group's proposals are closely related to a specific element of State practice (i.e. the Pacific Island States' proclamation), the debate in the Sixth Committee on the ILC report essentially proved to be the main forum for States to take a position on that specific element of practice (Vidas & Freestone, 2022, pp. 26–28). Consequently, while other forms of evidence may not be of particular assistance in ascertaining the legal convictions of States,[15] the debating in the UNGA[16] has afforded the most valuable and accessible evidence of the legal positions of States.

Second, the impact of the Study Group's work went beyond this institutional aspect. More importantly, the Study Group has worked as an 'accelerator' of *opinio juris*, pushing States to react to the 2021 Proclamation. To understand the importance of the Study Group's endorsement of the low-lying States' claim, one could compare the 2021 Proclamation with an even more famous proclamation for the international law of the sea, i.e. the 1945 Truman Proclamation on the Continental Shelf (Truman, 1945). The latter, which came from a Great Power and intercepted the political and economic interests of a large number of coastal States, had its own intrinsic power to reshape the customary international law of the sea. The 2021 Proclamation presents almost the opposite characteristics: it comes from a group of small States—both in size and in economic/political strength—and it responds to a phenomenon posing

---

[15] See ILC, Draft Conclusions on Identification of International Customary Law, Conclusion 10 (Forms of evidence of acceptance as law [*opinio juris*]), with related commentary (ILC, 2018b, pp. 140–142).

[16] Dialoguing with the ILC itself by submitting information and comments on the Study Group's papers has been less relevant, given that only a few States have been vocal in that dialogue.

an existential (immediate and serious) threat to only a few States particularly vulnerable to sea level rise. According to the Intergovernmental Panel on Climate Change, 'satellite measurements of the height of the ocean surface relative to the centre of the Earth (known as geocentric sea level) show differing rates of geocentric sea level change around the world. For example, in the western Pacific Ocean, rates were about three times greater than the global mean value of about 3 mm per year from 1993 to 2012' (IPCC, 2013, p. 173).[17]

However, the Study Group endorsed the 2021 Proclamation and introduced a number of arguments into the public debate to strengthen its legitimacy, which required States to take a position on them. Thus, the 'institutionalized' discussion in the Sixth Committee (and in the dialogue with the ILC) revolved around the 2021 Proclamation, as well as the Study Group's 'practical solutions' and the apparatus of values and principles that the latter placed at the basis of its proposals (climate justice, certainty of legal relations, certainty of maritime borders, and the vital interests of local populations).

Third, the just cause of the low-lying States ended up benefiting from the position taken by the Study Group. Indeed, the main takeaway from the debate in the Sixth Committee is the emergence of a broad block of States that consider the 2021 Proclamation and the resulting legislation of the atoll States legitimate either under international law or under equity and policy considerations.[18] This block includes both States parties to UNCLOS and States that have not ratified it yet. In particular, the United States not only recognized 'that trends are developing in the practice and views of States, but also noted its own commitment not to challenge established baselines and maritime zone limits that are not subsequently updated despite sea level rise caused by climate change' (UNGA, 2023b, p. 12). Furthermore, this block also includes States with strong economic interests that have fishing fleets operating in the Pacific and therefore need to purchase fishing licences from the claimant States. In addition to

---

[17] The Co-Chairs' First Issues Paper refers to the non-uniform impact of sea level rise, as scientifically ascertained by the IPCC (2013), p. 16.

[18] Some States referred to equity (see, among the others, the statements of Peru, South Africa, and Cameroon, (UNGA, 2023c, pp. 10/14/17; others to justice (UNGA, 2023b, p. 5, statement of Austria); others to necessity (UNGA, 2023c, p. 16, statement of Cuba); and others to the general principle of stability and certainty (see, among others, UNGA (2023c), pp. 13/15, statements of Thailand and Viet Nam)

the United States, consider Japan, which 'officially adopted the position that it is permissible to preserve the existing baselines and maritime zones established in accordance with UNCLOS, notwithstanding the regression of coastlines caused by climate change' (UNGA, 2023e, p. 4, Statement of Japan).

Fourth, the institutionalization of the debate and the acceleration of *opinio juris* among States could play a role also in the way in which the 'silence' held by some States (i.e. their failure to publicly react to the 2021 Proclamation) should be assessed. While the lack of objections to the State practice emerging among low-lying, Small Island Developing States is worth noting (Vidas & Freestone, 2022, p. 40), it is well known that the legal significance of silence is very difficult to assess.[19] However, it is also accepted that, depending on the circumstances, silence can take on the value of acquiescence. This can be said particularly when a State, which is the bearer of specific interests, has obtained knowledge of a practice of other States that is likely to affect its interests and has had sufficient time to react. Qualified silence can be relevant in the process of formation of customary international law, as evidence of *opinio juris*,[20] in the

---

[19] On the difficulty of evaluating silence when identifying the constituent elements of custom and on the importance to be given to the specific context, see, in general, ILC (2018b).

[20] See Conclusion No. 10, Para. 3: 'failure to react over time to a practice may serve as evidence of acceptance as law (*opinio juris*), provided that States were in a position to react and the circumstances called for some reaction'. The ILC explains that 'for such a lack of objection or protest to have this probative value, however, two requirements must be satisfied [...] First, it is essential that a reaction to the practice in question would have been called for. This may be the case, for example, where the practice is one that affects – usually unfavourably – the interests or rights of the State failing or refusing to act. Second [...] the State concerned must have had knowledge of the practice and [...] it must have had sufficient time and ability to act' (ILC, 2018c).

process of the formation of a non-written bilateral agreement, as evidence of a tacit consent,[21] or as evidence of acquiescence to historic titles.[22]

From this point of view, it seems indisputable that the work of the ILC influenced the issue of awareness with the consequence that, for example, it would not even be conceivable today for a State to claim that it had no knowledge of the national legislation adopted in the execution of the 2021 Proclamation. Similarly, it could be reasonably argued that the work of the ILC is also influencing the 'time element', as it speeds up the time within which a State can assert its dissent. We are not saying that the silence kept by States so far necessarily reflects a particular position on the Pacific Island States' claim. From this point of view, some States, such as China and the United Kingdom, have expressly suggested that the ILC should take a cautious approach to this.[23] However, it is not far-fetched to consider that an element related to time, that is, the conclusion of the work of the Study Group and the elaboration of a position of the ILC

---

[21] Under article 3 of the Vienna Convention on the Law of Treaties (United Nations, 1969), 'the fact that the present Convention does not apply to… international agreements not in written form, shall not affect: 3 (a) the legal force of such agreements'. The International Law Commission had already stated in its work on the law of treaties that, despite the exclusion of tacit agreements from the scope of its work, 'an international agreement not in written form or a unilateral declaration or any other form of international act is excluded from the application of the present articles [and] shall not be understood as affecting in any way such legal force as these agreements or acts may possess under general international law' (ILC, 1962, p. 35). The International Court of Justice, for its part, acknowledged the possibility of tacit agreements in a case concerning a maritime dispute between Nicaragua and Honduras, while requiring that the 'evidence of a tacit agreement must be compelling' (ICJ, 2007, p. 80). Moreover, in a subsequent case concerning a maritime dispute between Peru and Chile, the Court considered that the Parties 'acknowledged in a binding international agreement that a maritime boundary already exists' (ICJ, 2014, p. 39). For a discussion of the ICJ's jurisprudence on the topic see D'Aspremont (2015).

[22] For a concise and rather recent explanation of the problems posed by acquiescence to historic titles see Gioia (2018). For a discussion of the topic that is still exemplary in terms of clarity and depth, see Verhoeven (1975), pp. 340 ff.

[23] See the following statements: United Kingdom: 'The fact that the Study Group's preliminary observations in the First Issues Paper, or other points it had raised in various strands of its work had not been contested should not be interpreted as agreement with them' (UNGA, 2023a, p. 20). China: 'the silence of affected States on the issue of sea level rise did not necessarily reflect a particular position on the interpretation of the United Nations Convention on the Law of the Sea, or endorsement of or opposition to a particular rule' (UNGA, 2023d, p. 13). See also the summary of the exchange of views among the members of the Study Group reported in Report of ILC (2018a), p. 91.

on the Study Group's final report, can be relevant in assessing the legal meaning of a State's prolonged silence.

One might reasonably assume that once a final text has been submitted, silence, at least on the part of a State with specific counter-interests (such as, as noted earlier, the presence of a fishing fleet in the Pacific), would amount to its acquiescence to the claim of the Pacific Island States. It seems to us that this argument is confirmed by the statements of some States that have not expressed a clear and definitive position. These States asked that no particular legal significance be attached to their silence or imprecise position given the still interlocutory stage of the Study Group's work.[24] This kind of reasoning does not rule out that the presumption be rebutted if the State qualifies its silence with an opposite meaning. One might add that the reduction in the time available to react is justified by the urgency of responding to the sea level rise and is compensated for by: (1) the publicity and transparency of the procedure in the Sixth Committee[25]; (2) the completeness of the information made available to States; and (3) the possibility of discussing the issue simultaneously in other international fora. Suffice it to say that the Security Council has held meetings devoted to the issue of the relationship between rising sea levels and international peace and security.[26]

Finally, in terms of a more general plan, the Study Group succeeded in ensuring a certain level of institutionalization of the customary law formation process in this specific matter. In other words, the four points above

---

[24] The statements of the United Kingdom and China end with the following words: 'This was particularly the case in the light of the Study Group's mandate and the stage its work had reached' (United Kingdom); 'it might be that the State concerned had not yet developed the relevant opinio juris' (China). In the debate held at the Security Council, the Russian Federation emphasized 'that legal experts have not yet reached a consensus, and the results will be presented only after the thorough consideration of each of the three sub-topics, namely, issues pertaining to the law of the sea, statehood and the protection of persons affected by sea level rise. The topic is also being dealt with in the Sixth Committee of the General Assembly. We therefore believe that it is premature to discuss the issue, particularly in the Security Council, which is not a dedicated platform for the topic' (UNSC, 2023, p. 25).

[25] 'The more publicity is given to this practice, without causing any protests, the quicker the customary rule can come into being' (Soons, 1990, p. 225).

[26] The Security Council held open debates on the implications of sea level rise for international peace and security (See Chapter 3 in this volume). On the role that debating sea level rise in the Security Council can have, see the critical observations of Palchetti (2022).

must be read together *within* the process of the formation of a new norm of general international law. Although this perspective already emerges from what has been said so far, some clarification is still appropriate since a generalized consensus among States is not necessarily reflected in customary law, as the debate in the Sixth Committee shows. In the following section, the reasons for such a position will be explained, albeit briefly.

7 Looking for 'The Most Appropriate Form': Customary Law?

The question of which outlet this emerging consensus will take was clearly posed by some States during the debate in the Sixth Committee. Hungary, for example, while agreeing with the Study Group that there exists no obligation for regular updating of baselines, added that 'further thought should be given to finding the most appropriate form to reflect this conclusion' (UNGA, 2023b, p. 9). Similarly, India stated that 'all States should commence deliberations on finding the solution to this' (UNGA, 2023c, p. 7).

Four options have been considered in the debate. Firstly, some States advanced the idea of a meeting of the States parties to UNCLOS which would come to a 'shared interpretation' or an interpretative declaration.[27] The second possible solution mentioned in the debate in the Sixth Committee is the making of an international treaty establishing the principle of stability of baselines and/or outer limits despite sea level rise. As to its form such a treaty could be conceived as an interpretative agreement of UNCLOS.[28] The third possible solution is based on the consideration

---

[27] See, for example, the statements of Italy (UNGA, 2023a, p. 20); Sierra Leone (UNGA, 2023d, p. 6); and Argentina (UNGA, 2023e, p. 3). Other States declared that they favor a declaration on the LOSC, but it is not clear whether the author of such a declaration should be the States Parties or the ILC. See, among others, Romania (UNGA, 2023c, p. 3). Other States do not rule out unilateral declarations of States or small groups of States (see, in particular, Denmark speaking on behalf of the Nordic countries (UNGA, 2023a)).

[28] See, for example, the statement of Ireland calling for a 'decision' of States parties to UNCLOS that 'would constitute a subsequent agreement between the States parties regarding the interpretation of the Convention or the application of its provisions, as contemplated by article 31, Paragraph 3 (a), of the Vienna Convention on the Law of

that the growing consensus among States would produce an interpretative agreement by itself without the need for a written text reflecting its content. In this perspective, the emerging consensus would be relevant as a 'subsequent practice in the application of the treaty'.[29] However, States adopting this position generally do not clarify whether this practice should be conceived as establishing an 'authentic means of interpretation' under Article 31, Para. 3 (b) or as establishing a 'supplementary means of interpretation' under Article 32 of the 1969 Vienna Convention (United Nations 1969).[30] While the realization of the first sub-option would need a convergent practice of all States parties to UNCLOS, a practice of a significant group of States parties could be considered under Article 32.[31]

Treaties' (UNGA, 2023c, pp. 8–9). See also the statement of Jordan (UNSC, 2023, p. 28).

[29] The representative of Australia stated that its government thought it 'was encouraging to see that the Declaration had garnered support beyond the Pacific region, thus contributing to the progressive development of international law and State practice on the interpretation of the Convention' (UNGA, 2023a, p. 15). The Representative of the United Kingdom sees in the proposed 'adaptive interpretation' a possible 'change in interpretation' and a possible 'emergent consensus' (UNGA, 2023a, p. 19). The representative of Germany referred to an 'ever-increasing convergence of States' on a 'contemporary reading and interpretation of the Convention' (UNGA, 2023b, p. 10), while the USA 'recognizes that new trends are developing in the practice and view of States' (UNGA, 2023b, p. 12); and the Philippines 'noted that there was a growing consensus among Member States that the Convention did not forbid or exclude the option of fixing baselines' (UNGA, 2023e, p. 12).

[30] But see the representative of Chile, who stated that 'there seemed to be agreement among the parties as to the interpretation of the Convention; therefore, subsequent practice existed for the purposes of article 31 of the Vienna Convention on the Law of Treaties', and added that 'if that practice was not considered sufficiently uniform for the purposes of article 31 of the Vienna Convention, in accordance with the Commission's conclusions on subsequent agreements and subsequent practice in relation to the interpretation of treaties, it might meet the requirements of article 32 (Supplementary means of interpretation)' (UNGA, 2023b, p. 16).

[31] The difference between these two sub-options was emphasized by the ILC, *Draft Conclusions on Subsequent Agreements and Subsequent Practice in Relation to the Interpretation of Treaties*. See Conclusion 2, Para. 4 ('Recourse may be had to other subsequent practice in the application of the treaty as a supplementary means of interpretation under article 32') and Conclusion 4, Para. 1 ('A subsequent agreement as an authentic means of interpretation under article 31, Paragraph 3 (a), is an agreement between the parties, reached after the conclusion of a treaty, regarding the interpretation of the treaty or the application of its provisions') and par. 4 ('A subsequent practice as a supplementary means of interpretation under article 32 consists of conduct by one or more parties in the application of the treaty, after its conclusion') (ILC, 2018b).

However, none of these three propositions would provide a final and objective solution to the question. On the one hand, interpretative declarations are not binding and 'could only be considered as a supplementary means of interpretation under Article 32 of the 1969 Vienna Convention'. On the other hand, any new agreement would only bind the States parties to it and only an agreement between *all* the States parties would have a general scope of application. Finally, the interpretative practice solution presents the same limit as the first two: i.e. depending on the circumstances, either the non-binding character of an interpretative declaration, or the limited scope of application given the general rule regarding third States. Last but not least, we find it misleading, as discussed above, to approach the issue as if it were only a question of drawing baselines (see Sect. 6, note 12, Starita (2022)).

Like it or not, one cannot avoid passing through customary international law, i.e. the fourth possibility discussed by States in the Sixth Committee (UNGA, 2023b, p. 7, Statement of Poland; UNGA, 2023c, p. 9, Statement of Peru; UNGA, 2023e, p. 9, Statement of El Salvador).[32] In this perspective, shared declarations by States parties, resolutions by international organizations, and interpretative agreements could be useful, but more for their indirect relevance in the formation of such a rule than for their direct legal effects.

As far as this general law-making process is concerned, it cannot be said that the rule in question has already been formed. Although the situation is still fluid, the main elements for and against the formation of a new custom—as well as their weight—can be identified. Starting with the elements against the formation of the norm, we could mention, first of all the fact that, as stated before, the silence of some States cannot be interpreted as acquiescence at this stage of the Study Group's work.[33]

---

[32] Brazil stated that 'current State practice is not sufficient to identify a clear rule', adding that 'it is crucial that any future rule on the topic be established on the basis of State consent' (UNGA, 2023a, pp. 16-17), while Croatia underlined that the ILC 'should also take a cautious approach, as State practice and *opinio juris* on the issue were non-existent' (UNGA, 2023d, p. 11). See also China, which stated that '[t]he Study Group should also give adequate consideration to the rules of general international law when analysing the Convention' (UNGA, 2023d, p. 13). Finally, the statement of India that 'all States, whether they were currently affected or not, should commence deliberations on finding a solution' indirectly refers to general international law (UNGA, 2023c, p. 7).

[33] The representative of the Netherlands expressly stated that '[h]is Government had not yet taken a position on that question' (UNGA, 2023b, p. 11). If one only reads the

Moreover, one cannot deny that other States have taken a cautious stance, pointing out that equity is not an 'overarching principle' (UNGA, 2023b, p. 7, Statement of Poland), while still other States observed that a solution must be found that would take into account not only the principle of stability and legal certainty and the requirements of climate justice, but also the legitimate interests of third states (UNGA, 2023b, pp. 3/19–20, Statement of Islamic Republic of Belarus and Iran; UNGA, 2023c, pp. 5/6, Statement of Slovakia and Czechia; UNGA, 2023d, pp. 8/16, Statement of Colombia and Türkiye).

Furthermore, some States seem to favour the formation of a special regime for the Pacific,[34] while others would be open to a rule encompassing, in general, all cases of sea level-rise caused by climate change.[35] Finally, the increasing polarization and conflict in international society may reduce the chances of reaching a general consensus among States.

At the same time, other elements point in the direction of a consolidation of the customary rule—at least to the benefit of the Pacific Island States—in the near future.

---

statements made in the Sixth Committee, he/she could also count other States among those States that did not take a clear position on the issue, such as France (UNGA, 2023a, p. 18), Portugal, Israel (UNGA, 2023b, pp. 14/18), Malta and Malaysia (UNGA, 2023d, pp. 7/10).

[34] Many States referred to the particular situation of small island States in the light of their specific vulnerability to sea level rise and their underdeveloped economic conditions. See, among others, the statements (all cited above) of India, Indonesia, Singapore, and Türkyie. The Republic of Korea noted that 'as sea level rise posed substantially divergent challenges to different States, the Commission might wish to take a more flexible approach that considered States' differing circumstances', adding that 'in May 2023, acknowledging the special circumstances faced by Pacific islands and their related concerns, her Government had expressed its support for the Declaration on Preserving Maritime Zones in the Face of Climate Change-related Sea Level Rise' (UNGA, 2023e, p. 5).

[35] Malaysia warned of the risk of 'measures designed to enlarge coastlines under the pretext of sea level rise' (UNGA, 2023d, p. 10). Germany stated that it is supporting an interpretation of UNCLOS as allowing 'for the stabilization of baselines in coastal areas affected by climate-change induced sea level rise' (UNGA, 2023b, para. 54). The United Kingdom noted that 'any emergent consensus... should not apply... for reasons not connected with sea level rise' (UNGA, 2023a, para. 120). Germany, South Korea, and the USA limited their commitment not to challenge fixed baselines and maritime zone to situations of sea level rise caused by climate change. See also Romania (UNGA, 2023c, paras. 11 ff.). More generally, those States justifying their positions in favor of the 2021 Proclamation by referring to equity and justice seem to implicitly back solutions with a limited scope of application. However, it is not clear how limited this scope should be.

One is the fact—on which we have already dwelt—that several maritime powers with economic and political interests in the Pacific have already committed themselves not to challenge baselines and outer limits of maritime zones. A second, specular element is that the silence of some states can also be explained by the fact that they do not have particularly deep-rooted interests in the Pacific region (Conforti, 1957, p. 288).[36] Third, a prolonged silence, i.e. a prolonged failure to react to the 2021 Proclamation, could end up constituting the given States' acquiescence to it, with the Study Group having assured a certain degree of institutionalization of the debate (as discussed above in Sect. 6).

Finally, as far as one can tell from reading the reports of the debates in the Sixth Committee (as well as in the other institutional fora where the issue was discussed), no State has openly contested the international legality of the 2021 Proclamation. Even those States that declared that the consolidation of a new general norm is necessary tend to deny that the 2021 Proclamation is contrary to existing international law on the (explicit or implicit) grounds that there would be a vacuum in the legal system without it, and that sea level rise is a new issue and, as such, is not yet regulated.[37] From this point of view, it is difficult to deny that in the absence of an applicable rule, no State conduct in this regard can be considered unlawful. When there is a gap in the legal system, in short, it is the underlying principles of the system itself that come directly into play by providing the yardstick for assessing State behaviour. Among the various potentially relevant principles, those of legal certainty and security of maritime borders seem to dominate the debate, while principles that could push in a different direction have very rarely been mentioned in the

---

[36] In the 1950s, Conforti (1957) made a similar point in assessing the silence of some states regarding the spread of the continental shelf practice.

[37] An argument of a gap in international law in this regard has been developed or at least mentioned by some Governments. See, among others, Romania ('the legal dimension of sea level rise and its effects... many of which were unprecedented'), Ireland ('The Convention assumed stable sea level. Such an assumption is no longer valid'), Cuba ('The LOSC provides no answer to the questions raised by sea level rise'), and Türkyie ('The United Nations Convention on the Law of the Sea did not address the current challenges, as sea level rise had not been contemplated at the time of its negotiation'). Croatia warned that the principle of self-determination does not concern the sea level rise. More generally, it seems to us that the argument of a gap in international law is implicit in statements highlighting the need to find a balance between States' interests (or rights and obligations). The perspective of gaps in the international legal order had already been advanced in legal analysis (Boré Eveno, 2019).

debate in the Sixth Committee on the ILC reports on sea level rise. The principle that the territory dominates the sea, for instance, was recalled only rarely in the 2023 debate and was never used to clearly qualify the Pacific Island States' proclamation as unlawful, but was rather essentially seen as a principle to consider in a law-making process.

8 Concluding but Provisional Remarks

The ILC Study Group on Sea Level Rise in Relation to International Law has strongly championed the cause of the countries most vulnerable to sea level rise caused by climate change. It has done so by, among other things, promoting the idea that island States in danger of progressively losing their territory may legitimately maintain their rights over maritime zones that develop from baselines as they stand today. The only condition imposed is that the baselines and outer limits of the maritime zones must be determined in accordance with international law. The Study Group, like the ILA, took forward the idea that the desired result could be achieved without considering customary law, but it did so through an interpretation of the LOSC (in particular Articles 5 and 7), according to which each State would be free to consider its baselines fixed (and not ambulatory).

The argument developed in this chapter reverses this perspective. Although the debate in the Sixth Committee shows that the 'interpretative' perspective put forward by the Study Group did not gain a general acceptance by the States, the work of the Study Group has had an impact on the process of the formation of a general rule. By endorsing the Pacific Island States' Proclamation of 2021 and the reasons—legal and political—behind it, the Study Group pushed all States to react to this claim. The Study Group has also ensured for customary law—a source lacking a proper legal procedure—a certain level of institutionalization, while at the same time favouring an acceleration in its process of formation.

However, the consolidation of a customary rule allowing States to maintain their rights over maritime areas despite the effects of sea level rise caused by climate change cannot be said to be complete yet. Several elements, however, suggest that an evolution of the international law of the sea in this direction is likely. Out of these elements, the one that is especially worth mentioning is the widespread conviction that there is a vacuum of regulation in this area due to the novelty of the phenomenon. This means that the 2021 Proclamation and the State legislation adopted on its basis are not necessarily contrary to pre-existing law, and tend to be evaluated not so much within the context of a process of *changing* an existing norm, but rather within that of *creating* a new norm. In such a context, the role played by general principles and considerations of ethics or justice inevitably expands.

Should the customary process not be concluded, the legal situation of the maritime zones of the Pacific Island States would remain uncertain. The rights claimed by these States would not be the subject of a generalized acquiescence, and thus of an objective norm, but only of unilateral recognitions, albeit numerous ones. Some recognitions have already been formalized. This is the case of those States that have already officially committed not to challenge the 2021 Proclamation and of those States parties to UNCLOS that have expressly adhered to the interpretative approach.

References

Anggadi, F. (2022). What States Say and Do About Legal Stability and Maritime Zones, and Why It Matters. *International and Comparative Law Quarterly, 71*(4), 767–798. https://doi.org/10.1017/S002058932200032X

Bjorge, E. (2014). The Evolutionary Interpretation of Treaties. *Oxford University Press.* https://doi.org/10.1093/acprof:oso/9780198716143.001.0001

Boisson de Chazournes, L. (2021). The International Law Commission in a Mirror: Forms, Impact and Authority. In The United Nations (Ed.), *Seventy Years of the International Law Commission: Drawing a Balance for the Future* (pp. 133–153). Brill/Nilhoff. https://doi.org/10.1163/9789004434271_016

Boré Eveno, V. (2019) Les impacts de l'élévation du niveau de la mer sur les limites maritimes: Du flou juridique aux éclairages de la pratique. *Annuaire du Droit de la Mer Tome, 24,* 59–67.

Busch, S. V. (2021). Law of the Sea Responses to Sea Level Rise and Threatened Maritime Entitlements: Applying an Exception Rule to Manage an Exceptional Situation. In E. Johansen, S. V. Busch & I. U. Jakobsen (Eds.), *The Law of the Sea and Climate Change: Solutions and Constraints* (pp. 309–335). Cambridge University Press. https://doi.org/10.1017/9781108907118.014

Conforti, B. (1957). *Il regime giuridico dei mari.* Jovene.

D'Aspremont, J. (2015). The International Court of Justice and Tacit Conventionality. *Questions of International Law, 2,* 296–310.

De Sena, P., & Starita, M. (2023). *Corso di diritto internazionale.* Il Mulino.

Galvao Teles, P. (2016). The Work of the International Law Commission in the Present Quinquennium (2012–2016) and Possible Future Topics: How to Remain Relevant in the 21st Century. *Anuario de Direito Internacional*, *2014–2015*, 215–223.

Gioia, A. (2018, November). *Historic Titles*. Max Planck Encyclopedia of Public International Law. https://opil.ouplaw.com/display/https://doi.org/10.1093/law:epil/9780199231690/law-9780199231690-e705?rskey=LjpjPt&result=1&prd=OPIL

Intergovernmental Panel on Climate Change (IPCC). (2013). *Fifth Assessment Report: Summary for Policymakers*. https://www.ipcc.ch/report/ar5/wg1/

International Court of Justice (ICJ). (2007). Territorial and Maritime Dispute Between Nicaragua and Honduras in the Caribbean Sea (Nicaragua v. Honduras), Judgment of 8 October 2007. *ICJ Reports 2007*, 1–109. https://www.icj-cij.org/sites/default/files/case-related/120/120-20071008-JUD-01-00-EN.pdf

International Court of Justice (ICJ). (2009). Navigational and Related Rights (Costa Rica v. Nicaragua), Judgment of 13 July 2009. *ICJ Reports 2009*, 213–272. https://www.icj-cij.org/sites/default/files/case-related/133/133-20090713-JUD-01-00-EN.pdf

International Court of Justice (ICJ). (2014). Maritime Dispute (Peru v. Chile), Judgment of 27 January 2014. *ICJ Reports 2014*, 3–73. https://www.icj-cij.org/sites/default/files/case-related/137/17928.pdf

International Law Association (ILA). (2012). *Sofia Conference: Baselines under the International Law of the Sea*. https://ilareporter.org.au/wp-content/uploads/2015/07/Source-1-Baselines-Final-Report-Sofia-2012.pdf

International Law Association (ILA). (2018). *Sydney Conference Report*. https://www.ila-hq.org/en_GB/documents/conference-report-sydney-2018-7

International Law Commission (ILC). (1962). *Yearbook of the International Law Commission: Documents of the Fourteenth Session Including the Report of the Commission to the General Assembly, Volume 2 (A/CN.4/SER.A/1962/Add.1)*. United Nations. https://legal.un.org/ilc/publications/yearbooks/english/ilc_1962_v1.pdf

International Law Commission (ILC). (2006, April 13). *Fragmentation of International Law: Difficulties Arising from the Diversification and Expansion of International Law. Report of the Study Group of the International Law Commission (A/CN.4/L.682)*. United Nations. https://documents.un.org/doc/undoc/ltd/g06/610/77/pdf/g0661077.pdf

International Law Commission (ILC). (2011). *Yearbook of the International Law Commission. Report of the Commission to the General Assembly on the Work of Its Sixty-Third Session, Volume 2 (A/CN.4/SER.A/2011/Add.1 (Part 2))*. United Nations. https://legal.un.org/ilc/publications/yearbooks/english/ilc_2011_v2_p2.pdf

International Law Commission (ILC). (2018a). *Annex B: Sea Level Rise in Relation to International Law: First Issues Paper by Co-Chairs Bogdan Aurescu* and Nilüfer Oral. United Nations. https://legal.un.org/ilc/reports/2018/english/annex_b.pdf

International Law Commission (ILC). (2018b). *Report of the International Law Commission, Seventieth Session (A/73/10)*. United Nations. https://documents.un.org/doc/undoc/gen/g18/252/67/pdf/g1825267.pdf

International Law Commission (ILC). (2018c). *Yearbook of the International Law Commission. Report of the Commission to the General Assembly on the Work of Its Seventieth Session, Volume 2 (A/CN.4/SER.A/2018/Add.1 (Part 2))*. United Nations. https://legal.un.org/ilc/publications/yearbooks/english/ilc_2018_v2_p2.pdf

International Law Commission (ILC). (2020, February 28). *First Issues Paper: Study Group on Sea Level Rise in Relation to International Law (A/CN.4/740)*. United Nations. https://documents.un.org/doc/undoc/gen/n20/053/91/pdf/n2005391.pdf

International Law Commission (ILC). (2023, February 13). *Sea Level Rise in Relation to International Law: Additional Paper to the First Issues Paper. Study Group on Sea Level Rise in Relation to International Law (A/CN.4/76/Add.1)*. United Nations.

Kamto, M. (2021). The Working Methods of the International Law Commission. In The United Nations (Ed.), *Seventy Years of the International Law Commission: Drawing a Balance for the Future* (pp. 198–214). Brill/Nilhoff.

Murase, S. (2021). Concluding Remarks. In The United Nations (Ed.), *Seventy Years of the International Law Commission: Drawing a Balance for the Future* (pp. 215–223). Brill/Nijhoff.

Pacific Islands Forum. (2021, August 6). *Declaration on Preserving Maritime Zones in the Face of Climate Change-Related Sea Level Rise*. https://forumsec.org/

Palchetti, P. (2022). Débattre des changements climatiques au Conseil de sécurité: Pour quoi faire? *Questions of International Law, Zoom Out, 91*, 39–50.

Peters, A. (2021). The American Law Institute's Restatement of the Law: Bastion, Bridge and Behemoth. *European Journal of International Law, 32*(4), 1377–1397. https://doi.org/10.1093/eji/chab095

Soons, A. H. A. (1990). The Effects of a Rising Sea Level on Maritime Limits and Boundaries. *Netherlands International Law Review, 37*(2), 207–232. https://doi.org/10.1017/S0165070X00006513

Starita, M. (2022). The Impact of Sea Level Rise on Boundaries: A Question of Interpretation of the UNCLOS or Evolution of Customary Law? *Questions of International Law, Zoom Out, 91*, 5–21.

Truman, H. S. (1945, September 28). *Proclamation 2667: Policy of the United States of America With Respect to the Natural Resources of Subsoil and Sea Bed of the Continental Shelf*. The American Presidency Project. https://www.presidency.ucsb.edu

United Nations. (1945). *Charter of the United Nations*. https://www.un.org/en/about-us/un-charter

United Nations. (1969, May 23). *Vienna Convention on the Law of Treaties*. https://legal.un.org/ilc/texts/instruments/english/conventions/1_1_1969.pdf

United Nations. (1982, December 10). *United Nations Convention on the Law of the Sea*. https://www.un.org/depts/los/convention_agreements/texts/unclos/unclos_e.pdf

United Nations General Assembly (UNGA). (1947, November 21). *Resolution Adopted by the General Assembly: Hundred and Eighteenth Plenary Meeting (A/RES/174(II))*. https://documents.un.org/doc/resolution/gen/nr0/038/81/pdf/nr003881.pdf

United Nations General Assembly (UNGA). (1950, December 12). *Resolution Adopted by the General Assembly: Fifth Session (A/RES/485(V))*. https://documents.un.org/doc/resolution/gen/nr0/060/83/pdf/nr006083.pdf

United Nations General Assembly (UNGA). (1955, December 3). *Resolution Adopted by the General Assembly (A/RES/984(X)/A/RES/985(X))*. https://documents.un.org/doc/resolution/gen/nr0/104/64/pdf/nr010464.pdf

United Nations General Assembly (UNGA). (1981, November 18). *Resolution Adopted by the General Assembly: 62nd Plenary Meeting (A/RES/36/39)*. https://documents.un.org/doc/resolution/gen/nr0/406/65/pdf/nr040665.pdf

United Nations General Assembly (UNGA). (2018, December 22). *Resolution Adopted by the General Assembly: Seventy-Third Session (A/RES/73/265)*. https://documents.un.org/doc/undoc/gen/n18/464/93/pdf/n1846493.pdf

United Nations General Assembly (UNGA). (2023a, December 11). *Summary Record of the 23th Meeting: Seventy-Eighth Session (A/C.6/78/SR.23)*. United Nations. https://documents.un.org/doc/undoc/gen/n23/314/40/pdf/n2331440.pdf

United Nations General Assembly (UNGA). (2023b, December 11). *Summary Record of the 24th Meeting: Seventy-Eighth Session (A/C.6/78/SR.24)*. United Nations. https://documents.un.org/doc/undoc/gen/n23/314/46/pdf/n2331446.pdf

United Nations General Assembly (UNGA). (2023c, December 11). *Summary Record of the 25th Meeting: Seventy-Eighth Session (A/C.6/78/SR.25)*. United Nations. https://documents.un.org/doc/undoc/gen/n23/314/52/pdf/n2331452.pdf

United Nations General Assembly (UNGA). (2023d, December 11). *Summary Record of the 27th Meeting: Seventy-Eighth Session (A/C.6/78/*

SR.27). United Nations. https://documents.un.org/doc/undoc/gen/n23/314/64/pdf/n2331464.pdf

United Nations General Assembly (UNGA). (2023e, December 11). *Summary Record of the 28th Meeting: Seventy-Eighth Session (A/C.6/78/SR.28)*. United Nations. https://documents.un.org/doc/undoc/gen/n23/314/70/pdf/n2331470.pdf

United Nations Security Council (UNSC). (2023, February 14). *9260th Meeting: Seventy-Eighth Year (S/PV.9260)*. United Nations. https://documents.un.org/doc/undoc/gen/n23/044/11/pdf/n2304411.pdf

Verhoeven, J. (1975) *La reconnaissance internationale dans la pratique contemporaine*. Pedone.

Vidas, D. & Freestone, D. (2022). Legal Certainty and Stability in the Face of Sea Level Rise: Trends in the Development of State Practice and International Law Scholarship on Maritime Limits and Boundaries. *The International Journal of Marine and Coastal Law*, 37, 1–53. https://doi.org/10.1163/15718085-bja10106

Voulgaris, N (2022). The International Law Commission and Politics: Taking the Science Out of International Law's Progressive Development. *European Journal of International Law*, 33(3), 761–788. https://doi.org/10.1093/ejil/chac051

Open Access This chapter is licensed under the terms of the Creative Commons Attribution 4.0 International License (http://creativecommons.org/licenses/by/4.0/), which permits use, sharing, adaptation, distribution and reproduction in any medium or format, as long as you give appropriate credit to the original author(s) and the source, provide a link to the Creative Commons license and indicate if changes were made.

The images or other third party material in this chapter are included in the chapter's Creative Commons license, unless indicated otherwise in a credit line to the material. If material is not included in the chapter's Creative Commons license and your intended use is not permitted by statutory regulation or exceeds the permitted use, you will need to obtain permission directly from the copyright holder.

CHAPTER 4

The Role of the UN Security Council in Addressing and Providing Responses to Peace and Security Risks Resulting From Sea Level Rise

Giuseppe Nesi and Elisa Fornalé

1 INTRODUCTION

Over time, the climate-security debate has gradually evolved and contributed to our understanding of whether and how climate change-related sea level rise could affect sustainable peace and its diverse implications for law and policy (Palchetti, 2022).

Giuseppe Nesi has written Sect. 4; Sects. 2 and 3 is by Elisa Fornalé. The introduction and conclusions are the outcome of a common reflection.

G. Nesi (✉)
University of Trento, Trento, Italy
e-mail: Giuseppe.nesi@unitn.it

UN International Law Commission, Geneva, Switzerland

© The Author(s) 2026
E. Fornalé (ed.), *Sea Level Rise*,
https://doi.org/10.1007/978-3-031-89171-7_4

Insights from recent reports of the Intergovernmental Panel on Climate Change (IPCC) contributed to providing scientific evidence of how climate change is contributing to increasing conflict risks and how addressing these risks at international level could be beneficial for international peace (IPCC, 2022).[1] By building this shared knowledge it allowed for identifying security implications with legal, economic, and political consequences for exacerbating existing vulnerabilities as well as driving new risks at global level beyond a state-centric perspective (Schroeder, 2021). The UN Secretary-General highlighted that: '[i]t is clear that climate change and environmental mismanagement are risk multipliers. Where coping capacities are limited and there is high dependence on shrinking natural resources and ecosystem services, such as water and fertile land, grievances and tensions can explode, complicating efforts to prevent conflict and to sustain peace' (UNSG, 2021b).

Sea level rise is becoming one of the most serious threats to human and international security (Arias, 2022a and 2022b; Maertens, 2018). Not surprisingly, attempts to link security and climate change have been brought to the debate of the United Nations Security Council (hereinafter, the UN Security Council), one of the main organs of the United Nations with the primary responsibility in the preservation of peace and security (Dietz et al., 2016; Maertens & Trombetta, 2023; Webersik, 2012).

It held its first debate on this topic in 2007 and over 50 States participated and presented their statements (Maertens & Trombetta, 2023). The United Kingdom led the move to include this issue in the agenda by highlighting the role of climate change in 'exacerbating many

---

E. Fornalé
World Trade Institute, Faculty of Law, University of Bern, Bern, Switzerland
e-mail: elisa.fornale@unibe.ch

[1] 'Climate change undermines human livelihoods and security, because it increases the population's vulnerabilities, grievances and political tensions through an array of indirect—at times nonlinear—pathways, thereby increasing human insecurity and the risk of violent conflict' (Chapter 18, p. 2673) See also Chapter 8: 'Poor institutional responses [to climate change] can directly drive violence, and there is robust evidence that inequitable responses further exacerbate marginalization, exclusion or disenfranchisement of some populations, which are commonly recognized drivers of violent conflict' (p. 1190).

threats, including conflict' (UNSC, 2007a).[2] This topic remains quite contested: initially the general positioning of States was not unanimous; some States—notably China and Russia—strongly opposed the opportunity to add climate-security implications to the agenda of the UN Security Council, and there was not a consensus on discussing climate change as a security issue instead of a development issue, for instance.[3] But already at this meeting small island states stressed the need to consider the challenges raised by climate change as an 'existential threat' (Torres Camprubi, 2016).

In 2011 under the Presidency of Germany another meeting on the topic took place.[4] After four years, the President of the UN Security Council acknowledged that 'the Security Council expresses its concern that possible adverse effects of climate change may, in the long run, aggravate certain existing threats to international peace and security' (UNSC, 2011a).[5] Despite the growing recognition of this topic, the role and competence of the Security Council in this regard remain an issue of debate (Bonafé & Arcari, 2021; UNSC, 2017b).

This chapter aims to explore how the debate has evolved by focusing on: (1) which are the security risks raised by sea level rise; (2) what legal challenges need to be addressed; and (3) what role the UN Security Council could play in facing the increasing uncertainty because of climate-related impacts. To this end the first part provides an overview of the 2007–2021 debates held at the UN Security Council on climate change and their evolution to identify the key trends and the connection with sea level rise. The second part is based on official documents and the emerging literature and focuses specifically on the adverse impacts of sea level rise on international security and peace. The final part will consider the role of the UN Security Council in this respect.

---

[2] Open Debate on 'Energy, Security and Climate', 3 April 2007, chaired by the United Kingdom.

[3] The Security Council includes 15 members, 5 are permanent members, and the decisions adopted by the Security Council are binding for all UN Members (article 25 of the United Nations Charter).

[4] The Open Debate on 'Maintenance of International Peace and Security: Impacts of Climate Change?' was scheduled on 20 July 2011.

[5] This Declaration was the first official document adopted on this issue (Palchetti, 2022).

2 Securitization of Sea Level Rise: A Historical Overview

Over the years, the adverse impacts of climate change have become part of what Boas defines as the 'process of securitisation', which is seen as 'the process through which non-traditional security issues' are drawn into the security domain (Boas, 2015, p. 1). The way in which security is understood requires a theoretical reflection because several of the threats raised by climate change do not fit in the traditional security agenda (Scott & Ku, 2018).

Even though the climate-security nexus has been widely discussed, this connection raises tensions, particularly when it comes to its conceptualization and the legal implications of the complexity of the interests involved (Cusato, 2022; Maertens, 2018; Trombetta, 2023; UNSG, 2009). This link requires specific choices connected to clarifying 'whose security is at stake and prioritized'—thus protection needs as part of human security, who is supposed to act—in terms of national and international security, and the ways to provide responses—preventive or reactive approaches. Engaging with this 'indeterminate concept' raises crucial normative implications (Benton Heath, 2022, p. 291; Cusato, 2022).

For many, security is strictly connected with the 'survival of the State' and the legitimation of the use by States of 'emergency measures' (e.g. defensive and reactive measures). In particular the use of 'security' as an 'evocative term' can mobilize 'powerful emotions and [the term] belongs to a political tradition that uses it to justify exceptional measures' by creating a 'sense of urgency' (Cusato, 2022; Maertens & Trombetta, 2023, p. 190; Trombetta, 2023, p. 2). This could mobilize divergent approaches to security by raising disagreement in the development of policy measures at domestic and international levels (Davies & Riddell, 2017; Nasu, 2021).

At the same time, the debate has evolved to include human-related risks of climate change (Joyeeta & Hilmer, 2021). The major challenge will be to identify ways of re-focusing on how to make people more secure by including the needs of the most vulnerable in the measures. Different concepts of security could in fact promote different responses to climate risks, and the needs of the most vulnerable—who are the most insecure due to the adverse impacts of climate change—could not be prioritized by exceptional and reactive measures. It is important to translate climate security into achieving human security by increasing the enjoyment of fundamental rights. As suggested by Ban Ki-moon in 2009, the 'interdependence between the security of individuals and communities and the security of nation States' (UNGA, 2009a) requires us to interrogate our traditional analytical frameworks to meet competing interests by including other actors (e.g. individuals/IOs) in the process to foster a more interconnected approach that would be attentive to cooperative approaches (Elliot, 2023).

This is particularly problematic in the context of international security[6] and its relationship to international law (Torres Camprubi, 2016; White & Davies-Bright, 2021). To this end, it is useful to explore how the emergence of this topic in the agenda of the UN Security Council as a security issue could be framed by addressing some of the linkages between human and international security (Palchetti, 2022).

The UN Security Council, in line with Article 24 of the UN Charter, has the 'primary responsibility for the maintenance of international peace and security'. Under Chapter VII, the UN Security Council can define a 'threat to peace determination' (Wood, 2022).

Article 39 of the UN Charter in particular establishes that the UN Security Council has the authority to

> determine the existence of any threat to the peace, breach of the peace, or act of aggression and shall make recommendations, or decide what measures shall be taken in accordance with Articles 41 and 42, to maintain or restore international peace and security.

There is no definition of 'what constitutes a threat to the peace', and this made it possible for the fifteen members of the UN Security Council to address non-traditional security threats and include them in the agenda (Poku, 2013; Schroeder, 2021).[7] In the past decades, several examples

---

[6] The concept of 'collective security' has been defined by de Wet and Wood as a 'system, regional or global, in which each State in the system accepts that the security of one is the concern of all, and agrees to join in a collective response to threats to, and breaches of, the peace' (De Wet & Wood, 2012).

[7] See, for instance, the Ebola health crisis in 2014 (Day et al., 2023). See also the following statement made by Tunisia: 'During its term on the Council, my country has sought to include unconventional threats, such as climate change and pandemics, on the Council's agenda. That is because we are convinced that those phenomena pose a threat to collective security and therefore should not be overlooked by the Security Council. Since peace and security are interlinked with many factors, it is necessary to take an

confirmed the trend in which the role of the Security Council expanded so that it could affirm its authority in regard to diverse issues (e.g. the coronavirus pandemic, gendered violence, human trafficking).

For this, an overview of the ways in which the links between climate change and security have been established by States could provide some insights into the legal impacts of this understanding, as suggested by legal scholars (Palchetti, 2022; Tomuschat, 2022).

In the first debate on the topic held in 2007, low island states raised their concerns about how sea level rise was undermining the very basis of their existence (Torres Camprubi, 2016). Papua New Guinea highlighted how in the Pacific 'the people may have to abandon their traditional lands, their home and possibly their nations' (UNSC, 2007b, p. 27).

Then the turning point came in 2009 with the adoption by the General Assembly of Resolution 63/281 on 'Climate Change and its Possible Security Implications' (UNGA, 2009b, p. 6),[8] which led to the Report of the UN Secretary-General on the subject that included the notion of a 'threat multiplier' to define climate change (Cusato, 2022; UNSG, 2009a). This report identified a threat posed to the 'viability and even the survival' of states as the fourth pathway through which climate change could reduce security. The major concern was to examine how to support the affected States in terms of safeguarding their statehood, protecting displaced populations and dealing with territorial disputes over exclusive economic zones.

In 2011, the UN Security Council convened a new meeting to engage with the mid- and long-term implications of sea level rise, nurturing the policy and the academic attention to the current and potential impacts for affected States, and the security implications of loss of territory for the protection of the rights of their citizens and communities (UNSC, 2011b, 2011c). In this meeting, the Pacific small islands states started to be more vocal by raising attention to their concerns about the potential loss of their territory. Still disagreements were raised by China, India, and Russia.

---

evolving approach as we address those factors that fuel conflicts and violence' (UNSC, 2021b, p. 14).

[8] The resolution suggested 'relevant organs of the United Nations, as appropriate and within their respective mandates, to intensify their efforts in considering climate change, including its possible security implications and requested the UN Secretary General to report on climate change and its security implications' (UNGA, 2009b, p. 2).

In 2015, New Zealand held an open debate on 'Peace and Security Challenges Facing Small Island Developing States' that confirmed how small island states were already facing 'an increasing rate of domestic migration and relocation, with people from rural areas and outlying islands moving to urban centres as they lost their livelihoods and lands' (UNSG, 2009, p. 16).

Also, the two informal discussions held under the Arria Formula Meeting (UNSC, 2020) in 2017 and 2021, identified sea level rise as a threat multiplier which could affect global security and development challenges (Elliot, 2023). The UN Secretary-General highlighted that 'a concept that puts people at its center' is required in the context of climate change (UNSG, 2021).

In 2021, a suggestion to adopt a thematic Resolution[9] was co-sponsored by Niger (the then Security Council President) and Ireland but it was rejected.[10] This draft Resolution represented a crucial attempt to define a role for the UN Security Council (Maertens & Trombetta, 2023, p. 182). The draft Resolution highlighted in particular

> that Small Island Developing States are particularly vulnerable to the adverse effects of climate change, and [the Security Council] expresses deep concern that the impacts, including the loss of territory caused by the rise of the sea level, may have implications for international peace and security, in addition to humanitarian, economic, social, cultural and ecological consequences (UNSC, 2021, p. 3).[11]

A major aim expressed in the draft text was to take preliminary steps to 'integrate climate-related security risk as a central component into

---

[9] It was requested 'to integrate climate-related security risks as a component into comprehensive conflict-prevention strategies of the United Nations'.

[10] India voted against it, China abstained and Russia invoked its right to veto. '12 Council members voted in favour of it, and 113 non-members co-sponsored the resolution' (Buhaug et al., 2023).

[11] See also the statement of Viet Nam: 'Viet Nam is among the countries most vulnerable to the adverse impacts of climate change, including sea level rise . Combating climate change has consistently been our priority during and prior to our term on the Security Council, and it will continue to be so beyond our term in the Council and within other international forums in the framework of comprehensive efforts to support strong and meaningful climate action. We concur that the United Nations Framework Convention on Climate Change remains the primary and inclusive forum for negotiating climate action' (UNSC, 2021b, p. 11).

comprehensive conflict-prevention strategies' (UNSC, 2021, p. 4).[12] In particular the draft text requested that the UN Secretary-General submit a thematic report; and that peacekeeping operations and the UN explore the role of environmental concerns while providing adequate training to personnel involved. In addition, the text encouraged collaboration between the scientific community and Member States.

The negative outcome confirmed the absence of a unanimous consensus among the permanent Members and the elected Members.[13] This was so despite the draft text noting

> that an increasing number of Member States are recognising the adverse effects of climate change on their security in their national contexts, and stressing the primary responsibility of States to develop and implement measures to manage and address climate-related security risks, and, in this regard, [the Security Council] emphasises the need for stronger and sustained international cooperation and capacity-building (UNSC, 2021, p. 2).[14]

In addition, among the arguments invoked to oppose the adoption of this Resolution was the lack of sufficient scientific evidence for connections between climate and security implications (Arias, 2022a).

Despite the failure to adopt this Resolution, the constant evolution of the debate is in favour of broadening the UN Security Council's mandate by contributing to what some authors define as a progressive 'climatization' of its role (Maertens & Trombetta, 2023). States, in their statements, contributed to a progressive conceptualization of the climate

---

[12] See the statement of Ireland: 'Today, we come to the Council to ask it to take the modest first steps to strengthen its ability to begin to assume its own responsibility on the defining issue of this generation: climate change. The draft resolution is aimed at responding to the climate-related security risks affecting the conflicts on the Council's agenda—no more, no less' (UNSC, 2021a, p. 2).

[13] The opposition to the role of the Security Council in engaging with this topic came mainly from Brazil, China, and Russia (Palchetti, 2022).

[14] See also the statement by Niger: 'For the 113 countries that sponsored draft resolution S/2021/990, on the impacts of climate change on conflict situations, a link clearly exists in the sense that climate degradation serves to exacerbate security challenges, especially in the Sahel and Lake Chad basin regions. For landlocked countries, their very existence as geographical entities is at stake. That is a fact and the reality. The force of the veto can block the adoption of a text, but it cannot hide that reality—the truth' (UNSC, 2021b, p. 6).

change-security nexus by selecting specific issues of concern. As described above, the attempt to include climate change under the mandate of the UN Security Council is taking into account how one could '[answer] the security implications of climate change' without necessarily imposing 'securitizing moves' that rest on 'alarmist rhetoric' or the adoption of extraordinary measures (Cusato, 2022). As suggested by scholars, these new concerns require challenging the traditional understanding of security by adopting a pragmatic and cooperative approach (Murphy, 2019). At the same time, human security needs to be included in the debate in order to focus on possible future threats (Jcyeeta & Hilmer, 2021). The way in which the growing recognition of the adverse impact of sea level rise could drive security risks reflects this trend as it will be explored in the next section, dealing with the open debate held in 2023 on this topic.

3 The 2023 Open Debate: Sea Level Rise—Implications for International Peace and Security

The debates hosted by the UN Security Council confirm its progressive engagement on this pressing issue (Bonafé & Arcari, 2021; Day & Harper, 2023). Particularly prominent was its expression of the need to understand the legal implications of climate-induced sea level rise, namely clarifying the nature of the 'security' threats and who is affected by these risks.

This section focuses on the UN Security Council's open debate on 'Sea Level Rise—Implications for International Peace and Security' held on 14 February 2023. This event marked a significant stage by illustrating how new challenges continue to appear and by promoting a new engagement on these critical issues at domestic level. It aimed to share State concerns and their views on how to frame a coordinated approach in response to the security implications of this phenomenon. The open debate was held under the presidency of Malta and under the agenda item 'Threats to International Peace and Security'. Malta's Presidency shared a concept note (UNSC. 2023a) asking the participants to explore how the Security Council could be involved in addressing these risks. In its concept note,

Malta's Presidency formulated four key questions to guide the sharing of good practices and lessons.[15]

The following points were raised. It was noted that the accelerating path of sea level rise can subject island States and low-lying coastal communities to territorial loss. For certain States the adverse impacts of sea level rise represent a 'direct security threat'. This was highlighted by Bangladesh, a low-lying coastal country where 'just a one-metre rise in sea level could result in an inundation of a large area of [the country] and potentially displace more than 40 million people by the end of the century' (UNSC, 2023b, pp. 38–39). Also, 'the Maldives could become uninhabitable by 2050'. 'This means 27 years from now' (UNSC, 2023b, p. 35). Also, Singapore, 'a small and low-lying city State with more than 50 per cent of [its] population living within about two miles of the coast', clearly identified the threats posed by sea level rise and said that they are of 'existential proportions' (UNSC, 2023b, p. 25). The Philippines, like other island nations, has observed that sea level rise has reached 60 centimetres. In its statement, it defines the impacts of sea level rise as a major 'national security' concern because the 'nation's sovereignty and territorial integrity, its peoples' well-being, core values and way of life, along with the State and its institutions, could be affected. All those elements of security are at risk, and this will result in compromising the lives and livelihoods of people living in coastal areas (UNSC, 2023b, p. 2).

Sea level rise is also seen as a 'threat multiplier' that could exacerbate existing challenges (e.g. water scarcity, poverty, food security)[16] or create new risks (loss of statehood, threats to maritime delimitations).[17]

---

[15] The questions were as follows: '1. How can the Security Council, in synergy with other United Nations bodies, best contribute to driving actions to address risks and responses to peace and security resulting from sea level rise as well as preventing tensions and conflict risks resulting therefrom?; 2. How will sea level rise jeopardize statehood, including that of small island developing States, and how is the limited and unambitious action to address climate change impinging on their sovereignty and territorial integrity?; 3. What role can national Governments, regional organizations and local actors play in addressing the security and humanitarian risks related to sea level rise ?; 4. How can the Security Council respond to the triple nexus of gender inequality, State fragility and climate vulnerability, and what actions can be identified to strengthen women's leadership and inclusion in decision-making?' (UNSC, 2023a).

[16] These challenges apply to Singapore; Indonesia; Haiti; the Holy See; Bangladesh (water scarcity); Egypt (food security); Guyana; and Bangladesh.

[17] These risks apply to Costa Rica; Ghana; Malta; Switzerland; Denmark; Slovenia; Mexico; Guyana; Greece; Thailand; Ukraine; and Argentina.

As highlighted by the UN Secretary-General, who briefed the UN Security Council, 'for the hundreds of millions of people living in small island developing States and other low-lying coastal areas around the world, sea level rise is a torrent of trouble' (UNSC, 2023c, p. 2).

Sea level rise can increase instability. The Republic of Haiti gave an accurate account of how the country is already 'grappling with the many challenges of resource scarcity, exposure to natural disasters, ecological crises and vulnerability to external economic shocks' (UNSC, 2023b, p. 21). The United States mentioned the case of Lousiana, where fishers report that rising water is causing damages to livelihoods and forces local communities to move to higher areas (UNSC, 2023b, p. 12).

In addition, sea level rise undermines prosperity also in the Caribbean, where forced migration could cause social instability and demographic tensions. Besides the long-term implications it raises issues related 'to the integrity of maritime boundaries, the ownership of marine resources', and jurisdictions (UNSC, 2023c, p. 20). It is crucial to address in legal terms how human-security-related sea level rise will affect lives and livelihoods in the course of the long-term process (Krieger et al., 2020). As highlighted by Gabon, 'the strong correlation between the vulnerability caused by climate change and insecurity results in a negative feedback loop' (UNSC, 2023c, p. 17).

Finally, concerns have been raised by the States about how human insecurities—for women and climate migrants—could amplify threats to international security.

3.1 Women

At international level the gender-climate nexus is becoming extremely relevant (Fornalé, 2023). This requires to advance its conceptualization by addressing some of the protection needs faced by women and girls (Borras-Pentinat, 2022; Campbell, 2023; Fornalé, 2023; Rowena, 2019; UNSC, 2022). The UN Special Rapporteur on Violence against Women Reem Alsalem, in her 2022 report, raised the attention on how the climate change risks disproportionately affecting them by increasing their vulnerability to violence (UN Special Rapporteur, 2022; UN Special Rapporteur, 2023). As highlighted in the report on 'Women, Girls and the Right to a Clean, Healthy, and Sustainable Environment' (2023) by the UN Special Rapporteur on Human Rights and the Environment,

David R. Boyd, the climatic crisis 'affects everyone, everywhere, but not equally' (para. 2).

The Special Rapporteur warns that the 'effects of the climate crisis could deepen gender divides', 'which [is] exacerbated by discriminatory legal systems and governance structure and unequal power distribution' (UN Special Rapporteur, 2022, para. 23; Fornalé, 2023). Slow-onset climate events particularly exacerbate the inequalities women in SIDS face and lead to greater harm and violence due to ineffective and unavailable reporting mechanisms (UN Women, 2021; UNGA, 2019).[18]

It is not surprising that the adverse impacts of sea level rise on women and girls have been identified as critical during the open debate (Sea Level Rise: Implications for International Peace and Security) organized by the UN Security Council in February 2023.[19] As mentioned above, among the key questions raised by Malta in its concept note, the member States were requested to address the question 'how can the Security Council respond to the triple nexus of gender inequality, State fragility, and climate vulnerability and what actions can be identified to strength women's leadership and inclusion in decision-making?' (UNSC, 2023a).

In the last years, some measures have been adopted to address gender-based discrimination and climate change (ILA, 2024, p. 30; Asian Development Bank, 2022).

---

[18] For instance, environmental degradation could expose 'women and girls into sexual exploitation in exchange for food' (e.g. Mozambique, South Africa, Jordan) (Fornalé, 2023; UN Special Rapporteur, 2022).

[19] 'The humanitarian consequences will have disproportionate impacts on women and children, in particular girls, by further exacerbating instability in regions with already existing tensions over resources, such as food and water. The gender dimensions of these risks and vulnerabilities are to be analysed, also within the changing global context of advancing the women and peace and security agenda' (UNSC, 2023a).

Key statements have been made to highlight the urgent need to adopt gender-responsive frameworks (Albania,[20] Liechtenstein,[21] Chile[22]) and ensure women's empowerment, here understood as increasing women's participation and leadership in measures and decisions that need to be developed (Ecuador,[23] Denmark,[24] Kiribati,[25] Nauru,[26] the Maldives[27]).

In line with this, the Special Rapporteur called for adopting a 'gender-transformative change' that could enhance rights-based climate action

---

[20] 'Climate and environmental action and disaster risk reduction therefore need to be gender responsive, value and promote all women and girls as agents of change and directly address the specific risks that they face' (UNSC, 2023c, p. 15).

[21] 'Especially for people living in low-lying atoll States, sea level rise is the most pressing security threat. As with all questions of peace and security, women and girls undoubtedly shoulder a disproportionate burden. At the same time, they remain crucial agents of change. A gender-sensitive lens is therefore urgently needed in all climate and security responses' (UNSC, 2023c, p. 26).

[22] 'Fourthly, we underscore that women are disproportionately affected by climate-related disasters. The reasons underlying that disparity are multidimensional, ranging from economic to cultural factors. Furthermore, the persistence of a humanitarian response that lacks a gender perspective exacerbates the disproportionate impact of climate emergencies' (UNSC, 2023b, p. 7).

[23] 'Women's leadership and full and equal participation must be included at all levels of decision-making, especially in addressing the impacts of climate change' (UNSC, 2023c, p. 24).

[24] 'The participation of rights holders, especially women, youth and the local community, must be at the centre of those efforts in order to ensure that solutions are locally owned and led, that marginalized groups are empowered and that we do not inadvertently harm those we seek to protect' (UNSC, 2023b, p. 5).

[25] 'Kiribati strongly believes that the international community, including the Security Council, must invest in women and girls in order to help alleviate the effects of sea level rise on sustainable peace. They have the voice and agency to withstand multiple hazards. In that respect, we need to advance gender-responsiveness in climate processes by ensuring that global knowledge is shared and provide platforms for both intergovernmental agencies and non-governmental organizations to brief the Council on gender-responsive resilience policies' (UNSC, 2023b, p. 23).

[26] 'Measures must also be taken to fast-track women's full, equal and meaningful participation in preventing and addressing climate-affected conflicts in relation to sea level rise' (UNSC, 2023b, p 33).

[27] 'Maldives, we are working to empower women and provide inclusive decision-making environments. We believe it is imperative that the Council do more to increase the participation of women in all aspects of decision-making and peacebuilding processes, including as mediators, negotiators and leaders of security forces' (UNSC, 2023b, p. 36).

and robustly address the protection needs of people from Small Island Developing States in the Pacific region.

3.2 Climate Migration

The second prominent theme related to the climate-security nexus has been the issue of climate migration.[28] Since the first meeting on the subject held by the Security Council, climate migration was an issue of concern, and in the open debate, it has been increasingly recognized, particularly by small island States (Mègret & Mayer, 2018; Joyeeta & Hilmer, 2021). In fact, successive reports of the Intergovernmental Panel on Climate Change (IPCC) have illustrated that rising sea level will increase human mobility at domestic and international levels (IPCC, 2022).

As Verónica Nataniel Macamo Dlhovo, the Minister for Foreign Affairs and Cooperation of Mozambique, observed, 'small island developing States are some of the most peaceful nations in the world today. However, population displacement, loss of territory and possible threats to national identity may deeply affect their stability, peace and security' (UNSC, 2023c, p. 11). Yet the concerns expressed by small island states are connected with other human insecurities:

> Since rising sea levels destroy people's livelihoods, lead to displacement and conflicts over fresh water and fertile land and threaten their very existence, the international community must not ignore their fate and must demonstrate its solidarity.[29]

Other States, such as the Netherlands, highlighted how throughout history, conflicts over resources have been connected with disruption and increased vulnerability by influencing the movement of populations.

---

[28] See the statement of the President: 'It goes without saying that the displacement of hundreds of millions of people is a security risk. With a good part of global agriculture concentrated on coastal plains and low-lying islands, sea level rise is also bringing up long-term questions about humankind's survival' (UNSC, 2023c, p. 4).

[29] See the statement of Alexander Marschik (Austria) who encouraged the Security Council to address climate-security issues as a key priority (UNSC, 2023b, p. 10).

Several States include the protection needs of climate migrants at the core of their statements.³⁰ It is worth mentioning that some scholars highlighted how 'viewing forced migrants as a threat to international peace and security, however, does not automatically mean a genuine interest in the protection of the human rights of these migrants' (Mègret & Mayer, 2018, p. 91). They also highlighted the real risk that the security narrative could be used to 'promote a Western agenda of containing migrants in developing countries at the expense of the effective protection of the human rights [of] migrants and other populations affected by the adverse impacts of climate change' (Mègret & Mayer, 2018, p. 96).

Despite this growing interest, no concrete measures to address this issue have been identified within international law.³¹ It appears difficult for the UN Security Council to take action in this specific domain.

4 THE ROLE OF THE SECURITY COUNCIL AND ITS PROGRESSIVE CLIMATIZATION

This short overview illustrates how the impacts of sea level rise are creating new sources of instability and aggravating existing threats to peace, security, and the identities of States and people. The statements of States and their practice confirm this global trend by consolidating the need for global action. As highlighted by the literature, under international law, the recognition of climate change as a critical 'threat to peace' under Article 39 of the UN Charter could open the door to specific measures such as the adoption of specific sanctions (e.g. Article 41 of the United Nations Charter) or the adoption of specific resolutions (Conca, 2019; Scott & Ku, 2018). Legal scholars have tried to map potential options for the UN Security Council in terms of its taking action in this regard (Scott, 2015).

---

³⁰ Jordan; Denmark; Austria; Dominican Republic; Botswana; Papa New Guinea; EU; Ukraine; Bahrain; Niger; Netherlands; Bangladesh (UNSC, 2023b; UNSC, 2023c).

³¹ The UN Secretary General recalled that: '[t]he current legal regime must be forward-looking and fill the gaps in existing frameworks. That includes international refugee law. It also means putting in place innovative legal and practical solutions to address the impacts of sea level rise on forced displacement and on the very existence of the land territory of some States' (UNSC, 2023c, p. 3).

4.1 Knowledge Sharing for Supporting Cooperative Actions

The majority of the UN Security Council members support and encourage its role as a catalyst of concerted actions (Allen & Yuen, 2022). They consider the UN Security Council as the appropriate actor for this role by arguing that climate change that can increase insecurity.

Antigua and Barbuda made clear how the absence of concrete collective actions 'has made sea level rise a devastating and crippling reality' for many countries.

When looking closely at the suggestions formulated during the open debate, one can see that the role of the UN Security Council could be strengthened by focusing on sharing knowledge and building a common understanding. The States invited to the open debate suggested that the Council should play an influential role by raising awareness of threats to peace by building a shared and deeper understanding of the climate-related security risks and giving States the opportunity to contribute to promoting human security (UNSC, 2023b).[32] As described by Allen and Yuen, the UN Security Council could use a mix of formal and informal procedures together with diverse kinds of meetings to accommodate emerging challenges (Allen & Yuen, 2022, pp. 21–22).[33]

In order to deliver effective responses, Vietnam suggested the 'development of a comprehensive UN Database on the multidimensional impacts of climate change' that could facilitate global climate action (UNSC, 2023b, p. 6). And the Maldives urged the 'Council to monitor and consider the planetary boundary thresholds' 'for maintaining a safe operating space for humankind' (UNSC, 2023b, pp. 34–36). Finally, several States reiterated the need for the appointment of a UN Special Representative on Climate Change and Security.[34]

Overall, it is worth noting that this practice of States confirms that the UN Charter does not prevent the UN Security Council from contributing

---

[32] See the statements by Albania; the United Arab States; Denmark; Greece; Italy, Korea; Vietnam; Denmark; Lebanon; Georgia; the Marshall Islands; Kenya; Niger; Sierra Leone; and the Maldives (UNSC, 2023b, 2023c).

[33] There are, in particular, four types of meetings: public meetings, Arria-formula meetings, informal consultations and private meetings. Arria-formula meetings are informal meetings that involve UN delegations, stakeholders and non-State actors.

[34] See the statements by Albania; the Federated State of Micronesia; Palau; Papua New Guinea; the EU; and Nauru.

to the dialogue on these issues with other UN organs by strengthening the UN system (Allen & Yuen, 2022).

4.2 Facilitating Normative Developments

Furthermore, going forward, as 'an agenda setter' the UN Security Council could shape international responses to emerging security challenges (Day & Malone, 2021). The notions of prevention and cooperation need to become more powerful in the sense of their practical meaning via the Security Council acting unilaterally and concertedly according to commonly agreed rules. This could be beneficial in translating preventive diplomacy (as suggested by Japan) for present and future generations and developing a common agenda that would enable the full realization of the rights of the affected populations. However, it remains unclear how the UNSC could place input into a coordinated approach to addressing 'this common concern' (UNSC, 2023b).

The UN Security Council could contribute to this not only by adopting decisions of a recommendatory nature, but also by requiring States to act under Chapter VII of the UN Charter (Scott & Ku, 2018). The UN Security Council has the power to impose obligations on the UN member States, acting 'de facto' as a 'global legislator'. So even in the absence of an explicit power to legislate, in the last years the Council has adopted significant resolutions that have created new international obligations (Scott, 2015).

So far, in terms of States' obligations, many States expressed great expectations for the outcomes of three requests for advisory opinions that have been promoted before three major international courts (the International Court of Justice, the International Tribunal of the Law of the Sea[35] and the Inter-American Court of Human Rights). For the very first time, these courts were confronted with the extremely controversial questions of how to frame the obligations and duties of the international communities in response to global climate risks that disregard the borders of nation-States. In this vein, one could wonder whether the UN Security Council, through its decisions in the field of climate change, could work

---

[35] ITLOS (2024, May 21). Advisory Opinion. https://www.itlos.org/en/main/cases/list-of-cases/request-for-an-advisory-opinion-submitted-by-the-commission-of-small-island-states-on-climate-change-and-international-law-request-for-advisory-opinion-submitted-to-the-tribunal. For a detailed analysis see Chapter 9 by Doebbler.

in parallel with the climate and human rights regimes by supporting their implementation.

5 Conclusions

The article offered some preliminary insights into the attempts to 'climatize' the role of the UN Security Council by illustrating the unsettled meaning of climate security. To invoke this notion some key questions still need to be addressed: Whose security? Who is responsible for providing security? The literature highlights the need to shift from a traditional approach that focuses on the role of States as 'the referent of security' to bring into this field 'real human insecurities and vulnerabilities' (Elliot, 2023, p. 39). This requires redefining existing practices and calling for new roles for existing institutions.

The chapter allowed us to engage with the debate at the UN Security Council, which is still contested, to explore how States understand the linkages between climate change and security and frame their practices to address major issues of concern (e.g. human mobility, State disappearance, the impact on human rights protection). In a possible articulation of this field, a better understanding of how threats of climate change—namely induced sea level rise—raise direct and indirect security threats has been repeatedly suggested by States. Consequently, the legal dimension of the debate requires us to focus on responsibilities of those failing to adapt to them because it risks perpetrating traditional ways of dealing with these problems.

Significant concerns remain as to how global challenges such as the adverse impacts of climate change could be addressed at the international level. How can we foster coordinated approaches at different levels? The overview has sought to illustrate some of the critical questions that persist for the international community in its efforts to establish concerted efforts in the face of sea level rise/climate change, particularly with regard to its security implications.

As stressed by the UN Secretary General,

> [t]he Security Council has a critical role to play in mobilizing the political will to address the devastating security challenges posed by rising seas. We must all continue to give this issue the visibility it deserves and to support the lives, livelihoods and communities on the front line of this crisis (UNSC, 2023c, p. 4).

The UN Security Council could play a key role by building a closer cooperation by meeting existing, new, and emerging threats. More specifically, facing unprecedented challenges requires overcoming the limits of nationally driven legislation by designing and collectively implementing multilateral, cooperative responses.

References

Allen, S. H., & Yuen, A. (2022). Bargaining in the UN Security Council: Setting the Global Agenda. Oxford University Press. https://doi.org/10.1093/oso/9780192849755.001.0001

Arias, S. B. (2022a, October 11). *The UN Security Council Declined to Take Up Climate Change as a Security Problem. Why?* Columbia. https://multilateralism.sipa.columbia.edu/news/un-security-council-declined-take-climate-change-security-problem-why

Arias, S. B. (2022b). Who Securitizes? Climate Change Discourse in the United Nations. *International Studies Quarterly, 66*(2). https://doi.org/10.1093/isq/sqac020

Asian Development Bank. (2022, August). *Women's Resilience in Fiji: How Laws and Policies Promote Gender Equality in Climate Change and Disaster Risk Management*. https://www.adb.org/publications/women-resilience-fiji-laws-policies-gender-equality-climate-change-disaster

Benton Heath, J. (2022). Making Sense of Security. *American Journal of International Law, 116*(2), 289–339. https://doi.org/10.1017/ajil.2021.63

Boas, I. (2015). *Climate Migration and Security, Securitisation as a Strategy in Climate Change Politics*. Routledge Research in Environmental Politics. https://doi.org/10.4324/9781315749228

Bonafé, B., & Arcari, M. (2021). Climate Security: Global Issues v. Regional Responses. *Questions of International Law, Zoom-in, 84*, 1–2.

Borras-Pentinat, S. (2022). Gender Climate Justice in a Context of Intersectional Vulnerabilities. In M. Campins Eritja & R. Bentirou Mathlouthi (Eds.), *Understanding Vulnerability in the Context of Climate Change* (pp. 103–122). Atelier.

Campbell, M. (2023). A Greener CEDAW: Adopting a Women's Substantive Equality Approach to Climate Change. In C. Albertyn, M. Campbell, H. Alviar Garcia, & S. Fredman (Eds.), *Feminist Frontiers in Climate Justice, Gender Equality, Climate Change and Rights* (pp. 90–115). Edward Elgar Publishing.

Committee on the Elimination of Discrimination Against Women (CEDAW). (2018). *Summary Record of the 1578th Meeting* (UN Doc CEDAW/C/

SR.1578). https://documents.un.org/doc/undoc/gen/g18/049/76/pdf/g1804976.pdf

Conca, K. (2019). Is There a Role for the UN Security Council on Climate Change? *Environment: Science and Policy for Sustainable Development, 61*(1), 4–15. https://doi.org/10.1080/14781158.2013.787058

Cusato, E. (2022). Of Violence and (In)Visibility. The Securitisation of Climate Change in International Law. *London Review of International Law, 10*(2), 203–242. https://doi.org/10.1093/lril/lrac015

Davies, K., & Riddell, T. (2017). The Warming War: How Climate Change Is Creating Threats to International Peace and Security. *The Georgetown International Environmental Law Review, 30*(1), 47–74.

Day, A., & Harper, E. (2023, May). *Research Brief: Climate Change in the Security Council: Obstacles, Opportunities and Options*. https://www.geneva-academy.ch/joomlatools-files/docman-files/Resarch%20Brief%20Climate%20Cange%20in%20the%20Security%20Council.pdf

Day, A., & Malone, M. D. (2021). The Role of the United Nations in Shaping Global Security Law. In R. Geiss & N. Melzer (Eds.), *The Oxford Handbook of the International Law of Global Security* (pp. 1071–1088). Oxford University Press.

Day, A., Vivekananda, J., & Pacillo, G. (2023, January 30). *Climate Change in the Security Council: What New Council Members Can Achieve in 2023*. The Global Observatory. https://reliefweb.int/report/world/climate-change-security-council-what-new-council-members-can-achieve-2023

De Wet, E. & Wood, M. (2012). Collective Security. In R. Wolfrum (Ed.), *Max Planck Encyclopaedia of Public International Law*. Oxford University Press.

Dietz, T., von Lucke, F., & Wellman, Z. (2016). *The Securitization of Climate Change: Actors*. Routledge. https://doi.org/10.4324/9781315665757

Elliot, L. (2023). Climate Change and Human Security: Implications for International Security. In M. J. Trombetta (Ed.), *Handbook on Climate Change and International Security* (pp. 34–50). Edward Elgar Publisher.

Fornalé, E. (2023). Slow Violence, Gender Equality and Climate Agency in Times of 'Polycrisis'. *Revista General de Derecho Europeo, 61.* https://www.iustel.com/v2/revistas/detalle_revista.asp?id_noticia=426572&d=1

Fornalé, E., Bilkova, V., Burgorgue-Larsen, L., Cristani, F., De Vido, S., Doebbler, C., & Hertogen, A. (2023). *Amicus Curiae - Request for an advisory opinion on the Climate Emergency and Human Rights submitted to the Inter-American Court of Human Rights by the Republic of Colombia and the Republic of Chile of January 9, 2023*. https://doi.org/10.48350/190814

Intergovernmental Panel on Climate Change (IPCC). (2022). *Sixth Assessment Report: Impacts, Adaptation and Vulnerability*. https://www.ipcc.ch/report/ar6/wg2/chapter/chapter-7

Joyeeta, G., & Hilmer, B. (2021). Climate Change and Security. In R. Geiss Robin & N. Melzer (Eds.), *The Oxford Handbook of the International Law of Global Security* (pp. 548–565). Oxford University Press.

Krieger, T., Panke, D., & Pregernig, M. (2020). *Environmental Conflicts, Migration and Governance*. Bristol University Press. https://doi.org/10.56687/9781529202175

Maertens, L. (2018). Le changement climatique en débat au Conseil de sécurité de l'ONU. *Revue Internationale et Stratégique, 109*, 105–114. https://doi.org/10.3917/ris.109.0105

Maertens, L., & Trombetta, M. J. (2023). Climate Change at the United Nations Security Council: Securitization, Climatization and Beyond. In M. J. Trombetta (Ed.), *Handbook on Climate Change and International Security* (pp. 182–200). Edward Elgar Publisher.

Mègret, F., & Mayer, B. (2018). 'Climate Migration' and the Security Council. In S. Scott & C. Ku (Eds.), *Climate Change and the UN Security Council* (pp. 85–96). Edward Elgar Publishing.

Murphy, A. (2019). The United Nations Security Council and Climate Change: Mapping a Pragmatic Pathway to Intervention. *Carbon and Climate Law Review, 19*(1), 50–62. https://www.jstor.org/stable/26739643

Nasu, H. (2021). The Global Security Agenda: Securitization of Everything? In R. Geiss & N. Melzer (Eds.), *The Oxford Handbook of the International Law of Global Security*. Oxford Academic. https://doi.org/10.1093/law/9780198827276.001.0001

Palchetti, P. (2022). Débattre des changements climatiques au Conseil de sécurité: Pour quoi faire? *Questions of International Law, Zoom Out 1*. http://www.qil-qdi.org/wp-content/uploads/2022/05/03_Sea-Level-Rise_PALCHETTI_FIN.pdf

Poku, N. K. (2013). HIV/AIDS, State Fragility, and United Nations Security Council Resolution 1308: A View from Africa. *International Peacekeeping, 20*(4), 521–535. https://doi.org/10.1080/13533312.2013.854580

Rowena, M. (2019). Gender, Climate Change and the United Nations Framework Convention on Climate Change. In S. Harris Rimmer & K. Ogg (Eds.), *Research Handbook on Feminist Engagement with International Law* (pp. 63–80). Edward Elgar Publishing.

Scott, S. (2015). Implications of Climate Change for the UN Security Council: Mapping the Range of Potential Policy Responses. *International Affairs, 91*(6), 1317–1333. https://www.jstor.org/stable/24539058

Scott, S., & Ku, C. (2018). *Climate Change and the UN Security Council*. Edward Elgar Publishing.

Schroeder, U. (2021). The Transformation of Security Concepts: Beyond the State. In R. Geiss & N. Melzer (Eds.), *The Oxford Handbook of the International Law of Global Security* (pp. 54–68). https://doi.org/10.1093/law/9780198827276.003.0004

Tomuschat, C. (2022). The Changing Faces of the UN Security Council. *German Yearbook of International Law, 63*(1), 647–705.

Torres Camprubi, A. (2016). Securitization of Climate Change: The Inter-Regional Institutional Voyage. *Yearbook of International Environmental Law, 27*, 82–105. https://doi.org/10.1093/yiel/yvy079

Trombetta, M. J. (2023). Handbook on Climate Change and International Security. *Edward Elgar Publisher*. https://doi.org/10.4337/9781789906448

Webersik, C. (2012). Securitizing Climate Change: The United Nations Security Council Debate. In D. Gallagher (Ed.), *Environmental Leadership: A Reference Handbook*. SAGE Publications. https://doi.org/10.4135/9781452218601.n58

White, N., & Davies-Bright, A. (2021). The Concept of Security in International Law. In R. Geiss & N. Melzer (Eds.), *The Oxford Handbook of the International Law of Global Security* (pp. 19–36). Oxford University Press.

Wood, M. (2022). The UN Security Council and International Law. *Cambridge University Press*. https://doi.org/10.1017/9781108692373

Security Council—Documents

United nations Security Council (UNSC). (2007a, April 17). *Security Council Holds First-Ever Debate on Impact of Climate Change on Peace, security, Hearing Over 50 Speakers* (UN Doc SC/9000). https://press.un.org/en/2007/sc9000.doc.htm

United Nations Security Council (UNSC). (2007b, April 17). *Meeting Record* (UN Doc S/PV.5663). https://www.securitycouncilreport.org/atf/cf/%7B65BFCF9B-6D27-4E9C-8CD3-CF6E4FF96FF9%7D/s_pv_5663.pdf

United Nations Security Council (UNSC). (2007c, June 25). *Declaration of the President of the Security Council* (UN Doc S/PRST/2007/22). https://www.securitycouncilreport.org/atf/cf/%7B65BFCF9B-6D27-4E9C-8CD3-CF6E4FF96FF9%7D/NRC%20SPRST%202007%2022.pdf

United Nations Security Council (UNSC). (2011a, July 20). *Declaration of the President of the Security Council* (UN Doc S/PRST/2011/15). https://www.securitycouncilreport.org/atf/cf/%7B65BFCF9B-6D27-4E9C-8CD3-CF6E4FF96FF9%7D/CC%20SPRST%202011%205.pdf

United Nations Security Council (UNSC). (2011b, July 20). *Meeting Record* (UN Doc S/PV.6587). https://www.securitycouncilreport.org/atf/cf/%7B65BFCF9B-6D27-4E9C-8CD3-CF6E4FF96FF9%7D/CC%20SPV%206587.pdf

United Nations Security Council (UNSC). (2011c, November 23). *Meeting Record* (UN Doc S/PV.6668). https://www.securitycouncilreport.org/atf/cf/%7B65BFCF9B-6D27-4E9C-8CD3-CF6E4FF96FF9%7D/SPV6668.pdf

United Nations Security Council (UNSC). (2015, July 30). *Meeting Record* (UN Doc S/PV.7499). https://www.securitycouncilreport.org/atf/cf/%7B65BFCF9B-6D27-4E9C-8CD3-CF6E4FF96FF9%7D/s_pv_7499.pdf

United Nations Security Council (UNSC). (2017b, July 31). *In Hindsight: The Security Council and Climate Change: An Ambivalent Relationship. Report*. https://www.securitycouncilreport.org/monthly-forecast/2017-08/the_security_council_and_climate_change_an_ambivalent_relationship.php

United Nations Security Council (UNSC). (2020, December 16). *Arria-Formula Meetings: Report on Working Methods*. https://www.securitycouncilreport.org/un-security-council-working-methods/arria-formula-meetings.php

United Nations Security Council (UNSC). (2021a, December 13). *Draft Resolution on Climate and Security* (UN Doc. S/2021/990). https://www.securitycouncilreport.org/atf/cf/%7B65BFCF9B-6D27-4E9C-8CD3-CF6E4FF96FF9%7D/s_2021_990.pdf

United Nations Security Council. (UNSC). (2021b, December 13). *Maintenance of International Peace and Security. Climate and Security* (UN Doc S/PV.8926) https://docs.un.org/en/S/PV.8926

United Nations Security Council (UNSC). (2023a, February 2). *Letter Dated 2 February 2023 From the Permanent Representative of Malta to the United Nations Addressed to the Secretary-General* (UN Doc S/203/79). https://documents.un.org/doc/undoc/gen/n23/032/39/pdf/n2303239.pdf

United Nations Security Council (UNSC). (2023b, February 14). *Meeting Record (Resumption 1)* (UN Doc S/PV.9260). https://documents.un.org/doc/undoc/gen/n23/045/04/pdf/n2304504.pdf

United Nations Security Council (UNSC). (2023c, February 14). *Meeting Record* (UN Doc S/PV.9260). https://documents.un.org/doc/undoc/gen/n23/044/11/pdf/n2304411.pdf

Other UN Documents

United Nations General Assembly (UNGA). (2009a, December 25). *Climate Change and Its Possible Security Implications: Report of the Secretary-General* (UN Doc A/RES/66/290). https://daccess-ods.un.org/access.nsf/Get?OpenAgent&DS=A/RES/66/290&Lang=E

United Nations General Assembly (UNGA). (2009b, May 18). *Resolution* (UN Doc A/63/281). https://digitallibrary.un.org/record/656156/files/A_RES_63_281-EN.pdf

United Nations General Assembly (UNGA). (2019, May 1). *Analytical Study on Gender-Responsive Climate Action for the Full and Effective Enjoyment of the*

Rights of Women (UN Doc A/HRC/C/41/26). https://www.ohchr.org/Documents/Issues/ClimateChange/GenderResponsive/A_HRC_41_26.pdf

United Nations Secretary-General (UNSG). (2009, September 11). *Climate Change and Its Possible Security Implications* (UN Doc A/64/350). https://sdgs.un.org/documents/a64350-climate-change-and-its-possible-securit-18103

United Nations Secretary-General (UNSG). (2021a). *Remarks to the Security Council on Addressing Climate-Related Security Risks to International Peace and Security Through Mitigation and Resilience Building*. https://www.un.org/sg/en/content/sg/statement/2021-02-23/secretary-generals-remarks-the-security-council-addressing-climate-related-security-risks-international-peace-and-security-through-mitigation-and-resilience-building

United Nations Secretary-General (UNSG). (2021b). *Global Climate Crisis' Dire Impact on Peace, Security Calls for Bolder Collective Action, Secretary-General Tells Security Council*. https://press.un.org/en/2021/sgsm20926.doc.htm

United Nations Secretary-General (UNSG). (2022, January 4). *Achieving Gender Equality and the Empowerment of All Women and Girls in the Context of Climate Change, Environmental and Disaster Risk Reduction Policies and Programmes* (UN Doc E/CN/.6/2022/3). United Nations Economic and Social Council. https://digitallibrary.un.org/record/3956348?ln=en&v=pdf

United Nations Special Rapporteur on the Issue of Human Rights Obligations Relating to the Enjoyment of a Safe, Clean, Healthy and Sustainable Environment. (2023, January 5). *Report of the Special Rapporteur on Human Rights and the Environment* (A/HRC/52/33). United Nations Human Rights Council. https://www.ohchr.org/en/documents/thematic-reports/ahrc5233-women-girls-and-right-clean-healthy-and-sustainable-environment

United Nations Women. (2021). *Measuring the Nexus Between Gender Equality and Women's Empowerment and the Environment, Including Climate Change and Disaster Risk Reduction*. https://www.unwomen.org/sites/default/files/Headquarters/Attachments/Sections/CSW/66/EGM/Info%20Papers/UN%20Women_CSW66%20Informational%20Paper.pdf

UN Special Rapporteur on Violence Against Women. (2022, July 11). *The Report on Violence Against Women and Girls in the Context of the Climate Crisis, Including Environmental Degradation and Related Disaster Risk Mitigation and Response* (UN Doc A/77/136). https://documents.un.org/doc/undoc/gen/n22/418/07/pdf/n2241807.pdf

Open Access This chapter is licensed under the terms of the Creative Commons Attribution 4.0 International License (http://creativecommons.org/licenses/by/4.0/), which permits use, sharing, adaptation, distribution and reproduction in any medium or format, as long as you give appropriate credit to the original author(s) and the source, provide a link to the Creative Commons license and indicate if changes were made.

The images or other third party material in this chapter are included in the chapter's Creative Commons license, unless indicated otherwise in a credit line to the material. If material is not included in the chapter's Creative Commons license and your intended use is not permitted by statutory regulation or exceeds the permitted use, you will need to obtain permission directly from the copyright holder.

PART II

Sea Level Rise and the Rights of Affected Population

CHAPTER 5

Sea Level Rise and Human Rights

Veronika Bílková

1 Introduction[1]

Sea level rise, one of the phenomena accompanying climate change, has a negative effect on the enjoyment of human rights. It makes it impossible for people to earn a living in the way they have been used to for centuries, it limits their access to drinking water or food supplies, and it forces them to leave their homes and, sometimes, their State. Such a State may even cease to exist in the future or, more specifically, it may lose its territorial basis (Bílková, 2017). It is therefore no surprise that international bodies studying the effects that the rising sea level has had or might have on international law could not ignore the effects of this phenomenon on human rights. This is the case of the UN International Law Commission (ILC), which included the topic of *Sea Level Rise in relation to International Law* in its programme of work in 2019 and

---

[1] This chapter is based on the work carried out within the framework of the Horizon Europe HRJust project (States' Practice of Human Rights Justification: a study in civil society engagement and human rights through the lens of gender and intersectionality), GA No. 101094346 and the SERI (Swiss State Secretariat for Education, Research and Innovation—SERI) under grant agreement n. 23.00131/1,010,943,546.

V. Bílková (✉)
Institute of International Relations, Prague, Czech Republic
e-mail: bilkova@iir.cz

© The Author(s) 2026
E. Fornalé (ed.), *Sea Level Rise*,
https://doi.org/10.1007/978-3-031-89171-7_5

identified the protection of persons affected by sea level rise as one of the crucial aspects of this topic (ILC, 2019). It is also the case of the International Law Association (ILA) Committee on International Law and Sea Level Rise (ILA, 2024), established in 2012, which has persistently described human rights and forced migration as one of the three areas of international law most likely to be affected by sea level rise, with the other two being the law of the sea and issues of statehood.

That sea level rise has a negative effect on the enjoyment of human rights is therefore largely accepted and unquestioned. What exactly this effect is, whether and when it entails violations of human rights law, which human rights can get violated, who the victims of such violations are, which States (or other actors) bear the corresponding duties, and what the content of such rights and duties could be, is, however, much less clear and settled.[2] The aim of this paper is to consider these questions. This is done mainly in the second section of the paper. The first section shows how sea level rise has (or has not) been reflected in international human rights instruments and the case law of international human rights bodies. The paper builds on the work done by the ILC and the ILA Committee, as well as on books and articles dealing with the relationship between sea level rise and human rights, although such publications are much less numerous than one could expect (Cataldi, 2014). It also takes inspiration from more general texts focusing on the relationship between climate change and human rights, including international case law that progressively piles up in all the regions of the world.[3]

2 Sea Level Rise and Its Effects on Human Rights

Sea level rise is one of the consequences of climate change.[4] Since 1880, the sea level has risen by some 21–24 cm, with almost half of it gained in the last three decades. With the rate of some 4 mm per year and with

---

[2] See also UN (2009), para 70(*'While climate change has obvious implications for the enjoyment of human rights, it is less obvious whether, and to what extent, such effects can be qualified as human rights violations in a strict legal sense'*).

[3] See, for instance, Humphreys (2009); and the case-law cited in this paper.

[4] For more details, see the chapter by Vilane Gonçalves Sales (Sea Level Rise: Scientific Evidence, Socio-Economic Realities, and Adaptation Challenges for Coastal Communities) in this book.

this rate gradually accelerating, the sea level is likely to rise by a further 20–25 cm by 2050. Melting of ice sheets and glaciers, especially in Greenland and the Antarctic, and thermal expansion of water, both of which are caused by global warming, are the main factors accounting for these changes (Lindsey, 2023; Mimura, 2013). Whereas climate change and global warming affect all countries of the world, sea level rise has a disproportionate effect on smaller island States and other low-lying countries. The four countries at the most acute risk in this respect are the Maldives, Tuvalu, the Marshall Islands, and Kiribati, with 99–100% of their population living within 5 metres above sea level. Other countries that are likely to be seriously affected by sea level rise include Bangladesh, Vietnam, Indonesia, Haiti, Cuba, Ireland, the Netherlands and the UK. As the list shows, the phenomenon is not limited to one region of the world but is truly global in nature. It also hits both well-developed and less-developed countries.

Sea level rise affects the enjoyment of human rights. It does so, on the one hand, by exposing people to increasing scarcity of resources, depriving them of their traditional lifestyles, and forcing them to change where they live and how they live. It also does so, on the other hand, by weakening States and making it more difficult for some of them to abide by their obligations to respect, protect and fulfil the human rights of their inhabitants. International human rights bodies have confirmed that sea level rise, and climate change more broadly, can not only affect the enjoyment of human rights but even violate these rights (IACHR, 2017). The range of human rights which can be impaired in this way is a broad one, as it includes the right to life, the right to private and family life, the right to the freedom of movement, the right to property, the right to an adequate standard of living, the right to the enjoyment of the highest attainable standard of physical and mental health, the right to development, and the right to a clean, healthy, and sustainable environment. Sea level rise thus potentially impairs all categories of human rights—civil and political rights, as well as economic, social, and cultural rights.

International human rights instruments do not refer to sea level rise explicitly. Several of them, however, enshrine the right to a clean, healthy, and sustainable environment (UN, 2019).[5] The existence of this right

---

[5] See Article 24 of the African Charter of Human and Peoples' Rights, Article 11 of the Protocol to the American Convention on Human Rights in the Area of Economic, Social

was also recognized by the UN General Assembly in a landmark resolution titled *The Human Right to a Clean, Healthy, and Sustainable Environment* adopted in July 2022 (UNGA, para. 1).[6] The resolution followed a resolution with the same title and broadly similar content adopted in October 2021 by the UN Human Rights Council (UNHRCo, 2021). Although neither resolution explicitly mentions sea level rise, they both refer to climate change, noting that it constitutes one of the *'most pressing and serious threats to the ability of present and future generations to effectively enjoy all human rights'* (UNGA, para. 13 of the preamble; UNHRCo, 2021, para. 11 of the preamble) and that environmental damage *'has negative implications, both direct and indirect, for the effective enjoyment of all human rights"* '(UNGA, para. 10 of the preamble; UNHRCo, 2021, para. 9 of the preamble).

The right to a clean, healthy, and sustainable environment has already been invoked at the national level with respect to climate change, for instance in cases related to climate change impact assessment (High Court of South Africa, 2017; Montana First Judicial District Court, United States, 2023). It is certainly just a question of time when applications concerning sea level rise submitted at the national or international level will rely on this human right as well. The Advisory Opinions on climate change requested from the International Court of Justice (ICJ) (ICJ, 2023) and the Inter-American Court of Human Rights (IACHR, 2023) are also likely to cast light on the scope of the application and content of this right.

So far, however, other human rights have been more commonly relied on in case law related to sea level rise, especially at the international level. This case law has two main strands. One concerns climate change asylum claims, i.e. claims of individuals who have left their country of origin due to the problems stemming from sea level rise and have sought asylum, mostly unsuccessfully, in another country. These cases typically focus on the interpretation of the principle of non-refoulement, which prohibits States from expelling or returning a person to a country where that person

---

and Cultural Rights (Protocol of San Salvador), Article 38 of the Arab Charter on Human Rights and Article 28(f) of the Human Rights Declaration of the Association of Southeast Asian Nations (ASEAN). The right to a safe, healthy, and sustainable environment is also enshrined in many national constitutions.

[6] The resolution was adopted by a vote of 161 in favour to 0 against, with 8 abstentions (Belarus, Cambodia, China, Ethiopia, Iran, Kyrgyzstan, the Russian Federation, and Syria).

would face a real risk of persecution, of serious human rights violations, or of irreparable harm. The principle originated in international refugee law but has been gradually incorporated into international human rights law as well (Hamdan, 2016). The main question which arises here is whether and, if so, under what conditions and in which moment the effects of sea level rise in one country might trigger the application of the principle of non-refoulement by other countries.

The first case in which this question was discussed at the international level was the case *Teitiota v. New Zealand*, decided by the UN Human Rights Committee (HRC) in 2019 (UNHRC, 2019b; McAdam, 2020).[7] The applicant, Mr Ioane Teitiota, was a national of Kiribati who left his country and moved to New Zealand, where he applied for asylum. His application was not granted and together with his wife and children, he was deported back to Kiribati. He argued that by deporting him, New Zealand violated his right to life protected by Article 6 of the International Covenant on Civil and Political Rights (ICCPR) because sea level rise and other effects of climate change had rendered Kiribati almost uninhabitable due to environmental degradation, contamination of freshwater supplies and violent and disputes linked to the situation.

The HRC did not find any violation of Article 6 of the ICCPR. Yet, on a general level, it confirmed that *'environmental degradation, climate change and unsustainable development constitute some of the most pressing and serious threats to the ability of present and future generations to enjoy the right to life'* (UNHRC, 2019b, para. 9.4). It also noted that *'given that the risk of an entire country becoming submerged under water is / .../ an extreme risk, the conditions of life in such a country may become incompatible with the right to life with dignity before the risk is realized'* (UNHRC, 2019b, para. 9.4). Were this scenario to materialize—as was not the case in Kiribati at that time according to the HRC—the principle of non-refoulement would apply and the deportation of a person back to such a country could qualify as a violation of the right to life. The decision was heralded as *'a historic UN Human Rights case open/ing/ /a/ door to climate change asylum claims"* '(UN, 2020). Although no flood of such claims has yet reached international human rights bodies, cases involving the application of the principle of non-refoulement to asylum

---

[7] See also the chapter by Sara De Vido (Sea Level Rise as a Form of Gendered Climate Violence: International Legal Implications for Migration) in this book.

seekers leaving countries affected by sea level rise are likely to increase in number in the coming years.

The second strand of case law deals with the impact of sea level rise on human rights more broadly. It considers whether and, if so, how and in which context sea level rise, its effects and the failure by States to take adequate measures to prevent such effects interfere with human rights. Vulnerable individuals and groups of individuals, such as indigenous peoples, minorities, or women, have been at the centre of attention in this regard so far. This was so, for instance, in the case *Billy v. Australia* (UNHRC, 2022; Kahl, 2022; Cullen, 2018), considered by the HRC in 2022. The complaint was filed by eight Australian nationals and six of their children, who belonged to indigenous communities of the small low-lying islands in Australia's Torres Strait region. The applicants claimed that Australia violated their right to life, right to private, family and home life, minority rights and rights of children under Articles 6, 17, 24, and 27 of the ICCPR by failing to adopt measures mitigating the impact of climate change on vulnerable groups of populations, especially indigenous people, children, and, also, future generations.

The HRC did not find a violation of the right to life in this case. Relying on its position in the *Teitiota* case, it recalled that *'without robust national and international efforts, the effects of climate change may expose individuals to a violation of their rights under article 6 /of the ICCPR/"* '(UNHRC, 2022, para. 8.7). This, however, according to the HRC, was not the case here, first because Australia was *'taking adaptive measures to reduce existing vulnerabilities and build resilience to climate change-related harms on the islands'* (UNHRC, 2022, para. 8.7), and second because the situation was not so critical as to expose the applicants to adverse impacts on their health or a real and reasonably foreseeable risk of being exposed to a situation of physical endangerment or extreme precarity that could threaten their right to life. At the same time, the HRC found violations of the right to private, family and home life and of minority rights. It did so by stressing the special relationship of persons belonging to indigenous communities with their territory and the obligation of States to adopt positive measures ensuring the protection of this relationship. The HRC opined that although Australia sought to put such measures in place, there were unsubstantiated delays in their implementation, which amounted to a violation of Articles 17 and 27 of the ICCPR. Having found these violations, the HRC did not consider it necessary to examine the claims related to the rights of children and future generations (Article 24 of the ICCPR).

Billy v. Australia was rightly labelled as *'a case of many firsts for the advancement of environmental protection at the international law level, as well as for the advancement of Indigenous Persons' rights'"* (ESCR-Net, 2022). It is indeed the first case in which an international human rights body has not only generally confirmed that climate change may affect human rights but also found a violation of such rights. It is also the first case in which the special vulnerability of certain communities, mainly indigenous people, was recognized. That *'the damaging effects of climate /.../ impact most heavily on various vulnerable groups in society, who need special care and protection from the authorities"* (ECtHR, 2024, para. 410)[8] has also been declared since then, this time with respect to elderly persons, by the European Court of Human Rights (ECtHR) in *Verein KlimaSeniorinnen Schweiz and Others v. Switzerland*. Although this case did not refer to sea level rise specifically, it confirmed the readiness of international human rights bodies to use the concept of positive obligations stemming from various human rights to assess whether States were doing enough to protect their inhabitants against negative effects of climate change, including sea level rise. This approach, however, gives rise to questions relating to the scope of the obligations, their content as well as the States or other actors concerned by them (right-holders and duty-bearers). These questions will be examined in the next section.

3 Legal Questions Related to Human Rights and Sea Level Rise

During the debate on the protection of persons affected by sea level rise in the ILC Study Group on the topic, some of the members expressed doubts as to whether the international human rights law framework could be truly relevant for this protection. They noted that

> while States had human rights obligations towards individuals, the sea level rise phenomenon was not directly attributable to any particular State. Accordingly, it was unclear how human rights rules would operate within that context and, specifically, how and against whom claims related to sea level rise could be brought. Those questions were considered even more

---

[8] See also the chapter by Cristani and Fornalé in this book.

pertinent in the case of a State whose territory was completely submerged or rendered uninhabitable (ILC, 2022, para. 64).

Sea level rise, and climate change more generally, thus not only affect the enjoyment of human rights but also raise interesting and uneasy questions about how human rights actually operate. More specifically, they raise questions about the categories of human rights that can be violated in the context of sea level rise, the identity of potential victims, the range of potential duty bearers, and the content of relevant rights and duties. These four questions are discussed in the four parts of this section. It is important to stress that the questions are interrelated and the answer to one of them largely determines how the others shall be addressed.

3.1   Human Rights Potentially Violated in Connection with Sea Level Rise

The question of which human rights can be violated in the context of sea level rise has already been partially addressed in the previous section. Three points are worth highlighting and elaborating on here.

First, we saw that the range of human rights that can be violated in this context is very broad. It encompasses not only the right to a clean, healthy, and sustainable environment but also various other human rights belonging both to the category of civil and political rights (the right to life, the right to private and family life, the right to freedom of movement, the right to property, etc.) and to that of economic, social and cultural rights (the right to an adequate standard of living, the right to the enjoyment of the highest attainable standard of physical and mental health, etc.). Certain new rights, which are often considered collective in nature, such as the right to development, may be affected as well. Put the other way round, it might actually be difficult to identify a single human right which could not be possibly violated in connection with sea level rise. This shows the comprehensive nature of the challenges posed by sea level rise and climate change as such. It also attests to the interrelated and interconnected nature of all human rights. As rightly noted by the IACHR, *'the different categories of rights constitute an indivisible whole based on the recognition of the dignity of the human being'* (IACHR, 2017, para. 47).

In this way, sea level rise, and climate change more broadly, are similar to certain other emergency situations that also have a dramatic impact on a range of human rights and can seriously affect and limit the ability of

States to respect, protect and fulfil these rights. Such situations include armed conflicts, natural disasters, manmade disasters, or pandemics. In all these situations, human rights continue to apply, though States may be obliged to restrict some of them or even temporarily suspend their application (so-called derogation). Such restrictions and derogations must be required by the exigency of the situation, proportionate to the threat posed by it, and necessary for helping to overcome it and installing the normal legal regime. There is a rich case law of international human rights bodies related to such exceptional situations that could serve as a source of inspiration for sea level rise-related cases (Hafner-Burton et al., 2011), as could instruments related to these situations, such as, for instance, the *Draft Articles on the Protection of Persons in the Event of Disasters*, adopted by the ILC in 2016 (UN, 2016, pp. 13–17).

It is, however, important to keep in mind that sea level rise and climate change differ from most of the other exceptional situations in several important ways. Unlike armed conflicts, disasters, or pandemics, they are not so much an event limited in time and space but rather a process which unfolds gradually in time without any specific starting or ending point. They are also not limited to any particular area or region, though different areas and regions may be hit by them in different ways and in different periods. Such a process has a complex impact on societies and individuals and it might, in fact, be more similar to underlying causes of other emergency situations—for instance, tensions leading to an outburst of violence, or neglect in management leading to a manmade disaster—rather than to those situations as such. The specific features of sea level rise, especially its irreversibility and long-term nature (ILC, 2022, para. 69), have been acknowledged by the ILC Study Group on the protection of persons affected by sea level rise, which *'cautioned against drawing too many parallels, in particular with the draft articles on the protection of persons in the event of disasters'*(ILC, 2024, para. 18). While the standards applicable to other emergencies can serve as a source of inspiration, they cannot be mechanically transferred to the sea level rise context and used here without taking the specific features of this phenomenon into account.

The second point which deserves attention concerns the relevance of various human rights that can be violated in connection with sea level rise. What are the pros and cons of relying on certain rights rather than on others? Is it more appropriate to invoke the specific right to a clean, healthy, and sustainable environment than more general human rights (such as the right to life or the right to property)? These questions are very important for individuals who decide to submit an application to national courts or international human rights bodies. It is essential to stress that the answers to them are to a large extent context-specific, depending on the legal and jurisdictional basis for the operation of such courts and bodies. It is, for instance, impossible to rely on the right to a clean, healthy, and sustainable environment at the UN level or before the ECtHR because neither the two UN Covenants nor the ECHR enshrine such a right. The possibility of including it in the ECHR has been discussed for years (Council of Europe, Parliamentary Assembly, 2009) without, however, yielding any tangible effects so far.

The right to a clean, healthy, and sustainable environment has not, moreover, been actively relied on even in regional systems which recognize it. One exception is the *Ogoniland* case (AfCmHR, 2001; Coomans, 2003), decided by the African Commission on Human Rights (AfCmHR) in 2001. The applicant, representing the Ogoni community, alleged that Nigeria failed to prevent and, in fact, took part in the contamination of the environment of the Ogoni people through oil-extraction activities run by the State-owned oil company and that, by doing so, it violated, *inter alia*, the right to a general satisfactory environment favourable to the development enshrined in Article 24 of the African Charter of Human and Peoples' Rights. The AfCmHR shared this view, stressing that the right *'imposes clear obligations upon a government. It requires the State to take reasonable and other measures to prevent pollution and ecological degradation, to promote conservation, and to secure an ecologically sustainable development and use of natural resources'*(AfCmHR, 2001, para. 52). Although the case did not specifically refer to sea level rise, the obligations defined in it would undoubtedly apply in this context as well.

As Boyle (2012) noted in his article, relying on the specific right to a clean, healthy, and sustainable environment, rather than on general human rights, might have advantages. This right would *'address the environment as a public good'* (Boyle, 2012, p. 628) and would add *'a broader and more explicit focus on environmental quality'*(Boyle, 2012, p. 629). The case law of the HRC and the ECtHR, however, shows that the absence of such a right in the relevant instrument is not an obstacle to the efficient judicial protection against the effects of climate change. Obligations largely similar to those identified in the *Ogoniland* case have been read into various substantive human rights. The case law also shows that the obligations, and the threshold of violation, differ among various rights. Thus, in the *Billy* case, the failure by Australia to take timely measures to protect indigenous communities against climate change-related environmental degradation was found serious enough to constitute a violation of the right to private and family life and of minority rights, but not so critical as to give rise to a violation of the right to life. The positive obligations stemming from different human rights with respect to sea level rise are therefore not identical.

The third point to raise relates to the nature of human rights that can be violated in connection with sea level rise. More specifically, it deals with the interplay between substantive and procedural rights. The *Convention on Access to Information, Public Participation in Decision-making and Access to Justice in Environmental Matters,* also known as the Aarhus Convention, adopted in 1998 and ratified by almost 50 States and the EU, guarantees in its Article 1 *'the rights of access to information, public participation in decision-making, and access to justice in environmental matters'*. These rights can apply autonomously but they have also been repeatedly read into substantive rights, even with respect to States which have not ratified the Aarhus Convention. For instance, in *Taşkin and Others v. Turkey* (ECtHR, 2004b), the ECtHR noted that *'whilst Article 8 contains no explicit procedural requirements, the decision-making process leading to measures of interference must be fair and such as to afford due respect for the interests of the individual'* (ECtHR, 2004b, para. 118). It then specified that

> *where a State must determine complex issues of environmental /.../ policy, the decision-making process must firstly involve appropriate investigations and studies in order to allow them to predict and evaluate in advance the effects of those activities which might damage the environment and infringe individuals' rights /.../ The importance of public access to the conclusions of such studies and to information /.../ is beyond question. /.../ the individuals concerned must also be able to appeal to the courts against any decision, act or omission where they consider that their interests or their comments have not been given sufficient weight in the decision-making process* (ECtHR, 2004b, para. 119).

The ECtHR found a violation of the right to private and family life on account of the failure by Turkey to abide by these procedural obligations.

Overall, therefore, a whole range of human rights, both specific and general, and both substantive and procedural, can be violated in the context of sea level rise. These rights do not exist in clinical isolation from each other but rather influence each other and elements of some are often read into others. Which one of them it is appropriate to invoke depends on the specific situation as well as on the human rights system within which the invocation takes place. This conclusion also holds for the identification of potential victims of violations of human rights occurring in connection with sea level rise, to which we will turn now.

3.2 Victims of Human Rights Violations Occurring in Connection With Sea Level Rise

'The issue of victim status is one of the salient issues of the climate-change cases' (ECtHR, 2024, para. 459), the ECtHR held in *Verein KlimaSeniorinnen Schweiz*. This claim is not limited to the European human rights system or to any particular category of climate change cases but applies more broadly. Usually only individuals claiming to be victims of violations of one of the rights set forth in the relevant human rights instrument can file an application to an international human rights body (UN, 1966, Article 1; Council of Europe, 1950, Article 34). Most of such instruments do not allow *actio popularis*, i.e. they do not *'permit individuals or groups of individuals to complain about a provision of national law simply because they consider, without having been directly affected by it, that it may contravene /the relevant instrument/'* (ECtHR, 2024, para. 460). There are some exceptions to this rule, e.g. the procedure before the AfCmHR. The *Ogoniland* case discussed above was brought to the AfCmHR as a case of *actio popularis*.[9] This does not mean that the AfCmHR could ignore the victim issue altogether, as it is relevant not only at the procedural level (who is entitled to submit an application?) but also at the substantive level (who can be the victim of a human rights violation?).[10]

---

[9] *'The [African] Commission thanks the two human rights NGOs who brought the matter under its purview /.../ Such is a demonstration of the usefulness to the [African] Commission and individuals of actio popularis, which is wisely allowed under the African Charter'* (AfCmHR, 2001, para 49).

[10] Schmid (2022) rightly notes that no judicial body *'can be expected to neatly separate admissibility from merits if the applicants invoke allegations of violations by regulatory omissions. /.../ If the allegation is an omission to take sufficient measures required by a positive obligation, the legal assessment inevitably requires at least a preliminary engagement with the scope of the positive obligation when deciding whether the applicants have victim status'*.

International human rights bodies have taken a rather broad approach to the concept of victims. The HRC has held that to qualify as a victim of a human rights violation, an individual *'must show either that an act or an omission of a State /.../ has already adversely affected his or her enjoyment of such right, or that such an effect is imminent'* (UNHRC, 2005, para. 4.3). These cases may include situations in which a certain piece of legislation or a judicial or administrative act or practice has already been applied to the detriment of a certain individual, or in which it is applicable to them *'in such a way that the alleged victim's risk of being affected is more than a theoretical possibility'* (UNHRC, 2005, para. 4.3). The ECtHR has distinguished three categories of victims: direct victims, who have already been affected by the alleged violation of the ECHR; indirect victims, who have been indirectly affected by it; and potential victims, who might potentially be affected by it (ECtHR, 2024, para. 463). For the last category, the ECtHR has stressed that an applicant must *'produce reasonable and convincing evidence of the likelihood that a violation affecting him or her personally will occur; mere suspicion or conjecture is insufficient'* (ECtHR, 2004a). It also noted that the category may include both those *'who claim that they are at present, or have been, affected by the general measure complained of, and, in other circumstances, /.../ those who claim that they might be affected by such a measure in the future'* (ECtHR, 2024, para. 471).

Cases related to sea level rise, and climate change more broadly, constitute a challenge in this respect (Mariconda, 2023). There might be specific instances in which concrete individuals would become direct victims of human rights violations stemming from sea level rise, for instance when a State would construct a dam on land belonging to a person without providing any compensation to them. Yet, in most instances, such direct victims may not be available. As a long-term process, rather than a one-time event, sea level rise produces its effects gradually, simultaneously affecting many different people in many different ways. As the ECtHR rightly held, *'in the climate-change context, everyone may be, one way or another and to some degree, directly affected, or at a real risk of being directly affected, by the adverse effects of climate change'* (ECtHR, 2024, para. 483). This creates a dilemma. Accepting, on the one hand, that anyone is a potential victim of sea level rise would mean not only de facto introducing *actio popularis* into systems that do not know it but also turning courts into the final arbitrators of policies regarding climate change. Limiting, on the other hand, the status of victims to those who have been directly affected by specific effects of sea level rise might make individuals powerless even in the face of a blatant failure by a State to take any relevant measures to protect its inhabitants against sea level rise (ECtHR, 2024, para. 484).

To solve this dilemma, the ECtHR has suggested that *'in order to claim victim status /…/ in the context of complaints concerning harm or risk of harm resulting from alleged failures by the State to combat climate change, an applicant needs to show that he or she was personally and directly affected by the impugned failures'* (ECtHR, 2024, para. 487). To do so, two cumulative conditions need to be met. First, the applicant needs to be subject to a high intensity of exposure to the adverse effects of climate change, i.e. the level and severity of the adverse consequences of the governmental action or inaction affecting the applicant must be significant. Second, there must be a pressing need to ensure protection for the individual owing to the absence or inadequacy of any reasonable measures to reduce harm (ECtHR, 2024, para. 487). The victim status would thus be attained only in *'highly exceptional circumstances'* (ECtHR, 2024, para. 470). Whether such circumstances have materialized needs to be assessed on a case-by-case basis and, due to the general nature of the two-part test, it is highly probable that it would often be subject to disagreements (ECtHR, 2024, Conclusion and the Partly Concurring Partly Dissenting Opinion of Judge Eicke).

Other international human rights bodies have not considered the victim status in climate change cases in such detail but at least some of them seem to adhere to a similar approach as the ECtHR, reserving the status for those individuals for whom the effects of sea level rise have been particularly dramatic. These would be individuals who have either been exposed to very serious consequences of sea level rise or who are increasingly vulnerable to such consequences. The two HRC cases introduced in the previous section illustrate these two categories. In *Teitiota v. New Zealand*, the applicant attained victim status (though he was not found to be a victim of any human rights violations in the end) due to the fact that the country to which he had been deported, Kiribati, was seriously affected by sea level rise (UNHRC, 2019b, para. 9.3). In *Billy v. Australia*, the applicants attained victim status (and were finally declared victims of human rights violations) due to their belonging to a vulnerable group of indigenous persons.[11] The two categories are, however, defined in rather general terms and it may thus be expected that their interpretation and application will give rise to controversies.

3.3 States Having Human Rights Obligations in Connection With Sea Level Rise

Determining who victims of human rights violations occurring in connection with sea level rise are, is difficult. So is determining which State shall bear responsibility for these violations. As was noted during the debate in the ILC Study Group, *'while States /have/ human rights obligations towards individuals, the sea level rise phenomenon /is/ not directly attributable to any particular State. Accordingly, it /is/ unclear how human rights rules would operate within that context and, specifically, how and against whom claims related to sea level rise could be brought'* (ILC, 2022, para. 64). Similarly, in *Billy v. Australia,* Australia argued that *'/i/t is not possible under international human rights law to attribute climate change to the State party. It is not possible as a legal matter to trace causal links between the State party's contribution to climate change, its efforts to address climate change and the alleged effects of climate change on the enjoyment of the authors' rights'* (UNHRC, 2022, para. 4.3). The specificities of cases related to climate change, which *'are not concerned with single-source local /.../ issues but with more complex global problems'* (ECtHR, 2024, para. 424), were acknowledged by the ECtHR in *Verein KlimaSeniorinnen Schweiz* as well.

International human rights bodies have found a way to overcome this difficult situation by stressing that sea level rise, and climate change more broadly, give rise to positive obligations aimed at protecting individuals against the negative effects of these phenomena and that every State must *'do its part to ensure such protection'* (ECtHR, 2024, para. 545). That entails that all States have human rights obligations in this context, and they all have to 'do their part' by adopting measures to respond to environmental challenges. How big the respective parts of individual States would be and which concrete measures each State shall adopt is already a

---

[11] *"The Committee therefore considers that the authors are among those who are extremely vulnerable to experiencing severely disruptive climate change impacts intensely'* (UNHRC, 2022, para. 7.10).

question of the substance, i.e. it relates to the content of the legal regulation, not to the identity of the duty bearers. Concerning this identity, the principle of common but differentiated responsibilities is often invoked, suggesting that *'all States have common responsibilities to protect the environment and promote sustainable development, but with different burdens due to their different contributions to environmental degradation and to their varying financial and technological capabilities'* (OHCHR, 2021, p. 63).

Two further questions arise in this context. The first one is whether States are the only actors having human rights obligations in connection with sea level rise. Traditionally, this question was answered in the affirmative, as States were seen as the *exclusive* duty bearers under human rights law. In recent years, however, it has been suggested that States are not the exclusive but the *primary* duty bearers and that other actors, such as private business companies, could have obligations stemming from this law as well (OHCHR, 2021, para. 30). Whether such obligations truly exist and what their nature would be is, however, still being discussed.

The second question is whether the human rights obligations (negative or positive) that States have in connection with sea level rise, arise only with respect to their own inhabitants or whether they could be invoked by individuals living in other countries as well. This question touches upon many other substantive and procedural issues (the concept of victims, the jurisdiction of human rights bodies, etc.). As we saw in the previous section, international human rights bodies have already discussed some aspects of this question, mainly those related to the non-refoulement principle. Other aspects, for instance the relevance of the concept of transboundary pollution, still remain to be explored in more detail both in case law and in scholarly literature (Boyle, 2012; IACHR, 2017, pp. 32–43).

3.4 Content of Human Rights Obligations Arising in Connection With Sea Level Rise

As shown previously in this text, States, and potentially other actors, may violate the human rights of their inhabitants, and potentially other individuals, if they fail to meet obligations stemming from human rights law in connection with sea level rise. What these obligations are is probably the most difficult and comprehensive question. International human rights bodies, in line with their mandate, have addressed this question mostly in regard to specific individuals.

Thus, in the *Teitiota* case, the HRC held that New Zealand had the obligation to assess the risks involved in the deportation of the applicant to a country hit by negative effects of sea level rise and that it would have the obligation not to carry out the deportation if there was a risk to the applicant's right to life. In the *Billy* case, the HRC found that Australia had the obligation to take adaptative measures that would reduce existing vulnerabilities and build resilience to climate change-related harms on the islands inhabited by indigenous communities. It indicated some examples of such measures, such as the construction of sea walls on the islands. In the *Verein KlimaSeniorinnen Schweiz* case, the ECHR stated that each State must '*undertake measures for the substantial and progressive reduction of their respective GHG emission levels, with a view to reaching net neutrality within, in principle, the next three decades*' (ECtHR, 2024, para. 548). Although States have a margin of appreciation to decide what these measures should be, the ECtHR set certain criteria in the light of which it would assess whether this margin has been respected (ECtHR, 2024, para. 550).

On the one hand, all these obligations identified by international human rights bodies are case-specific and they might also differ depending on the concrete human rights invoked by the applicant and the human rights system within which the case is discussed. On the other hand, the obligations can be generalized, albeit with caution, as it is highly probable that they would arise in similar cases as well. There have indeed been attempts to define human rights obligations arising in connection with sea level rise, and climate change as such, in more general terms. For instance, in its General Comment No. 36 on the right to life (Article 6 of the ICCPR), the HRC noted that:

> *[i]mplementation of the obligation to respect and ensure the right to life, and in particular life with dignity, depends, inter alia, on measures taken by States parties to preserve the environment and protect it against harm, pollution and climate change caused by public and private actors. States parties should therefore ensure sustainable use of natural resources, develop and implement substantive environmental standards, conduct environmental impact assessments and consult with relevant States about activities likely to have a significant impact on the environment, provide notification to other States concerned about natural disasters and emergencies and cooperate with them, provide appropriate access to information on environmental hazards and pay due regard to the precautionary approach* (UNHRC, 2019a, para. 62).

The obligations are defined in very general terms and would need to be translated into more concrete duties. They would moreover only arise under the right to life, where the threshold for violation is particularly high.

The 2017 Advisory Opinion of the IACHR is also primarily focused on the obligations arising under the right to life and also those related to the right to personal integrity (Articles 4 and 5 of the American Convention on Human Rights) (IACHR, 2017, paras. 123–243). It, however, deals with the issue in a more detailed way. The IACHR distinguishes four obligations arising in the face of potential environmental damage. The first is the obligation of prevention, which encompasses the duties to regulate, supervise and monitor, require and approve environmental impact assessments, prepare a contingency plan, and mitigate if environmental damage occurs. The second obligation relies on the principle of precaution, under which States must act with due caution to prevent severe and irreversible damage to the environment if there are plausible indications that an activity could result in such a damage, even in the absence of scientific certainty. The third is the obligation to cooperate, which includes the duty to notify other States about the potential impact of one's activities, the duty to consult and negotiate with potentially affected States, and the duty to exchange information. The fourth obligation is procedural in nature and consists of ensuring for individuals' access to relevant information, public consultations, and access to justice. Although these obligations relate to climate change in general, they are relevant in the context of sea level rise as well.

This specific context has been considered by the ILC and the ILA Committee. The ILC has also sought to identify certain leading principles or, rather, elements of the legal regulation applicable to the protection of persons affected by sea level rise (ILC, 2024, paras. 68–83). Despite the emphasis placed by the ILC on the particular nature of sea level rise, the discussion has been clearly inspired by the 2016 *Draft Articles on the Protection of Persons in the Event of Disasters*. It thus takes human dignity as the starting point, seeks to combine a needs-based and a rights-based approach, and stresses the importance of the protection of the vulnerable and of international cooperation. The elements have so far been only outlined and will require further specification. The ILA Committee for the moment largely focuses on the protection of human rights of individuals who have been forced to leave their homes in connection with sea level rise (ILA, 2024, pp. 33–41). Human rights obligations are divided between the affected State, the host State, and the international community, and they differ depending on whether the affected State is able to secure substantial parts of its habitable territory, maintains only small parts of its territory, or loses all of its habitable territory. The analysis is useful but it only deals with one set of human rights obligations, concerning individuals on the move from the countries hit by sea level rise.

The overview of several initiatives aimed at defining the content of human rights obligations arising in connection with sea level rise, or climate change as such, demonstrates the difficulties involved in this task. First, sea level rise and climate change are new and, in many ways, unprecedented challenges. Using analogy when identifying general principles applicable in this area or when deciding specific cases, warrants caution, as the approaches used with respect to other challenges, e.g. natural disasters, may not necessarily be appropriate here. Second, sea level rise and climate change are global phenomena which concern all States. Consequently, all States have the obligation to protect their inhabitants, and possibly other individuals, against the negative effects that these phenomena can have on their human rights. When implementing this obligation, States might be guided by a uniform set of general principles but the concrete measures to be implemented need to be defined for each of them individually, taking into account their specific situation in a given moment. Third, the content of human rights obligations also differs depending on the identity (and vulnerability) of the victims, the human rights at stake, and the parameters of the human rights system.

Fourth, it is important to make a distinction between human rights obligations arising with respect to sea level rise as such and those linked to the forced migration from countries hit by effects of sea level rise. The two issues are interrelated but whereas the former has to do more with the primary protection of individuals against activities harmful to the environment that States either engage in or fail to prevent, the latter is more about ensuring subsidiary protection for individuals who can no longer rely on their own State. Fifth, lawyers, including judges, are not necessarily the best placed to decide which concrete measures States shall take in connection with sea level rise. Or, put differently, they cannot take this decision based solely on their legal knowledge. What needs to be done and by whom to prevent or mitigate the negative effects of climate change is a factual question to be answered by experts from various non-legal fields. While this is not specific for this area, here the degree of scientific uncertainty might be more extensive, and the expert consensus more limited. Due to that, moreover, quite a large margin of appreciation needs to be left for States and their political representatives to decide which measures to adopt. Legal institutions, including courts, have an important role to play in this area but their work must not substitute for the work of other institutions.

4 Conclusions

Sea level rise affects the enjoyment of human rights. It makes people change their way of life and, sometimes, forces them to leave their homes or even their country. States, and possibly other actors, have the obligation to 'do their part' to ensure the protection of their own inhabitants, and possibly other individuals, against such effects. A failure to abide by these obligations and adopt the appropriate measures might constitute a violation of human rights. This general scheme is relatively clear. Yet, many aspects thereof remain to be further discussed and clarified. One of these aspects concerns the concrete human rights that can be violated in connection with sea level rise. Does it make a real difference whether the human rights system contains the right to a clean, healthy, and sustainable development or whether it is necessary to rely on general human rights (such as the right to life or the right to property)? And what is the relationship between substantive and procedural human rights here? Another aspect pertains to the victims. Since sea level rise, and climate change more broadly, affect everyone, the number of victims of human rights violations occurring in this context is potentially unlimited. Is it possible to limit this number somehow, for instance by focusing on the special vulnerability of certain individuals?

The third aspect is also related to actors, this time to duty bearers. Provided that sea level rise is a global phenomenon that is not based on easy causation links, which States shall be held responsible for its human rights effects? Moreover, do only States have human rights obligations in this area or should the category of duty bearers include non-State actors, such as private business companies? And do States have obligations with respect to their inhabitants only? Finally, the fourth, most comprehensive aspect concerns the content of human rights obligations. Which measures are individual States expected to adopt to meet their obligation to respect, protect and fulfil human rights? What should the role of lawyers, including judges, be in defining these measures? And to what extent should they defer to experts or political decision-makers? All these questions, which are closely intertwined, have already been addressed by international human rights bodies and expert bodies such as the ILC or the ILA Committee. Yet, the answers provided to them are neither comprehensive nor identical. They also need to be constantly checked, and occasionally revised, as the effects of sea level rise progressively unfold and our understanding of this phenomenon improves. The relationship between sea level rise and human rights will thus certainly remain on the agenda of legal and other institutions for many years to come, raising many more questions than those addressed in this paper.

References

African Commission on Human and Peoples' Rights (AfCmHR). (2001). *The Social and Economic Rights Action Center, et al. v. Nigeria (Communication No. 155/96)*.

Bílková, V. (2017). A State Without Territory? *Netherlands Yearbook of International Law*, *2016*(47), 19–47. https://doi.org/10.1007/978-94-6265-207-1

Boyle, A. (2012). Human Rights and the Environment: Where Next? *European Journal of International Law*, *23*(3), 613–642. https://doi.org/10.1093/ejil/chs054

Cataldi, G. (2014). Human Rights of People Living in States Threatened by Climate Change. *Questions of International Law*, *1*, 23–39

Coomans, F. (2003). The Ogoni Case Before the African Commission on Human and Peoples' Rights. *International and Comparative Law Quarterly*, *52*, 749–760. https://doi.org/10.1093/iclq/52.3.749

Council of Europe. (1950). *European Convention on Human Rights*. https://www.echr.coe.int/documents/convention_eng.pdf

Council of Europe, Parliamentary Assembly. (2009). *Recommendation 1885: Drafting an Additional Protocol to the European Convention on Human Rights Concerning the Right to a Healthy Environment*. https://assembly.coe.int/nw/xml/XRef/Xref-XML2HTML-en.asp?fileid=17751&lang=en

Cullen, M. (2018). Eaten by the Sea: Human Rights Claims for the Impacts of Climate Change Upon Remote Subnational Communities. *Journal of Human Rights & the Environment*, *9*(2), 171–193. https://doi.org/10.4337/jhre.2018.02.03

ESCR-Net. (2022). *Daniel Billy et al. vs Australia (Torres Strait Islanders petition)*. https://www.escr-net.org/caselaw/2022/daniel-billy-et-al-vs-australia-torres-strait-islanders-petition/

European Court of Human Rights (ECtHR). (2004a). *Senator Lines GmbH v. Fifteen Member States of the European Union (Application No. 56672/00, Decision (Grand Chamber))*.

European Court of Human Rights (ECtHR). (2004b). *Taşkin and Others v. Turkey (Application No. 46117/99, Judgment)*.

European Court of Human Rights (ECtHR). (2024). *Verein KlimaSeniorinnen Schweiz and Others v. Switzerland (Application No. 53600/20, Judgment (Grand Chamber))*.

Hafner-Burton, E. M., Helfer, L. R., & Fariss, C. J. (2011). Emergency and Escape: Explaining Derogations From Human Rights Treaties. *International Organization*, 65(4), 673–707. https://doi.org/10.1017/S0020818311000229

Hamdan, E. (2016). *The Principle of Non-Refoulement Under the ECHR and the UN Convention Against Torture and Other Cruel, Inhuman or Degrading Treatment or Punishment*. Brill. https://doi.org/10.1163/9789004319394

High Court of South Africa (2017). *South Africa, Earthlife Africa Johannesburg v Minister of Environmental Affairs and Others (Case No. 65662/16)*.

Humphreys, S. (2009). *Human Rights and Climate Change*. Cambridge University Press. https://doi.org/10.1017/CBO9780511770722.002

Inter-American Court of Human Rights (IACHR). (2023, January 9). *Climate Emergency and Human Rights, Request for an Advisory Opinion Submitted by the Republic of Colombia and the Republic of Chile*.

Inter-American Court of Human Rights (IACHR). (2017). *The Environment and Human Rights (Advisory Opinion OC-23/17)*.

International Court of Justice (ICJ). (2023). *Request for Advisory Opinion: Obligations of States in Respect of Climate Change (Resolution 77/276)*.

International Law Association (ILA). (2024, July 1). *Athens Conference Report: Committee on International Law and Sea Level Rise*.

International Law Commission (ILC). (2019, December 10). *Report on the Work of the Seventy-First Session*. United Nations. https://legal.un.org/ilc/reports/2019/

Kahl, V. (2022, November 21). Rising Before Sinking: The Landmark Decision of the UN Human Rights Committee in *Daniel Billy et al. v. Australia*: Völkerrechtliche Tagesthemen: Spotlight (Episode 31). *Völkerrechtsblog*. https://voelkerrechtsblog.org/rising-before-sinking-the-landmark-decision-of-the-un-human-rights-committee-in-daniel-billy-et-al-v-australia/

Lindsey, R. (2023, August 22) Climate Change: Global Sea Level. *Climate.gov*. https://www.climate.gov/news-features/understanding-climate/climate-change-global-sea-level

Mariconda, A. (2023). Between Old Certainties and New Challenges. *Italian Review of International and Comparative Law*, 3(2), 260–282.

McAdam, J. (2020). Protecting People Displaced by the Impacts of Climate Change: The UN Human Rights Committee and the Principle of Non-Refoulement. *American Journal of International Law, 114*(4), 708–725. https://doi.org/10.1017/ajil.2020.31

Mimura, N. (2013). Sea level Rise Caused by Climate Change and Its Implications for Society. *Proceedings of the Japan Academy, Series B Physical and Biological Sciences, 89*(7), 281–301. https://doi.org/10.2183%2Fpjab.89.281

Montana First Judicial District Court, United States (2023). *Rikki Held, et al. v. State of Montana, et al.* (CDV-2020-307).

Office of the United Nations High Commissioner for Human Rights (OHCHR). (2021). *Frequently Asked Questions on Human Rights and Climate Change (Fact Sheet No. 38)*. United Nations.

Schmid, E. (2022, April 30). Victim Status Before the ECtHR in Cases of Alleged Omissions: The Swiss Climate Case. *EJIL: Talk!*. https://www.ejiltalk.org/victim-status-before-the-ecthr-in-cases-of-alleged-omissions-the-swiss-climate-case/

United Nations (UN). (1966). *Optional Protocol to the International Covenant on Civil and Political Rights*. https://www.chchr.org/en/instruments-mechanisms/instruments/optional-protocol-international-covenant-civil-and-political-rights

United Nations Human Rights Committee (UNHRC). (2005). *Beydon et al. v. France (CCPR/C/85/D/1400/2005)*. https://undocs.org/CCPR/C/85/D/1400/2005

United Nations (UN). (2009). *Report of the Office of the United Nations High Commissioner for Human Rights on the Relationship Between Climate Change and Human Rights (A/HRC/10/61)*. https://undocs.org/A/HRC/10/61

United Nations (UN). (2016). *Report of the International Law Commission Sixty-Eighth Session (2 May-10 June and 4 July-12 August 2016) (A/71/10)*. https://undocs.org/A/71/10

United Nations (UN). (2020, January 21). *Historic UN Human Rights Case Opens Door to Climate Change Asylum Claims*. https://www.ohchr.org/en/press-releases/2020/01/historic-un-human-rights-case-opens-door-climate-change-asylum-claims

International Law Commission (ILC). (2022). *Study Group on Sea Level Rise in Relation to International Law: Report (A/CN.4/L.972)*. https://undocs.org/A/CN.4/L.972

International Law Commission (ILC). (2024). *Study Group on Sea Level Rise in Relation to International Law: Report (A/CN.4/L.1002)*. https://undocs.org/A/CN.4/L.1002

United Nations (UN). (2019). *Right to a Healthy Environment: Good Practices. Report of the Special Rapporteur on the Issue of Human Rights Obligations Relating to the Enjoyment of a Safe, Clean, Healthy and Sustainable Environment (A/HRC/43/53)*. https://undocs.org/A/HRC/43/53

United Nations Human Rights Council (UNHRCo). (2021). *The Human Right to a Clean, Healthy and Sustainable Environment (A/HRC/RES/48/13)*. https://undocs.org/A/HRC/RES/48/13

United Nations General Assembly (UNGA). (2022). *The Human Right to a Clean, Healthy and Sustainable Environment (A/RES/76/300)*. https://undocs.org/A/RES/76/300

United Nations Human Rights Committee (UNHRC). (2019a). *General Comment No. 36 Article 6: Right to Life (CCPR/C/GC/36)*. https://www.ohchr.org/en/issues/development/Pages/GeneralComments.aspx

United Nations Human Rights Committee (UNHRC). (2019b). *Ioane Teitiota v. New Zealand (CCPR/C/127/D/2728/2016)*. https://undocs.org/CCPR/C/127/D/2728/2016

United Nations Human Rights Committee (UNHRC). (2022). *Daniel Billy et al. v. Australia (CCPR/C/135/D/3624/2019)*. https://undocs.org/CCPR/C/135/D/3624/2019

Open Access This chapter is licensed under the terms of the Creative Commons Attribution 4.0 International License (http://creativecommons.org/licenses/by/4.0/), which permits use, sharing, adaptation, distribution and reproduction in any medium or format, as long as you give appropriate credit to the original author(s) and the source, provide a link to the Creative Commons license and indicate if changes were made.

The images or other third party material in this chapter are included in the chapter's Creative Commons license, unless indicated otherwise in a credit line to the material. If material is not included in the chapter's Creative Commons license and your intended use is not permitted by statutory regulation or exceeds the permitted use, you will need to obtain permission directly from the copyright holder.

CHAPTER 6

Sea Level Rise as a Form of Gendered Climate Violence: International Legal Implications for Migration

Sara De Vido

1 INTRODUCTION

Climate change is 'an issue of inequality towards women and injustice' (Albertyn et al., 2023). An only apparently neutral phenomenon, climate change exacerbates situations of discrimination, including gender discrimination, which are already present in societies. Like other phenomena of environmental deterioration caused by (a part of) humanity, it contributes to determining migration flows, constituting a reason for why groups of individuals decide to leave their country of origin.[1]

---

[1] On migration as a consequence of climate change, see, *inter alia*, McAdam (2012), Fornalé (2020), Atapattu (2020), Behrman and Kent (2022) and Caracciolo et al. (2022). Already in 2016, the New York Declaration for Refugees and Migrants recognized that migration also occurs in response to climate change, natural disasters or other environmental factors (Declaration of 19 September 2016, A/71/L.1). This chapter draws on and further develops the ideas in De Vido (2023a), with a focus on human rights justifications.

S. De Vido (✉)
Ca' Foscari University of Venice, Venice, Italy
e-mail: Sara.devido@unive.it

© The Author(s) 2026
E. Fornalé (ed.), *Sea Level Rise*,
https://doi.org/10.1007/978-3-031-89171-7_6

Sea level rise, which is one of the major negative effects of climate change, is also a gendered phenomenon: it does not cause discrimination *per se*, but the effects of a human-caused environmental phenomenon differentiate based on gender. Several studies have shown that particularly in coastal areas, women and girls are extremely vulnerable to sea level rise (MacGregor, 2010; Pratiwi, 2023).

The aim of this chapter is to conceive sea level rise as a form of what we have defined, in broader research, as environmental 'chronic emergency',[2] and a form of climate violence, and to analyse, using an ecofeminist legal approach, gendered migration *because of* the effects of climate change. After explaining the notions of 'climate violence' and 'gendered migrations', the contribution will rely on the work of several UN bodies and rapporteurs to identify the connection between women, climate, and migration, and will explain what an ecofeminist method in international law means for this connection. It will then critically read the *Teitiota* case decided by the Human Rights Committee in 2019, and will argue that human rights justifications invoked by States in migration and climate adaptation and mitigation policies conceal patterns of discrimination and oppression on the basis of gender.

---

[2] The chapter is one of the outcomes of the project *Gendering International Legal Responses to Chronic Emergencies* (GenREm) PRIN 2022—P2022XYHPTC—CUP H53D23002940006 (Ca' Foscari University of Venice, the University of Florence and the University of Palermo), financed by the EU and NextGenerationEU (https://genrem.wordpress.com/). The project will lead to the publication of the book *Gendering International Legal Responses to Chronic Emergencies,* co-edited by Sara De Vido, Deborah Russo and Enzamaria Tramontana, forthcoming in 2026, and to be published by Elgar Publishing. The book conceptualizes and analyses from a gender perspective the concept of environmental chronic emergencies, which refers to emergencies that are not mere disasters, and that put into question time, space and legal obligations in international law. The concept of environmental chronic emergencies derives from the innovative idea of slow violence and elaborates further the concept of slow onset events to appreciate situational vulnerabilities and the disproportionate impact of these emergencies on women and girls. The adjective chronic gives the idea of something that is rooted, often silent, and surely unseen, but has an impact for present and future generations, for the environment and for women at the intersection of different grounds of discrimination. The book offers a pioneering gendered analysis of chronic emergencies and their effects on actors, the legal obligations involved and remedies for them.

2 SEA LEVEL RISE AS CLIMATE VIOLENCE

This chapter argues that sea level rise, just as much as climate change, is a form of climate violence that disproportionately affects women and girls. Violence is not 'natural' in the sense of a manifestation of a vital process, but instead it is an instrument of perpetuation of power relations that have taken root and reproduce themselves in society (Arendt, 1970). That is the case, for example, with the 'structural' and systemic violence against Iranian women, who are the object of a form of 'State' violence whose civil rights are constantly trampled upon for the mere fact of 'being a woman' (UN Women, 2022b) Violence also occurs when the State 'tolerates' its manifestations in private relationships, for example when the authorities do not respond immediately to victims' complaints or when they do not carry out an adequate risk assessment to prevent the escalation of violent behaviour (De Vido & Frulli, 2023). Violence also expresses itself through the domination of human beings over other species, thus leading to repeating intra-species patterns (Jones, 2023). Law in itself can be argued to be violence, as the latter provides 'the occasion and method for founding legal systems, gives law (as the regulator of force and coercion) a reason for being' and identifies 'the means by which law acts' (Hamzic, 2018; Sarat & Kearns, 1995). Why then is it important to speak of violence in the 'State'—and 'inter-State'—sense to delineate the boundaries of what we have called 'climate violence' and its (disproportionate) impact on women?

The concept that this chapter proposes, that of climate violence, captures, on the one hand, the fact that climate change and its correlative effects, such as sea level rise, are a consequence of a set of situations *caused by* humanity—*rectius*, a part of humanity (Gaard, 1993)—including the violence exercised by humans on other species. On the other hand, the term climate violence identifies the forms of violence which are exacerbated *due to* climate change, including gender-based violence against women. It is argued here that climate change contributes to the exacerbation of already existing forms of discrimination in societies and to the materialization of violent phenomena. Understanding the roots of the disproportionate impact of climate change on women allows for the formulation of a gender-sensitive and trauma-sensitive response to migration, a response that is not universalizing and one-size-fits-all but considers individual experiences (Lambrou, & Piana, 2006). As has been observed, climate is never the sole cause of migration (or non-migration) but is 'socially mediated' (Lama et al., 2021). Nicole Rogers (2023, p. 144) explained that 'climate crisis' is an 'assemblage of synergistic violences, encompassing violence to human, non-human and more-than-human beings, violence to Earth systems, and violence to all forms of property including, but not limited to, the annihilation of the entire territory of individual nation States. These violences are both direct and structural in nature'.

Climate violence as understood here goes beyond climate change to embrace other forms of 'slow violence' (Nixon, 2011), which often escape legal understanding. Violence is indeed 'often conceived as an event or action that 'erupts' at a particular moment in history, such as a natural disaster' (De Vido, 2024). International disaster management law 'focuses on legal issues arising from the prevention [of], response [to] and recovery [from] various natural catastrophic events, but also from human-made disasters, such as large-scale industrial accidents' (De Guttry et al., 2016; De Vido, 2023a; Zorzi Giustiniani, 2021).

However, 'this conception of violence is only able to capture a part of the situations produced by human activity and fails to identify other forms that are 'neither spectacular nor instantaneous', but rather 'incremental" (De Vido, 2023a),. Slow violence is an 'apparently invisible [phenomenon], although its effects occur [just] as much [in relation to] human beings, in an intra-generational and intergenerational perspective, as [in relation to] nature' (De Vido, 2024).

Some examples of this are 'climate change (Fornalé, 2023), thawing permafrost, ocean acidification, deforestation, [...] sea [level rise], pesticides, and the use of substances such as mercury' (OHCHR, 2022; De Vido, 2023a),.[3]

In Kiribati, e.g. seawater floods have polluted wells, limiting rural women's access to water, but also to firewood that is needed to prepare food and to medicinal plants (CEDAW, 2020). A new notion, that of environmental 'chronic emergency', has been conceptualized by the author in a recent project. It has its roots in Nixon's slow violence but describes the phenomenon in a possibly less evocative but more thought-provoking way.[4] The concept of chronic emergencies derives from the

---

[3] Special Rapporteur on the implications for human rights of the environmentally sound management and disposal of hazardous substances and wastes 2022.

[4] See note 2 above.

innovative idea of slow violence and elaborates further the concept of slow-onset events to appreciate situational vulnerabilities and the disproportionate impact of these emergencies on women and girls. The adjective *chronic* gives the idea of something that is rooted, often silent, and surely unseen, but has an impact for present and future generations, for the environment, and for women at the intersection of different grounds of discrimination. Chronic emergencies can result in an environmental disaster—and probably 'all the examples cited will, sooner or later, [and] in the short, medium, long, [or] very long term, result in a disaster of [major] proportions—but the peculiarity of [this kind of disaster] is that it occurs gradually and, […] for this [very] reason, is [poorly] considered, even in legal terms, [as] stuck in the 'trap' of the present or at least the imminent' (De Vido, 2023a),. It has been shown that sea level rise produces negative effects on women's reproductive health (Al Hasnat & Teirstein, 2024) and is a driving force for migration, which is often neglected.

Thus—and this is a paradigmatic example, as there are international obligations stemming from, *inter alia*, the Paris Agreement—climate change debates were sidelined during the COVID-19 pandemic, and overtaken by the urgency of responding to a health emergency of global proportions without grasping the relationship between environmental destruction and the spread of pandemics. Chronic emergencies can be legally read from multiple perspectives: in terms of obligations under international conventions—see, for example, the Paris Agreement with regard to climate change, the Montego Bay Convention with regard to marine pollution, the 2001 Stockholm Convention on Persistent Organic Pollutants, in terms of enforceable principles, including those of international environmental law, as well as in terms of violations of fundamental human rights. The chapter will elaborate on this topic in the light of international human rights law and international migration law.

3 'Gendered' Migration

It should be noted that the recognition of a gender dimension in climate change-induced migration has been affirmed by numerous international bodies (IOM, 2014; UNHCR, 2022; UN Women, 2022a); however, the response from a gender perspective to this phenomenon has not been as decisive: what is lacking is a response capable of capturing the impact not only of large and devastating disasters but also of the progressive deterioration of the environment, the so-called slow violence (Nixon, 2011). Climate change-induced migration for women is often closely linked to the reduction of natural resources of which they are the main users. There is thus a generally disproportionate effect of climate change, including its effects such as sea level rise, on women. This aspect does not derive from the biological differences between women and men, but from the social, economic and political barriers women face in societies. For example, in countries where women are primarily responsible for the water supply, they suffer the greatest effects of climate change and may be induced to migrate to another country or become displaced within the original country. Women might also be denied migration during these phenomena, as they are in a subordinate position in many societies, to the extent that they might become the main direct and indirect victims of natural disasters. The two examples of the disproportionate impact of climate change on women presented here—the water supply and the ban on migration—call for a non-essentialist reading of the phenomenon (Lambrou & Piana, 2006): a reading, in other words, that does not automatically place women in the category of 'vulnerable subjects', but is able to grasp the complexity of intersectional forms of discrimination and the context in which they occur (UN Women, 2022a).

Regarding gendered migration, much has been written and said about this topic in legal (and non-legal) scholarship.[5] Here, to contextualize the analysis, the chapter will merely observe that the gender approach is almost absent in treaty law pertaining to forced migration. Without claiming to exhaust the topic, suffice it to point out that the 1951 UN Convention relating to the Status of Refugees defines the term 'refugee' in an apparently neutral way, but it is conceived, as a feminist jurist pointed out, 'in the male form', as it reproduces legal categories of persecution that reflect male experiences (Kelly, 1993). The Convention, as formulated in the immediate post-war period, was incapable of responding to cases of refugee status claims due to gender-based violence understood as a form of persecution. The absence of the gender dimension is due not only to the socio-political context in which the negotiations were conducted, but also to the then entrenched public/ private contraposition, by virtue of which the State had no international legal obligations with regard to the so-called private dimension of

---

[5] As an example of the rich scholarship on the topic, see Mora and Piper (2021).

violence. Women's rights and the recognition of gender discrimination only emerged in the 1970s with the adoption in 1979 of the Convention on the Elimination of All Forms of Discrimination against Women (CEDAW) and the subsequent General Recommendation No. 19 of the Committee established by the CEDAW Committee, which brought violence against women within the scope of the UN legal instrument. Women were absent and invisible in the international arena, and relegated to the private sphere. Remaining in this private sphere, women's activities could not obtain the character of 'politics', which was traditionally conceived as belonging to the male experience and, therefore, subject to protection even under international law. International soft-law instruments recovered the gender dimension, though. Hence, for example, the 2002 UNHCR Gender Guidelines indicated that women are an example of a 'social subset defined by innate and immutable characteristics [...] and who are frequently treated differently than men' (OHCHR, 2002).

4 Climate Change and Sea Level Rise Are not Gender-Neutral Phenomena: From CEDAW's Recommendation No. 38 to the Report of the UN *Special Rapporteur* on Violence Against Women

At first glance, it could be argued that climate change affects all of humanity and does not distinguish according to, for example, gender, age, ethnicity, geographical origin or the social and economic status of individuals. Yet, if one goes beyond the surface, one grasps the problematic nature of this Statement. Climate change, like the other forms of chronic emergencies mentioned above, is a human, social, and environmental phenomenon. It is a(n) '(un)natural disaster', causing more pronounced effects where there are already situations of vulnerability that are exacerbated by climate change itself. In other words, climate change becomes a human rights issue for all, but especially for those whose human rights are already compromised (Atrey, 2023). This is as true for the situation of people living in Third World countries as it is for women, especially those at the intersection of different grounds of discrimination. UN bodies have already explained the impact of climate change and its effects, including sea level rise, on women.

Thus, in its General Recommendation No. 37 (GR37) of 2018 on the gender dimension of disaster risk reduction and climate change, the

CEDAW Committee noted that structural inequalities faced by women make them more exposed to risks from disasters and risks of the loss of their livelihoods. Mortality and disease levels are higher among women in disaster situations as a consequence of their already precarious living conditions (those related to access to health services, water, and food, for example), but also as a result of environmental emergency response mechanisms that do not take into account the specific needs of different groups of women. Gender-based violence against women, including sexual violence, which is widespread in humanitarian crises, is exacerbated in situations of disaster and destruction of natural resources. Access to education, which is already limited for girls and young women as a result of social, cultural and economic barriers, becomes even more complex following disasters as a result of the destruction of infrastructure, the lack of teachers, and economic difficulties. Even with reference to the right to work and social protection, the situation of social inequality between women and men, with the former often employed in precarious and informal jobs, is exacerbated during disasters (Fornalé et al., 2023). The burden of domestic and care work increases as a result of disasters, namely due to the destruction of food stocks, housing and infrastructure, including water and energy supply. Women's right to health is disproportionately restricted because of disasters and climate change, again due to structural inequalities in society with regard to access to food, health and care.

The CEDAW Committee mentioned access to reproductive health and health services, but did not mention that, for instance, high levels of pollution from certain substances, including mercury, impact women's reproductive rights. Women, especially those living in rural areas, are also directly affected by climate change, which causes reduced food security, land degradation and reduced access to water and other natural resources. In many areas, even though they do not have the title to own land, women are left alone in farming as the male component of society emigrates. The movement of women is at risk after disasters and because of climate change, particularly due to violence that operates through trafficking, which occurs in the fields, at the border and in the country of destination. It has also been noted that some societies prevent women from emigrating, due to discriminatory laws, gender stereotypes and care responsibilities.

In applying CEDAW to disaster risk reduction and climate change, the Committee emphasized three general principles of the legal instrument: equality and non-discrimination; participation and empowerment; and access to justice. Despite the important step forward in considering environmental issues as issues of inequality under CEDAW, as has been noted, the Committee focuses on disaster management, particularly mitigation and adaptation, but refrains from taking a 'more radical and transformative approach' (Atrey, 2023).

The UN Commission on the Status of Women, at its sixty-sixth session from 14 to 25 March 2022, identified the following as its priority theme: 'achieving gender equality and the empowerment of all women and girls in the context of climate change, and environmental and disaster risk reduction policies and programme' (UN Commission on the Status of Women, 2022). In its conclusions of 29 March 2022, the Commission confirmed the disproportionate impact of climate change, environmental degradation and disasters on women, who are more exposed to risks and loss of livelihoods, but also highlighted the importance of recognizing the role of women as agents of change (agency). The following phenomena create the conditions for the migration of women, and constitute 'drivers and factors that compel women and girls to leave their countries of origin' (Fornalé et al., 2023): 'loss of their homes, water [shortages] and/or interruption [of water] supply, [and] destruction [of] and damage to schools and health facilities' (Fornalé et al., 2023). It is precisely in displacement caused by climate change and environmental degradation that women face 'specific challenges', including increased risks of violence and reduced access to employment, education, and essential health services, including reproductive health and psychosocial support services. With specific regard to migration, the Commission emphasized the importance of recognizing the positive contribution of migrant women in promoting a gender perspective in migration policies and programmes aimed at responding to situations of vulnerability experienced by migrants, and in responding to all forms of violence that may arise as a result of displacement.

A few months later, in July 2022, the Special Rapporteur on Violence against Women, its Causes and Consequences, Reem Alsalem, published a report on violence against women in the context of the climate crisis, including the environmental degradation and the related disaster response and mitigation (UN General Assembly, 2022). The Special Rapporteur noted that 'the combined impacts of sudden-onset natural disasters and slow-onset events, environmental degradation and forced displacement seriously affect women's and girls' rights to life, access to food and nutrition, safe drinking water and sanitation, education and training, adequate housing, land, decent work and labor protection' (UNGA, 2022). The report emphasized, in line with the definition of violence provided above, that climate change exposes those affected, especially women, to human rights violations that constitute forms of violence and persecution for the purposes of them obtaining a refugee status. The category of vulnerable women, because of intersectional elements of discrimination, must include those who defend, conserve and denounce the condition of resources and ecosystems, particularly human rights defenders (those dealing with women's rights and environmental rights). Migration and internal displacement emerge at various points in the Special Rapporteur's analysis. Thus, for example, the report highlighted how the likelihood of experiencing violence is multiplied when 'women are displaced or [are] in emergency shelters, [or] when they migrate to [countries], cities and peri-urban areas as a [result] of forced displacement or planned relocation, encountering difficulties in accessing adequate housing, employment and social protection mechanisms' (Fornalé et al., 2023). Women are also at risk of being trafficked for sexual exploitation or domestic work following disasters: this happened, for example, in the Philippines after Typhoon Haiyan in 2013.

Similarly to the CEDAW Committee's findings, the Special Rapporteur pointed out that climate change may cause outward migration by the male component of society, leaving women to provide for the survival of their families and enter a labour market that may be characterized by a gender pay gap and economic disempowerment. With the aim of avoiding forced migration or human trafficking, and thus escaping violence, girls are forced into early marriages, which are themselves forms of gender-based violence but are perceived by families bent on avoiding insecurity and economic instability as forms of protection. Climate change and environmental degradation result in the forced displacement of women, who are thus forced to face high risks of violence, especially sexual violence.

Although it does not specifically address the situation of displaced women, it is also worth mentioning the report published in 2020 by the Special Rapporteur on the Human Rights of Internally Displaced Persons (SRIDP), which highlighted the 'particular challenges' that displaced persons face 'in the context of the adverse slow-onset effects of climate change' (UN General Assembly, 2020). Referring to a report by the

Task Force on Displaced Persons under the UN Framework Convention on Climate Change (Task Force on Displacement, 2018), the SRIDP recalled four ways in which the adverse slow-onset effects of climate change can turn into a disaster and increase the risks of displacement. The first is the fact that climate change gradually reduces the availability of vital resources such as water and food (e.g. through soil drying), leading to food insecurity phenomena that cause displacement and migration. Second, a slow-onset phenomenon can turn into a disaster due to a sudden-onset event, such as a flood or fire. Third, slow-onset events weaken the ability of communities to cope with future risks, increasing their vulnerability. Fourth, slow-onset events constitute an aggravating factor that multiplies economic, social, cultural and political crisis factors. For the purposes of this chapter, it is important to note that movement, understood in the report as obviously movement to another area of the country of origin, is not necessarily an element that increases the vulnerability of those affected, but is also seen as an adaptation strategy which may take the form of, for example, seasonal or temporary migration. The SPIDP has grasped the complexity of what we define here as environmental chronic emergency, as she understands its 'human' nature, and the fact that it occurs in times of both peace and conflict.

Unfortunately, the International Law Commission has been so far silent on the gendered effects of sea level rise, apart from some references to 'vulnerable groups', including pregnant women (ILC, 2024a, 2024b). Even though the presence of women in a commission does not ensure gender sensitivity or a gendered approach, it should be noted that in October 2023, two co-Chairs of the International Law Commission, Patrícia Galvão Teles (Portugal) and Nilüfer Oral (Türkiye), were the first women to address the Sixth Committee as Commission Chairs (United Nations, 2023).

5 Ecofeminism as a Method in International Law

Migration because of climate change has already received adequate attention from internationalist scholarship, beginning with the recognition of the historically inevitable, but today no longer justifiable silence of the 1951 Convention relating to the Status of Refugees on this point. However, the traditional approach commonly used to identify who qualifies for refugee status, which is based on this very Convention, is that of answering the following questions: a. whether the individual belongs to a particular group, e.g. women belonging to a community affected by climate change; and b. whether the individual is specifically at risk by virtue of belonging to that group—namely whether he/she fails to grasp the complexity of the challenge posed not by disasters (the sudden-onset events mentioned above), but by slow-moving phenomena, such as climate change, resource depletion and pollution by specific substances.

Before proceeding to the critical reading of a case submitted to the UN Human Rights Committee, the chapter will endorse ecofeminism as a method that has never been fully explored in international legal scholarship. It should be first acknowledged that 'patterns of oppression and domination are not only *intra*-species but also *inter*-species[; they are in the relationship] between humans and [...] nature' (De Vido, 2024), and can be looked at from an intergenerational perspective. The 'human/nature dichotomy has been used to determine patterns of oppression and discrimination [that go] beyond human[beings to] includ[e] non-human animals and natural objects (De Vido, 2021; Grear, 2015).

Law 'has been theorized in [terms of a specific] human/nature [dichotomy], where the former dominates the latter' (De Vido, 2021).

Ecofeminism, despite not being a uniform concept, 'has played a [crucial] role in denouncing [...] patterns of oppression [among] humans and [between] a part of humanity [and] nature' (De Vido, 2021). The word was 'coined by Françoise d'Eaubonne in a 1974 work, *Le féminisme ou la mort*, [in which] she highlighted the environmental costs of development and argued that the overpopulation of the planet was caused by the patriarchal rejection of women's right to self-determination [with regard to] their bodies' (d'Eaubonne, 1974; De Vido, 2021).[6]

However, the roots of ecofeminism can be traced back to Rachel Carson's thought: she was 'a pioneer in [uncovering] the disconnections between human[s...] and the environment and paved the way for [...] ecofeminism' (De Vido, 2021)—or, to put it better, ecofeminisms as a plural—in the world' (Lear, 1998).[7]

---

[6] On ecofeminism, see, among others, Warren (1990), Mies and Shiva (1993), Warren (1997), Mellor (1997), Mallory (2009), Bianchi (2012), and Vakoch and Mickey (2018).

[7] Carson published *Silent Spring* in 1962, where she highlighted the toxic relationship between pesticides, environmental degradation and inter-species health. In the same period, Yayori Matsui, a journalist, worked on the Minamata disaster and reflected on the impact of colonialism on women's rights and environmentalism (Yoshida, 2023).

As the philosopher Val Plumwood argued, ecofeminists differ 'on how and even whether women are connected to nature, whether this connection can be shared with men, how to deal with the exclusion of women from culture, and how to re-evaluate the connection to nature' (Plumwood, 1993). Ecofeminism does not simply unify the struggles of environmentalists and feminists. Puleo explained that it is 'an attempt to hypothesize a new utopian horizon, one that addresses environmental problems arising from the categories of patriarchy, androcentrism, care, sex and gender' (Puleo, 2017). This author also emphasized 'how environmentalism is not always feminist, and how, in turn, feminism does not necessarily demonstrate 'great ecological sensitivity'" (Puleo, 2017). The 'dialogue between feminism and environmentalism is [decisive in emphasizing] the impact of environmental degradation on gender and the contribution women' can make (De Vido, 2021). A detailed description of ecofeminism is beyond the scope of this contribution, but it is worth pointing out how this approach has been undervalued by the international legal scholarship, even in recent writings that have valorized post-human methods and theories.[8] One of the reasons for the underestimation of the potential of this school of thought is the accusation of essentialism ecofeminists received in the 1990s.[9] This critique casts a shadow on the richness of a debate that has not simply equated women with nature (Gaard, 2011).[10] Ecofeminism expresses a great potential for application as a legal method of international law because it questions monolithic legal categories and reconsiders them in light of the interdependence between all species (human and non-human) and with nature, prioritizing the needs of marginalized communities that are most affected by climate change and, as this contribution argues, by chronic emergencies, including sea level rise.

To reply to an obvious objection commonly addressed to feminist lawyers (Charlesworth, 1999)—why they are using ecofeminism and not exploring other methods and theories, such as deep ecology or

---

[8] See the interesting post-human analysis by Jones (2023).

[9] For a reflection on the scarce legal scholarship on ecofeminism, see Malone (2015).

[10] See also the interesting ecofeminist analysis of whaling provided by this author (Gaard, 2001) and an examination of this article from a legal perspective (De Vido, 2023b).

Marxism[11]—it should be acknowledged that there are indeed several theories that endorse non-anthropocentric perspectives on law. Nonetheless, 'revitalizing ecofeminism and using it as a method of international law opens new interesting paths of research that can better look at schemes of oppression within and across species' (De Vido, 2024). It is not necessarily the best method—each method has points of strength and weakness, and mirrors the given author's background and sensitivity—or one that necessarily has to replace the others, but it is a 'method that combines an understanding of persistent discriminations in our societies with the knowledge that humans belong to the environment and not *vice versa*' (De Vido, 2024). It is a 'method that unravels dynamics of power and oppression within and between different species' (De Vido, 2024), expanding the view of these dynamics beyond humans. Karen Morrow, one of the leading theorists of ecofeminism in international law, argued that 'when ecofeminism is pursued in an international context, its goal is to import this paradigm shifting approach to accommodating alternative/additional procedural and substantive expertise and experience, from the periphery to the mainstream' (Morrow, 2023, p. 194). Since international law 'embraces and even embeds dualism at its very core, with entrenched power holders enjoying the advantages of othering non-privileged humans and the environment', an ecofeminist approach 'could offer a corrective […] entailing at base a radical, relational process of (re)connection—as humans with one another and as humans with the non-human environment' (ibid.).[12]

6 An Ecofeminist Reading of the *Teitiota* Case

The views issued by the UN Human Rights Committee (hereinafter the Committee) in the case *Ioane Teitiota v. New Zealand* in 2019 (Human Rights Committee, 2019) represent an interesting 'test' for the argument

---

[11] One could use Marxism, for example, which examines the relationship between the base and the 'superstructure', and concepts of ideology and hegemony, but not the exploitation of nature by (a part of) humanity. 'Marxist approaches are committed to grounding the law in its wider material context: understanding the ways in which political-economic relationships—and their attendant conflicts—shape and are manifested within (international) law' (Knox & Tzouvala, 2022). One could also use radical naturalism, which is also underdeveloped in international law (see an application of Spinoza's thought on international environmental law in De Lucia Dahlbeck, (2018)).

[12] See the discussion of ecofeminism in private international law in De Vido (2024).

presented in these pages. In the case under analysis here, the applicant, Ioane Teitiota, a resident of the island of Tarawa in Kiribati in the Central Pacific, applied for asylum status in New Zealand, claiming that the situation in his home country had become increasingly unstable and precarious because of rising sea levels caused by climate change. The authorities and the courts in New Zealand denied him the status and rejected his appeals with the consequence that Teitiota was expelled back to Kiribati along with his wife and their children that were born in New Zealand in the meantime

Without delving into the details of the case,[13] it is useful to remember that the Committee found that: (a) there was no generalized risk of violence in Kiribati such as to create a real risk of irreparable harm to Teitiota; (b) the appellant had not shown obvious arbitrariness or error in the verification carried out by the national authorities as to the existence of a real, personal or reasonably foreseeable risk to his right to life as a result of the climate of violence on the island; (c) the appellant had not provided sufficient information on the inaccessibility, insufficiency or insecurity of drinking water such as to pose a reasonable threat to his health and thus a violation of his right to life or a risk of his premature death; and (d) the applicant had not shown that the local authorities' assessment of the State of his crops, his major source of income—which is difficult but not impossible—was arbitrary, erroneous or such as to result in a denial of justice. It follows that according to the Committee, there had been no violation of the applicant's right to life.

It is also interesting to quote a passage in which the Committee recognized that both slow-onset processes, which produce a gradual deterioration of resources, and sudden-onset events, such as storms and flooding, 'can propel cross-border movement of individuals seeking protection from climate change-related harm' (UN Human Rights Office of the High Commissioner, 2022). The Committee also concluded that without national and international efforts, the effects of climate change in some countries may expose individuals to a violation of their rights guaranteed by Articles 6 and 7 of the Covenant. Kiribati was taking adaptation measures to reduce the risk of existing vulnerabilities and therefore the deportation decision of the New Zealand authorities, which had taken into consideration multiple elements, should not be considered arbitrary.

---

[13] See, *inter alia*, Foster and McAdam (2022), Behrman and Kent (2020).

The expert Vasilka Sancin disagreed with the majority opinion that the deportation of the appellant to Kiribati did not cause an imminent, or likely, risk of arbitrary deprivation of the applicant's life. In particular, the expert noted that it was not for the applicant but for New Zealand to prove that the author and his family would in fact enjoy access to 'safe drinking' (and not merely potable) water in Kiribati. The burden of proof would be onerous if imposed on the person claiming refugee status.

The re-reading from an ecofeminist perspective sheds light on the potential of UN bodies to unveil discriminatory patterns and identify legal obligations for States stemming from legally binding instruments in force. Kiribati, home to around 120,000 people, lies in the middle of the Pacific Ocean, north of New Zealand and south-west of Hawaii. Of its 33 far-flung islands, none are more than four metres above sea level at their highest point (WHO, 2024). Saltwater intrusion affects the quality of water in wells, and floods taro patches and gardens, and this has a disproportionate impact on women and girls (Kiribati National Statistics Office, 2019; UN Women, 2013; Vaughan et al., 2022). However, we know about the existence of Teitiota's wife only through her testimony (the testimony of the 'author's wife'), which was reported by the Committee. Her situation, which was not simply the situation of a woman within a family but rather that of a woman as part of a community of vulnerable women dramatically affected by sea level rise, was not taken into consideration at all by either the lawyers that filed the case or the experts of the Committee (because of how the facts were presented).

According to a study by the Asian Development Bank (2021), women, children, the elderly and the most disadvantaged households 'often bear a disproportionate share of the burden of inadequate WASH (water, sanitation and hygiene) infrastructure and behaviours'. In the southern part of Tarawa, women are responsible for all water-related household tasks, such as washing, cleaning, and caring for children and the elderly. The intermittent supply of water is a problem for many households and especially for women, who have menstrual hygiene needs: clean water and appropriate materials are often not available in homes, schools, and workplaces, resulting in absenteeism and an inability to handle this biological phenomenon with dignity. Although children suffer more than adults from diseases related to the use of unhealthy water, in Tarawa girls suffer more than boys from diarrhea and dysentery due to gender inequalities structured in a strongly patriarchal society, which provides for different feeding practices for the two sexes (to the detriment of girls) (Lal, 2015).[14]

Then there is the natural element, i.e. the impact of climate change on non-human species and natural objects. However, as some investigations in recent years have shown, rising sea levels have a significant effect on ecosystem functions and biodiversity.

A World Bank study (2021), for example, reports that phenomena such as flooding and inundation destroy important refuges for migratory bird species. Even when it comes to ocean water, the situation must be assessed by considering not only humans, but also non-human animals and natural species. Ocean acidification and sea level rise damage ecosystems, resulting in a decline in fishing capacity for residents.

The environment, of which human beings are a part, plays a key role in defining the imminence and predictability of risk when considered in the light of the ecofeminist method. It is true that the Committee specified that the situation in Kiribati must be monitored and updated before it could decide on future cases of deportation pertaining to it, thus 'opening up' to ever quasi-legal developments; however, it can be assumed that the Teitiota case already contained additional elements to be valued in the legal reasoning.

The communication to the Committee, submitted only by the 'head' of the family, raises a procedural question about the role of the applicant, as it does not consider the impact that phenomena such as climate change and other forms of slow violence have on women and children. Some scholars correctly argued that in the case of climate change, what matters in the analysis is not a strict imminence, but rather the foreseeability of harm (Foster & McAdam, 2022). Foreseeability entails a consideration of the impact on present and future generations of both human and non-human species of phenomena that can be brought under the concept of 'chronic emergency'. The concept of 'imminence' should consider the effects of environmental degradation and climate change on human and non-human beings, determining whether there are grounds to grant international protection to avoid violations of the human rights of

---

[14] As noted by the CEDAW Committee in its 2020 Concluding Observations ('Concluding observations on the combined initial, second and third periodic reports of Kiribati'), there is a patriarchal cultural legacy in Kiribati that manifests itself through customs and practices that account for the highest levels of gender-based violence against women in the Pacific.

the individual (thus not losing the individual dimension, and the assessment of the fear of persecution, which is pivotal for the recognition of the individual's refugee status), who lives in an 'environment' composed of multiple interconnected elements that stretch across time.

7 (Unjustifiable) Human Rights Justifications: The Responsibility of States for not Considering the Gendered Impact of Sea Level Rise

In this final part, the chapter briefly reflects on the concept of human rights justifications[15] as related to the phenomenon of sea level rise seen in a gendered perspective. The analysis challenges the binary system State compliance/violation of human rights from a feminist perspective and argues that in the case of sea level rise and migration, the human rights justifications States use as a governance tool are unjustifiable without a gendered and ecological perspective. Leaving aside the matter of a feminist analysis of human rights,[16] which is not within the scope of this section, it should be stressed that the legal point here is not what kind of justifications human rights have or should have, but rather how to evaluate the human rights-based justifications States utilize for their actions.[17] Feminist lawyers have highlighted how the State promotes human interests and, in particular, interests of 'dominant groups claim[ing] universality' (Chinkin & Charlesworth, 2022; Young, 1990, p. 97). The concept of the State, as traditionally conceived in international law, can be criticized and deconstructed from a feminist, queer, and postcolonial perspective to unpack its inherent contradictions, and reveal how it perpetuates existing inequalities in society by promoting a unified, hierarchical structure (Jones, 2023, p. 33). The sovereign State, as a non-human actor, is actually 'all-too-human', being 'deeply embedded in constructions of power, gender, and race in international

---

[15] See the Introduction. Additional information is available at: https://hrjust-intersect-observatory.eu/.

[16] On how international human rights law has disregarded women's rights, see Chinkin and Charlesworth (2022).

[17] On the right to justification, see the interesting analysis in Herlin-Karnell and Klatt (2020).

law' (Jones, 2023, p. 37). In that sense, the dichotomy State compliance/State violation of human rights can be disrupted by unravelling the fact that a State, even when formally undertaking actions to respect human rights, de facto violates them. The example of *Teitiota* is particularly relevant here. As the UN Human Rights Committee argued—even though with some relevant dissenting opinions—New Zealand did not violate human rights in the case. However, is this compliance not in itself a violation of human rights when read using a feminist and ecological perspective? As this chapter showed in the previous section, researchers should read the concept of imminence in a perspective that is less State-centric and, using the hermeneutic technique, explore new (legally sound) directions in this respect. Laws and policies that are adopted to help communities to adapt to climate change so as to help them enjoy their right to survival might be a cause of human rights violations. In the case of New Zealand, for example, feminist research has highlighted that in climate adaptation policies in Aotearoa 'women's experiences and knowledges are not adequately reflected in the policy', and 'technocratic, masculinist, and top-down approaches to climate adaptation have been prioritized over knowledges and approaches from diverse feminist, indigenous, queer, and broader social-science perspectives' (Macpherson et al., 2024, p. 4).

8 Conclusions

This chapter has shown the importance of studying sea level rise, one of the major consequences of climate change, from a gendered perspective, while considering its disproportionate impact on women, especially those at the intersection of different forms of discrimination. Sea level rise is a form of climate violence and falls under those phenomena that are only apparently natural because they are entrenched in a societal discriminatory structure: it is 'chronic' like an illness that it is difficult to cure. The re-reading of the Teitiota case, focusing on one aspect of the climate crisis there—sea level rise—and the effects on women, has demonstrated the potential of using an ecofeminist method in such an analysis, as an ecofeminist method does not consider women as vulnerable subjects per se, but empowers them and makes their often neglected voices heard. There is a long way to go towards the incorporation of a gendered perspective in the official debate on sea level rise, as the works of the International Law Commission demonstrate. From an academic and political point of view, however, it is important to consider how fundamental the debate on a *de facto* gender equality, along with women's access to climate justice (De Vido & Fornalé, 2023), is and will be in the near future, while dealing with these unnatural, (part of) humanity-made phenomena.

REFERENCES

Al Hasnat, M., & Teirstein, Z. (2024, May 30). *The Bizarre Link Between Rising Sea Levels and Complications in Pregnancy*. Vox. https://www.vox.com/climate/351534/salt-water-pregnancy-rising-sea-levels

Albertyn, C., García, H. A., Campbell, M., Fredman, S., & Rodriguez de Assis Machado, M. (2023). Introduction. In *Feminist Frontiers in Climate Justice: Gender Equality, Climate Change and Rights* (pp. 1–16). Edward Elgar Publishing. https://doi.org/10.4337/9781803923796

Arendt, H. (1970). *On Violence*. Harcourt, Brace & World.

Asian Development Bank. (2021, August). *Climate Change, Water Security, and Women: A Study on Water Boiling in South Tarawa, Kiribati*. https://www.adb.org/publications/climate-change-water-security-women-kiribati

Atapattu, S. (2020). Climate Change and Displacement: Protecting 'Climate Refugees' Within a Framework of Justice and Human Rights. *The Journal of Human Rights and the Environment*, 11, 86–113. https://doi.org/10.4337/jhre.2020.01.04

Atrey, S. (2023). The Inequality of Climate Change and the Differences It Makes. In C. Albertyn, H. A. García, M. Campbell, S. Fredman, & M. Rodriguez de Assis Machado (Eds.), *Feminist Frontiers in Climate Justice: Gender Equality, Climate Change and Rights* (pp. 17–39). Edward Elgar Publishing. https://doi.org/10.4337/9781803923796

Behrman, S., & Kent, A. (2020). The Teitiota Case and the Limitations of the Human Rights Framework. *Questions of International Law*, 75, 25–39. http://www.qil-qdi.org/the-teitiota-case-and-the-limitations-of-the-human-rights-framework/

Behrman, S., & Kent, A. (2022). *Climate Refugees: Global*. Cambridge University Press. https://doi.org/10.1017/9781108902991

Bianchi, B. (2012). Ecofemminismo: il pensiero, i dibattiti, le prospettive. *Deportate, Esuli e Profughe*, 20, I–XXVI. https://www.unive.it/pag/fileadmin/user_upload/dipartimenti/DSLCC/documenti/DEP/numeri/n20/02_20_-_numero_completo.pdf

Caracciolo, I., Cellamare, G., Di Stasi, A., & Gargiulo, P. (2022). *Migrazioni internazionali: Questioni giuridiche aperte*. Editoriale Scientifica.

Charlesworth, H. (1999). Feminist Methods in International Law. *The American Journal of International Law, 93*(2), 379–394. https://doi.org/10.2307/2997996

Chinkin, C., & Charlesworth, H. (2022). *The Boundaries of International Law: A Feminist Analysis*. Manchester University Press. https://doi.org/10.7765/9781526163592

Committee on the Elimination of Discrimination Against Women (CEDAW). (2020). *CEDAW/C/KIR/CO/1–3: Concluding Observations on the Combined Initial, Second and Third Periodic Reports of Kiribati by the Committee on the Elimination of Discrimination Against Women*. https://www.ohchr.org/en/documents/concluding-observations/cedawckirco1-3-concluding-observations-combined-initial-second

D'Eaubonne, F. (1974). *Le féminisme ou la mort*. Verso Books.

De Guttry, A., Gestri, M., & Venturini, G. (2016). *International Disaster Law Response*. Asser. https://doi.org/10.1007/978-90-6704-882-8

De Lucia Dahlbeck, M. (2018). *Spinoza, Ecology and International Law: Radical Naturalism in the Face of the Anthropocene*. Routledge. https://doi.org/10.4324/9781315177267

De Vido, S. (2021). A Quest for an Eco-centric Approach to International Law: the COVID-19 Pandemic as Game Changer. *Jus Cogens, 3*, 105–117. https://link.springer.com/article/https://doi.org/10.1007/s42439-020-00031-0

De Vido, S. (2023a). 'Violenza climatica' e migrazioni di genere nel diritto internazionale. In A. Di Stasi, R. Cadin, A. Iermano, & V. Zambrano (Eds.), *Migrant Women and Gender-Based Violence in the International and European Legal Framework* (pp. 137- 170). Editoriale Scientifica. https://doi.org/10.20318/cdt.2024.8479

De Vido, S. (2023b). L'ecofemminismo di Greta Gaard e la caccia alle balene: una riflessione giuridica. *Deportate, esuli, profughe, 52*, 93–109. https://www.unive.it/pag/fileadmin/user_upload/dipartimenti/DSLCC/documenti/DEP/numeri/n.51/n._52/07_De_Vido.pdf

De Vido, S. (2024). The Privatisation of Climate Change Litigation: Current Developments in Conflict of Laws. *Jus Cogens, 6*, 65–88. https://doi.org/10.1007/s42439-023-00084-x

De Vido, S., & Fornalé, E. (2023). *Achievements and Hurdles Towards Women's Access to Climate Justice*. In E. Fornalé, & F. Cristani (Eds.), *Women's Empowerment and Its Limits* (pp. 33–51). Palgrave Macmillan. https://doi.org/10.1007/978-3-031-29332-0

De Vido, S., & Frulli, M. (2023). Preventing and Combating Violence Against Women and Domestic Violence: A Commentary on the Istanbul Convention. *Edward Elgar Publishing*. https://doi.org/10.1007/s10991-024-09364-y

Fornalé, E. (2020). A l'envers: Setting the Stage for a Protective Environment to Deal with 'Climate Refugees' in Europe. *European Journal of Migration and Law, 22*(4), 518–540. https://doi.org/10.1163/15718166-12340088

Fornalé, E. (2023). Slow Violence, Gender Equality and Climate Agency in Times of 'Polycrisis'. *Revista General de Derecho Europeo, 61*. https://www.iustel.com/v2/revistas/detalle_revista.asp?id_noticia=426572&d=1

Fornalé, E., Bilkova, V., Burgorgue-Larsen, L., Cristani, F., De Vido, S., Doebbler, C., & Hertogen, A. (2023). *Amicus Curiae - Request for an advisory opinion on the Climate Emergency and Human Rights submitted to the Inter-American Court of Human Rights by the Republic of Colombia and the Republic of Chile of January 9, 2023*. https://doi.org/10.48350/190814

Foster, M., & McAdam, J. (2022). Analysis of 'Imminence' in International Protection Claims: Teitiota v. New Zealand and Beyond. *International Comparative Law Quarterly, 71*(1), 22–31. https://ssrn.com/abstract=4220389

Gaard, G. (1993). Living Interconnections with Animals and Nature. In *Ecofeminism, Women, Animals, Nature* (pp. 1–12). Temple University Press.

Gaard, G. (2001). Tools or a Cross-Cultural Feminist Ethics: Exploring Ethical Contexts and Contents in the Makah Whale Hunt. *Hypatia, 16*, 1–26. https://doi.org/10.1111/j.1527-2001.2001.tb01046.x

Gaard, G. (2011). Ecofeminism Revisited: Rejecting Essentialism and Re-Placing Species in a Material Feminist Environmentalism. *Feminist Formations, 23*(2), 26–53. https://doi.org/10.1353/ff.2011.0017

Grear, A. (2015). Deconstructing Anthropos: A Critical Legal Reflection on 'Anthropocentric' Law and Anthropocene 'Humanity.' *Law and Critique, 26*(2), 225–249. https://doi.org/10.1007/s10978-015-9161-0

Hamzic, V. (2018). International Law as Violence. In D. Otto (Ed.), *Queering International Law: Possibilities, Alliances, Complicities, Risks* (pp. 290–315). Oxford University Press. https://doi.org/10.1093/ejil/chy046

Herlin-Karnell, E., & Klatt, M. (2020). *Constitutionalism Justified*. Oxford University Press. https://doi.org/10.1093/oso/9780190889050.001.0001

Human Rights Committee. (2019). *Ioane Teitiota v. New Zealand: Communication No. 2728/2016*. United Nations.

International Law Commission (ILC). (2024a). *Sea level Rise in Relation to International Law: Elements in the Previous Work of the International Law Commission That Could Be Particularly Relevant to the Topic. Memorandum by the Secretariat A/CN.4/768* (26 January 2024). United Nations.

International Law Commission (ILC). (2024b). *Study Group on Sea level Rise in Relation to International Law: Report A/CN.4/L.1002* (15 July 2024). United Nations.

International Organization for Migration (IOM). (2014). *A Gender Approach to Environmental Migration: Brief No. 15*. https://www.iom.int/sites/g/files/tmzbdl486/files/about-iom/gender/Gender-Approach-to-Environmental-Migration.pdf

Jones, E. (2023). *Feminist Theories and International Law: Post-human Perspectives*. Routledge. https://doi.org/10.4324/9781003363798

Kelly, N. (1993). Gender-Related Persecution: Assessing the Asylum Claims of Women. *Cornell International Law Journal*, *26*, 625–674. https://scholarship.law.cornell.edu/cilj/vol26/iss3/5

Kiribati National Statistics Office. (2019). *Kiribati Social Development Indicator Survey 2018–19: Survey Findings Report*. National Statistics Office.

Knox, R., & Tzouvala, N. (2022). Marxist Approaches to International Law. In C. Binder, M. Nowak, J. A. Hofbauer, & P. Janig (Eds.), *Elgar Encyclopedia of Human Rights* (pp. 447–450). Edward Elgar Publishing. https://doi.org/10.4337/9781789903621

Lal, P. N. (2015). *The Economic Burden of Inadequate Water and Sanitation in South Tarawa, Kiribati: Who Bears the Cost?* Global Water Forum. Retrieved September 12, 2024, from https://www.globalwaterforum.org/2015/05/25/the-economic-burden-of-inadequate-water-and-sanitation-in-south-tarawa-kiribati-who-bears-the-cost/

Lama, P., Hamza, M., & Wester, M. (2021). Gendered dimensions of migration in relation to climate change. *Climate and Development*, *13*(4) 326–336. https://doi.org/10.1080/17565529.2020.1772708

Lambrou, Y., & Piana, G. (2006). *Gender: The Missing Component of the Response to Climate Change*. Food and Agriculture Organization of the United Nation. https://www.fao.org/4/i0170e/i0170e00.pdf

Lear, L. (1998). *Lost Woods*. Beacon Press.

MacGregor, S. (2010). Gender and Climate Change: From Impacts to Discourses. *Journal of the Indian Ocean Region*, *6*(2), 223–238. https://doi.org/10.1080/19480881.2010.536669

Macpherson, E., Masselot, A., Jefferson, D., & Gunn, J. (2024). A Critical Feminist Evaluation of Climate Adaptation Law and Policy: The Case of Aotearoa New Zealand. *Climate Law*, *14*, 1–35. https://doi.org/10.1163/18786561-bja10050

Mallory, C. (2009). Val Plumwood and Ecofeminist Solidarity: Standing with the Natural Other. *Ethics and the Environment*, *14*, 3–21. https://doi.org/10.2979/ete.2009.14.2.3

Malone, L. A. (2015). Environmental Justice Reimagined Through Human Security and Post-Modern Ecological Feminism: A Neglected Perspective on Climate Change. *William & Mary Law School Scholarship Repository*, *38*, 1445–1472. https://ir.lawnet.fordham.edu/ilj/vol38/iss5/4

McAdam, J. (2012). *Climate Change, Forced Migration, and International Law*. Oxford University Press. https://doi.org/10.1093/acprof:oso/9780199587087.001.0001

Mellor, M. (1997). *Feminism and Ecology*. New York University Press.

Mies, M., & Shiva, V. (1993). *Ecofeminism*. Zed Books.

Mora, C., & Piper, N. (2021). *The Palgrave Handbook of Gender and Migration*. Palgrave Macmillan. https://doi.org/10.1007/978-3-030-63347-9

Morrow, K. (2023). Towards an Ecofeminist Critique of International Law? In V. Chapaux, F. Mégret, & U. Natarajan (Eds.), *The Routledge Handbook of International Law and Anthropocentrism* (pp. 183–197). Routledge. https://doi.org/10.4324/9781003201120-8

Nixon, R. (2011). *Slow Violence and the Environmentalism of the Poor*. Harvard University Press. https://psycnet.apa.org/. https://doi.org/10.4159/harvard.9780674061194

Office of the High Commissioner for Human Rights (OHCHR). (2022). *Report of the Special Rapporteur on the implications for human rights of the environmentally sound management and disposal of hazardous substances and wastes*. United Nations.

Plumwood, V. (1993). *Feminism and the Mastery of Nature*. Routledge.

Pratiwi Misbahul, A. (2023, October 2). *How Women's Environmental Action Across the Global South Can Create a Better Planet*. The Conversation. https://theconversation.com/how-womens-environmental-action-across-the-global-south-can-create-a-better-planet-214083

Puleo, A. H. (2017). What is Ecofeminism? *Quaderns de la Mediterrània*, 25, 27–34

Rogers, N. (2023). Climate Violence and the Word. *Journal of Human Rights and the Environment*, 14(2), 144–168. https://doi.org/10.4337/jhre.2023.02.02

Sarat, A., & Kearns, T. R. (1995). Introduction. In *Law's Violence*. University of Michigan Press. https://doi.org/10.3998/mpub.13488

Task Force on Displacement. (2018, September 17). *Report*. https://unfccc.int/sites/default/files/resource/2018_TFD_report_17_Sep.pdf

UN Commission on the Status of Women. (2022). *Achieving Gender Equality and the Empowerment of All Women and Girls in the Context of Climate Change, Environmental and Disaster Risk Reduction Policies and Programmes* (E/CN.6/2022/L.7). United Nations.

UN General Assembly (UNGA). (2020). *Report of the Special Rapporteur on the Human Rights of Internally Displaced Persons* (A/75/207). United Nations.

UN General Assembly (UNGA). (2022). *Report of the Special Rapporteur on Violence Against Women and Girls* (A/77/136). United Nations.

UN Woman. (2013). Climate change, Disasters and Gender-Based Violence in the Pacific. https://www.uncclearn.org/wp-content/uploads/library/unwomen701.pdf

UN Women. (2022a, February 28). *Explainer: How Gender Inequality and Climate Change Are Interconnected*. United Nations. https://www.unwomen.org/en/news-stories/explainer/2022/02/explainer-how-gender-inequality-and-climate-change-are-interconnected

UN Women. (2022b, September 27). *Statement on Women's Rights in Iran*. United Nations. https://www.unwomen.org/en/news-stories/Statement/2022/09/un-women-Statement-on-womens-rights-in-iran

United Nations (UN). (2023, October 23). *'We Must Look to International Law as a Beacon of Light', Commission Co-Chair Tells Sixth Committee, as Review of Annual Report Begins*. https://press.un.org/en/2023/gal3698.doc.htm

United Nations High Commissioner for Refugees (UNHCR). (2002). *Gender-Related Persecution Within the Context of Article 1A(2) of the 1951 Convention and/or Its Protocol Relating to the Status of Refugees* (HCR/GIP/02/01).

United Nations High Commissioner for Refugees (UNHCR). (2022). Gender, Displacement and Climate Change. https://www.unhcr.org/media/gender-displacement-and-climate-change

Vakoch, D. A., & Mickey, S. (2018). Women and Nature? Beyond Dualism in Gender, Body, and Environment. *Routledge*. https://doi.org/10.4324/9781315167244

Vaughan, C., Moosad, L., & Rowe, J. (2022). *Sexual and Reproductive Health and Gender-Based Violence in Kiribati: A Review of Policy and Legislation*. University of Melbourne.

Warren, K. (1990). The Power and the Promise of Ecological Feminism. *Environmental Ethics, 12*, 125–146. https://doi.org/10.5840/enviroethics199012221

Warren, K. (1997). *Ecofeminism*. Indiana University Press.

World Bank Group (2021). *Climate Risk Country Profile: Kiribati*. https://climateknowledgeportal.worldbank.org/sites/default/files/country-profiles/15816-WB_Kiribati%20Country%20Profile-WEB.pdf

World Health Organization (WHO). (2024, March 8). *Countering the Climate Crisis in Kiribati*. https://www.who.int/westernpacific/news-room/feature-stories/item/countering-the-climate-crisis-in-kiribati

Yoshida, K. (2023). Yahori Matsui: Challenging the Silences of International Law Through Pan Asian Feminist Solidarity. In I. Tallgren (Ed.), *Portraits of Women in International Law* (pp. 160–169). Oxford University Press. https://doi.org/10.1111/1468-2230.12903

Young, I. M. (1990). Justice and the Politics of Difference. *Princeton University Press*. https://doi.org/10.2307/j.ctvcm4g4q

Zorzi Giustiniani, F. (2021). *International Law in Disaster Scenarios: Applicable Rules and Principles.* Springer. https://doi.org/10.1017/ajil.2022.79

Open Access This chapter is licensed under the terms of the Creative Commons Attribution 4.0 International License (http://creativecommons.org/licenses/by/4.0/), which permits use, sharing, adaptation, distribution and reproduction in any medium or format, as long as you give appropriate credit to the original author(s) and the source, provide a link to the Creative Commons license and indicate if changes were made.

The images or other third party material in this chapter are included in the chapter's Creative Commons license, unless indicated otherwise in a credit line to the material. If material is not included in the chapter's Creative Commons license and your intended use is not permitted by statutory regulation or exceeds the permitted use, you will need to obtain permission directly from the copyright holder.

CHAPTER 7

Rethinking Sustainable Migration for the Anthropocene

Samuel Ballin and Sandra Mantu

1 Introduction

In March 2024 scientists voted against recognizing the Anthropocene as a new geological epoch, a decision that highlights the controversies that abound around this notion (Witze, 2024). As a cultural concept, the Anthropocene is generally understood as the age of irreversible human impacts on the planet leading to, among others, climate change and biodiversity loss. Dipesh Chakrabarty argues that at the heart of debates about the Anthropocene is the failure to reconcile two different time scales: on the one hand, geological time, and, on the other hand, human-historical time, which would force us to engage with the Anthropocene as produced not by an amorph humanity but by specific systems, such as capitalism

---

Authors have contributed equally to this chapter.

S. Ballin · S. Mantu (✉)
Radboud University, Nijmegen, The Netherlands
e-mail: sandra.mantu@ru.nl

S. Ballin
e-mail: samuel.ballin@ru.nl

© The Author(s) 2026
E. Fornalé (ed.), *Sea Level Rise*,
https://doi.org/10.1007/978-3-031-89171-7_7

or the global economic system, and the moral choices they represent (Chakrabarty, 2018).

Besides systemic implications, the Anthropocene has profound effects on people, who may find themselves contemplating the degradation or loss of their physical environments and livelihoods, leading to supposed 'climate migrants'[1] (Baldwin & Fornalé, 2017; ILA , 2022, 2024). In the debate on the origins, nature, effects, and responses to the Anthropocene, environmental change either because of slow-onset events, such as sea level rise, or sudden-onset extreme weather is thus understood to have important consequences for states, their citizens, and international relations more broadly. It is generally accepted that the disappearance of physical territory because of sea level rise, as in the case of the so-called sinking or disappearing states, raises important questions about how persons affected by such changes can continue to benefit from rights and protections (McAdam, 2010; Oliver, 2009; Stewart, 2023). The role of law in dealing with the changes brought about by the Anthropocene has been examined from the perspective of how different branches of international law relate and may need to adapt and develop new lexicons to capture the new conditions of the Anthropocene (Vidas, 2014; Vidas et al., 2015). Safeguarding the human rights of persons affected by environmental change and degradation raises new dilemmas around State sovereignty, State cooperation, and legal personhood (Vidas, 2014).

In this chapter, we interrogate how migration law, and more specifically study and labour migration, engages with the Anthropocene. Since migration has been addressed as an adaptation strategy that predicts increased migration flows from territories affected by climate change and environmental degradation (Baldwin & Fornalé, 2017; Boas et al., 2022; Farbotko, 2022; Ferris, 2020), the climate-migration nexus has become one of the major foci of interest in migration governance. There is no consensus on whether 'climate migrants' should be categorized as forced migration, voluntary or economic migration. There is nonetheless agreement that the traditional legal categories of refugee, human rights, and migration laws do a poor job in accommodating the human rights needs of those affected, as illustrated by discussions on 'climate migrants' versus so-called climate refugees (Aleinikoff, 2019; Scott,

---

[1] There is much variety and debate regarding the terminology used by scholars and policymakers on the climate-migration nexus. The authors use the term 'climate migrants'.

2016; Yamamoto & Esteban, 2017). The interlinked failures to accommodate forms of 'climate migration' within existing legal frameworks, to open up the possibility of designing new international law rules, and more broadly, to respond to the human rights issues that they raise have contributed to growing social and political anxieties about large influxes of climate migrants (Bettini, 2019; ILA, 2022, 2024; see also Chapters 5 and 7 by De Vido and Cristani and Fornalé).

The debate on climate and migration is heavily influenced by how States in the Global North perceive migration more broadly, namely as a source of ontological insecurity that they seek to manage and contain, while sticking firmly to the script of territorially bounded sovereignty. The latter implies a State's right to decide who enters its territory and under what conditions. At the same time, the need to future-proof migration and respond to the multiple crises faced by humanity in the late Anthropocene is equally taking centre stage, as illustrated by the UN's linking of migration, climate, and sustainable development in its 2030 Agenda for Sustainable Development. The 2030 Agenda is meant to offer a 'shared blueprint for peace and prosperity for people and the planet, now and into the future', thereby highlighting the need to think about the ways in which climate mobilities (Baldwin et al., 2019; Boas et al., 2022) need to be included in our collective future.

In this chapter, we discuss State responses to migration and climate change as sources of ontological insecurity, that are appeased by the mobilization of notions such as sustainable migration and environmental sustainability. This strategy minimizes urgent questions around the use of study and labour migration as adaptation strategy available for persons affected by slow-onset events. We examine questions of social and environmental sustainability in relation to migration law and policy in the EU, with a focus on the Netherlands, one of the lowest-lying countries in the world. The Netherlands relies heavily on EU migrant workers, while seeking to attract highly skilled (potential) workers, including from States affected by slow-onset events, such as India and Bangladesh. We argue that much of the future-proofing of migration remains within a capitalist and postcolonial mode of viewing power relations between States that limits the possibility to frame study and labour migration as an avenue to respond to climate migration. We illustrate this with the examples of EU migrant workers and third country national (TCN) students and researchers, which allow us to explore how migration and sustainability fit together within a 'knowledge-based economy' that seeks to meet its obligations as an EU Member State while also living up to its image as a champion of 'green politics' and environmental progress.

We proceed in the following section by discussing the problem of ontological security as it is experienced by the State. Section 3 explores the notion of environmental sustainability and its links with migration, followed by a discussion of socially sustainable migration in the EU in Sect. 4. Section 5 presents two case studies on EU migrant workers in the meat industry and non-EU migrant students and researchers in higher education. Section 6 concludes by considering how we might reorient migration law and policy in ways that promote sustainability and ontological security in the Anthropocene.

2 Setting the Stage: The State and the Problem of Ontological Security for Human Rights Protection of Climate Migrants

The world of the Anthropocene is one in which interactions between social and natural systems are taking novel forms as well as reaching new levels of intensity. According to Farbotko, scholars of the Anthropocene consider environmental change as an 'ontological threat', raising the question as to whether the very experience of being human changes in the context of catastrophic alterations in global systems (Farbotko, 2019, p. 252). Ontological security describes 'the need of humans to feel as if they are whole, continuous, and stable over time which requires an ongoing effort to hold at bay the hard uncertainties of Being, such as meaning and mortality and the awareness of the fundamental contingency of the social order' (Mitzen, 2018, p. 1374). The argument is that profound shifts in the worlds of people living with the threat of their homes being rendered uninhabitable may lead to ontological insecurity, inasmuch as those threats cannot be effectively reduced or managed (Farbotko, 2019, pp. 253–254).

In the Anthropocene, migration interacts with environmental change to produce new forms and sources of ontological insecurity at both the personal and the State level. At the personal level, migrants and citizens may experience migration and environmental change as generating multiple sources of interconnected uncertainty and insecurity. This raises the question of how law responds to these interconnected and multiple sources of insecurity to safeguard the human rights of persons affected.

The concept of ontological security has been deployed in international relations (IR) literature to articulate relationships between identity and security, and between identity and political outcomes in relation to States as actors in the international community and as part of regional integration projects, such as the EU. It is, of course, a familiar and well-trodden story that many states in the Global North perceive migration as a source of such ontological insecurity (Boas, 2015; Browning, 2017; Mitzen, 2018). Politically, collective organizations of states such as the UN and the EU view migration related to environmental change as among the greatest security challenges faced by States in the Anthropocene (Boas, 2015).

This explains why migration and environmental change have become major foci for scholarship ranging from political science to law, sociology, development studies, and IR. These fields all question the sustainability and 'future-proofing' of existing migration law and policy in the light of unprecedented social and environmental changes in the Anthropocene. The fitness of current State approaches to migration governance in dealing with such phenomena is generally assessed as problematic. Debates have focused on both the need and the desirability of rethinking existing categories in international law to accommodate the diverse and novel realities of migration in the Anthropocene, but the widening of existing legal frameworks to include forms of climate migration is resisted on the grounds that it may blur or dilute the special protections currently afforded to other groups such as refugees (McAdam, 2011), and that it may also function to increase migration flows to States in the Global North.

This picture is complicated by the view of migration itself as a source of perceived ontological insecurity for many States in the Global North, as already mentioned above. Discourses on the governance and containment of migration through managed and regulated pathways, as in the Global Compact for Safe, Orderly, and Regular Migration (UN, 2019), co-exist with a growing acknowledgement at the international level of the need to future-proof migration law and policy to respond to the many challenges faced by humankind in the late Anthropocene (Kälin, 2018). The UN has linked migration, climate, and sustainable development in the Global Compact as well as in the 2030 Agenda for Sustainable Development, which sets out the Sustainable Development Goals (the SDGs) and offers a 'shared blueprint for peace and prosperity for people and the planet, now and into the future' (UN, 2025).

In the EU context, climate migration is part of a complex discourse that treats migration in general as a stressor to the EU's post-national political project, the core of which is the internal market as an area of free movement without internal border controls. To safeguard the idea of free movement, EU migration policy is formulated based on the premise that internal freedom is underpinned by secure and controlled external borders, so that only those migrants who have a right to be present in the EU can actually enter. This has resulted in EU migration policy moving towards the greater securitization and externalization of border controls, coupled with increased policing and surveillance, which has had a detrimental impact on the EU's commitments to asylum, non-refoulement, and human rights (Guild, 2020a).

In this setting, rethinking legal migration categories is further frustrated by a political discourse that emphasizes the right to select and actively recruit economically 'desirable' migrants. The decision-making processes of such migrants may also be influenced by environmental changes, but this is not considered relevant for policymakers and legislators. When it comes to groups such as tourists and migrant workers, there has therefore been far less emphasis on environmental sustainability compared climate migrants who are essentialized as displaced or forced migrants (for exceptions see Salazar, 2022; Loxa, 2023).

In this chapter, our aim is to look away from displacement and forced migration to consider instead the ways in which ontological insecurity is perceived and managed at the intersection of climate change and regular migration as potential adaptation strategy—something which has been strikingly underdiscussed in public, political and academic discourse. We examine how sustainability is perceived in study and labour migration law and policy in Europe, both as a present reality and as a future-oriented policy objective and ask how European States are grappling with the problem of ontological insecurity in the Anthropocene.

3 Environmental Sustainability and Migration as Adaptation

Ontological insecurities and challenges associated with migration and environmental sustainability are closely interlocked in the Anthropocene. Climate change threatens the stability and viability of States within their own borders, impacting the availability of energy and other resources, livelihoods, and economic production (see chapter by Nesi and Fornalé in this book). Many States and other actors are responding to these challenges with policies and strategies that aim to ensure environmental sustainability within their own borders or regional spaces. In the EU and Netherlands, as in many countries, there has been great emphasis on the phasing out of fossil fuels and the transition to 'greener' sources of energy. These strategies are exemplified at the European level in the Green Deal (European Commission, 2019). The Green Deal presents the EU Commission's vision of 'green growth', meaning the pursuit of economic growth together with environmental sustainability. It aims for the EU to achieve climate neutrality by 2050, largely focusing on emissions reductions, and to become a world leader in environmentally sustainable technologies in the same period. The Green Deal makes no reference to the role of migration in this transition, despite the EU's wider interest in competing to attract and retain international talent. There is perhaps an underlying tension between the economic and environmental goals of the Green Deal (Filipović et al., 2022). An emphasis on domestic emissions can mean 'offshoring' emissions from unsustainable industries to third countries instead of pursuing more meaningful climate mitigation policies, sometimes (though not always) with an increase in overall emissions at the global level (Dussaux et al., 2023; Cole et al., 2021).

It is clear that States and the EU cannot really address the source of their ontological insecurity without addressing global environmental sustainability. Failure to do so also exacerbates the fear of large-scale climate migration as a further source of ontological insecurity, reinforcing the link between these two areas of policy in the Anthropocene. Migration is considered as one response to environmental changes and ruptures in people's ontological worlds, although how people will respond remains unclear, and voluntary or involuntary immobilities are also a possibility for many people (Farbotko, 2019, p. 255; Rikani et al., 2023). The Intergovernmental Panel on Climate Change has consistently predicted that migration will be the greatest impact of climate change on human populations, underlined in its Sixth Assessment Report (Pörtner et al., 2022). It is difficult to quantify this impact or to draw direct links to specific migration flows, but it is now relatively clear that climate change leads to higher overall levels of global migration (Boas et al., 2019; Piguet, 2022). People experiencing slow-onset changes and future insecurities—such as disappearing territory and increased flood risks due to sea level rise—may respond by moving away from affected areas (Czaika & Münz, 2022).

None of this implies that a large-scale future influx of climate migrants into Europe is necessarily likely, or that climate migration in fact presents some kind of threat to destination countries. While displacement has tended to dominate public and policy discourse, the overwhelming majority is projected to occur domestically or regionally within the worst affected areas, with relatively very few displaced people expected to enter the EU (Czaika & Münz, 2022; Pörtner et al., 2022; Rikani et al., 2023). Nonetheless, there are very real effects associated with continued environmental degradation. These are perceived (accurately or erroneously) as sources of ontological insecurity and met by the State with individual and collective governance responses. This can be seen in academic and policy discourses that amalgamate migration as adaptation with other notions, such as 'environmental displacement' and the 'climate refugee' (Baldwin, 2013; Baldwin & Fornalé, 2017; Honarmand Ebrahimi & Ossewaarde, 2019; Remling, 2020, 2023). A truly sustainable migration policy for the Anthropocene thus needs to account for global environmental sustainability and the social impacts of actual and anticipated environmental risks.

Because few climate migrants are actually expected to enter Europe as a result of sudden displacement (Czaika & Münz, 2022), migration as adaptation to slow-onset changes, including study and labour migration, may therefore have greater value for sustainable migration law and policy in the EU. Migration is seen as a way to diversify incomes and generate remittances, and thus to help affected communities cope with the economic effects of climate change and to invest in more climate-resistant infrastructure (Dun et al., 2023; Klöck & Fink, 2019; Remling, 2020, 2023)—including in the EU context (Geddes & Jordan, 2012). More planned and gradual forms of climate migration, such as study and labour migration, may also help to alleviate states' fears of a sudden influx of climate migrants, and successful labour market integration can perhaps further reduce the sense of ontological insecurity that States associate with climate migration. As highlighted in the Global Compact for Migration, planned and managed migration is seen as a valuable social and economic resource for receiving states. It can also leave people vulnerable to exploitation, however, and places the burden on climate migrants to safeguard their own sustainability (Kitara & Farbotko, 2023; Farbotko et al., 2022). One of the criticisms made of 'adaptation' strategies more generally is that this approach may also feed back into the systems of unsustainable economic growth and consumption that are interacting unsustainably with the climate in the Anthropocene (Felli, 2021). They may even distract and at times conflict with efforts to mitigate the root causes of environmental unsustainability—focusing narrowly on well-being 'here and now' at the expense of 'there and for future generations'.

In the Anthropocene, environmental sustainability and ontological security 'here' cannot be achieved without also taking 'there' into consideration. It is increasingly clear that the State cannot afford to pursue short- and medium-term sustainability 'now' that does not fit into a longer-term plan 'for future generations'. European States and the EU are keenly aware of the vital importance of environmental sustainability in addressing the symptoms and sources of ontological insecurity, as policies like the Green Deal clearly illustrate. However, it is not yet fully clear that these policies take a sufficiently long-term and global approach demanded by the realities of the Anthropocene. It is likewise impossible to separate environmental sustainability from social and economic sustainability, and to scrutinize the existence of such holistic approaches to law and policymaking—particularly in the linking of climate and regular migration. The unprecedented levels of connection and feedback that exist today between the world's natural and social systems present a new challenge for these States, one facet of which is that it gives a new meaning to environmentally sustainable migration law and policy.

4 Socially Sustainable Migration

Franco Gavonel et al. (2021) have proposed the notion of the 'migration-sustainability paradox' to capture the ambiguous role that migration plays as both a factor in economic globalization driving the unsustainability crisis, and as potential force for transnational social and environmental change. Sustainability includes social, economic, and environmental dimensions, and migration can have a positive and negative impact in each of these regards. Despite its ambiguity, the authors argue that migration can make a positive contribution; they emphasize the potential of forms of migration that increase material well-being, promote diversity, reduce insecurity and lower the overall environmental burden (Franco Gavonel et al., 2021). A social sustainability framework for assessing migration would need to be comprehensive and wide ranging, encompassing well-being, integration, the distribution of power and resources, employment, education, the provision of basic infrastructure and services, freedom, justice, and access to influential decision-making.

The notion of sustainable migration has entered the lexicon of policymakers, with the EU being no exception (Loxa, 2023). Besides being something of a buzzword, sustainable migration is discursively approached as a solution to the many ills of current regimes of migration governance that are designed to safeguard the state's sovereign right to exclude and select desirable migrants and, ultimately, (re)produce inequality. The COVID-19 pandemic and its effects for migrants and global mobility regimes have brought a new impetus to the need to rethink migration regimes to make them sustainable, crisis proof, and resilient (Triandafyllidou & Yeoh, 2023). Consequently, scholars have called for grounding the notion of sustainable migration in migrants' human rights and State obligations towards them. Sustainable and resilient migration systems would require States to move away from temporary migration, offer more secure forms of migration and integration opportunities, and perhaps to invest in technological substitutions for some forms of low-skilled labour (Triandafyllidou & Yeoh, 2023).

Applying a sustainability lens to migration law requires first an understanding of these two notions. Sustainability is not a legal notion, but it has been linked to migration via development. Development is mainly an economic and policy-driven field, that seeks to promote the achievement of better living conditions for people; it may or may not rely on laws to achieve this. International migration on the other hand is very much a legal project that relies on the legal categorization of persons into citizens and various kinds of foreigners (see the IOM definition of 'migrant'). Moreover, migration is managed through the nation State and increasingly at the regional and international levels with the aim to regulate the international flow of people. The relationship between migration as a legal category and sustainability as a policy buzzword and desiderate of the international community remains to be clarified, not least in terms of how sustainable migration should be operationalized.

The issue of operationalization is a case in point for the EU. The Treaty on the European Union (TEU) contains references to sustainable development and the environment in connection to both the EU's internal and external dimensions as part of the values upon which the EU is built and which it seeks to promote in its relations with other countries. First, Article 3(3) TEU clarifies that the EU internal market shall work for the sustainable development of Europe which is 'based on balanced economic growth and price stability, a highly competitive social market economy, aiming at full employment and social progress, and a high level of protection and improvement of the quality of the environment'. Second, in the EU's relations with the wider world, the Union 'shall contribute to peace, security, the sustainable development of the Earth, solidarity and mutual respect among peoples, free and fair trade, eradication of poverty and the protection of human rights, […] and the strict observance and respect for international law' (Article 3(5) TEU). Article 21 TEU further elucidates that the EU's external action should foster the 'sustainable economic, social and environmental development of developing countries, with the primary aim of eradicating poverty' see Article 21(2)(d) TEU. Furthermore, in its external policies, the EU should foster cooperation and help develop international measures to preserve and improve the quality of the environment and the sustainable management of global natural resources, in order to ensure sustainable development (Article 21(2)(f) TEU).

While these legal provisions are relevant for orienting the development of policies, they fail to answer a number of important questions. What elements make the migration regime sustainable? For whom should migration be sustainable? How do migrants fit into a sustainable migration regime? How can or should the migration regime deal with perceived tensions between sustainability for countries of origin and destination? It is thus instructive to look closer at migration and development to understand some of the tensions mentioned above.

Starting with the 1980s migration has been discussed as a process for development. Most recently, the links between migration and development have been recognized at the UN level by the adoption of the UN Sustainable Development Goals. Elspeth Guild argues that by linking migration and development, the international community sends a clear message that migration can be a way to aid development and that moving from one State to another to find employment can assist in poverty reduction (Guild, 2020b). The UN Global Compact on Migration reflects this view when stating that 'migration is a multidimensional reality of major relevance for the sustainable development of countries of origin, transit and destination, which requires coherent and comprehensive responses'. At the same time, the inclusion of migration as a development-related issue into the UN SDGs, which paved the way for the adoption of the global compacts on refugees and migrants, is understood to have allowed for development, border control, and migration management to be lumped together, potentially to the detriment of enforceable rights for migrants at the international level (Guild, 2021). This framing reflects the politically contentious nature of migration as topic of international relations among States and the emphasis on safe, orderly, and regular migration.

The impact of migration on development is usually discussed in relation to brain drain, circular migration, and remittances. In this context, Baubock and Ruhs (2021) conclude that the triple win (for sending States, for receiving States, and for individual migrants) promised by many temporary labour migration programmes is elusive. Countries of origin and migrants do not always reap the benefits of such programmes because of uneven bargaining power and because the programmes do not always match the preferences of migrants (Rahim et al., 2021). Critical of the triple win discourse, scholars have suggested that for migration to be a tool for development there should be a greater emphasis on expanding legal pathways that match migrants' aspirations of migration and on the need to ensure the protection of migrants' labour rights (Bisong, 2024; Stoyanova, 2023; Vankova, 2022). Simply put, there are not enough safe migration channels for people to migrate in an orderly and regular fashion. Furthermore, migrant labour is often seen by governments of receiving countries and commercial operators as temporary and 'disposable', which leads to precarious employment, social cleavages, and economic inequalities, diminished contacts between migrants and locals (social exclusion), and the securitization of certain forms of migration. While most countries in the Global North depend on migrant labour to fuel economic growth, there is little political appetite for openly acknowledging this dependency or its implications for migrants and their rights.

Similar tensions are present in the EU. The 2020 EU Pact on Migration and Asylum published by the European Commission to orient future EU policy and laws on these topics was criticized among others for failing to create legal pathways for people to migrate, while prioritizing border controls and the fight against irregular migration (Farcy & Sarolea, 2022). The European Commission's 2022 Communication on attracting skills and talent seeks to remedy the weak emphasis on regular migration in the 2020 Pact by introducing an 'ambitious and sustainable EU legal migration policy' (European Commission, 2022, p. 1). This policy presents legal migration as a tool for development and as supporting the EU's green and digital transition, while contributing to making European societies more cohesive and resilient. It may be more accurate to say that sustainability is now enmeshed in the EU migration governance debate, but without adding enforceable elements. One critical point is that while sustainability is part of the proposal on legal migration policy, it is used as a buzzword without any clarity on how the Commission's proposals 'on legal migration contribute to sustainability or to the necessary transitions towards resilient and sustainable economies' (De Lange et al., 2022, p. 87). Legal scholars agree that for migrants, sustainability needs to be understood in terms of rights and future prospects (De Lange et al., 2022, p. 87; Loxa, 2023). Ensuring the rights of migrants, either via human rights or specialized legal instruments addressing migrants, remains relevant for understanding what sustainable migration can mean (De Lange, 2021; Guild, 2020b). Tesseltje De Lange (2021) argues in favour of mainstreaming the notion of social sustainability in the debate on the future of the EU's migration regime. This would require aligning the narrative on highly skilled migration with the environmental crisis we face; decouple study and labour migration from economic growth and recouple it to sustainable innovations working towards degrowth or, at least, green growth.

5 Case Studies: EU and Third Country National Migration in the Netherlands

As discussed above, we can speak of an emerging consensus among (legal) scholars that sustainable migration is linked to and requires respect for migrants' rights. The linkage between sustainability and migration via migrants' rights needs to be understood in the light of the existing legal framework at the EU level whereby migration management and control are emphasized in relation to migration from third countries to the EU, with the result that certain forms of migration are securitized. This understanding of migration is underpinned by normative assumptions about who moves and to where (Guild, 2020a) and the movement is always imagined as from poor, underdeveloped countries to rich, developed countries in the EU. This results in few legal pathways for migrants seeking to enter the EU as labour migrants and in differentiated rights for those migrants who are allowed entry depending on their legal categorization (for example, highly skilled migrants enjoy more rights than seasonal migrant workers).

The underlying issue remains that in a paradigm that interprets climate and environmental change as push factors for migration, the

EU retains its preoccupation with borders, security, and externalization of migration control while failing to provide tailored legal pathways for people managing their own environmental insecurities. This leaves people needing to fit themselves into limited existing channels for regular migration. We illustrate this point with a discussion of two groups of migrant workers in the Netherlands. Migrant workers, both EU citizens and third country nationals, play an important role in the Dutch economy. It is estimated that half of labour migrants in the Netherlands are EU citizens exercising their right to free movement, with some economic sectors being highly dependent on EU labour. When it comes to third country nationals (TCN), immigration policy in the Netherlands is divided between restricting access for most, while competing to recruit and retain certain other workers whose higher education qualifications and skills are essential to its 'knowledge-based economy'. We seek to illustrate some of the questions of environmental and social sustainability by examining the situation of two different categories of migrant workers: EU nationals employed in the Dutch meat industry and TCN students and researchers in higher education.

The Dutch 'brede welvaart' (general well-being) policy reflects the objective of sustainability as well as their contradictions. Prior to the UN 2030 Agenda, the Netherlands were interested in finding better ways to measure well-being beyond GDP. This preoccupation goes back to discussions that started in the 1960s concerning the measurement of GDP and economic performance, on the one hand, and social progress via social indicators on the other. Dutch policymaking was inspired and influenced by the work of the Club of Rome (established in 1968) which can be seen as an early frontrunner in the global sustainability movement. The Club of Rome put on the agenda the relationship between economic growth and the pursuit of economic expansion and the ecological footprint of humanity. In 2016, the Dutch Temporary Committee for a Broad Concept of Well-being recommended that the Dutch Central Statistics Bureau (CBS) develop a conceptual framework to report annually on well-being, and in 2017, the Dutch government adopted the recommendations and commissioned the CBS to publish an annual Monitor of Well-being. Because there is a clear thematic overlap, as well as a certain overlap between the SDG indicators and the indicators used for the Monitor of Well-being, in 2018 it was decided to combine the two reports. Thus, the Dutch 'brede welvaart' policy can be seen as the national implementation of the UN 2030 Agenda (Centraal Bureau voor de Statistiek, 2023). The Dutch policy understands general well-being as an umbrella term that combines economic and non-economic concerns within one framework (Boelhouwer, 2016). The importance of the economy and of economic growth within Dutch society is a recurrent theme in debates about the general well-being policy, as is the need to better accommodate social and ecological aspects. As discussed above, these questions are more than mere details—they are at the core of sustainability in the Anthropocene.

Research measuring the Dutch well-being policy takes the 'citizen' as the main focal point, with most questions seeking to understand well-being from the objective and subjective position of the 'citizen'. How migration fits into the Dutch general well-being policy is less straightforward. In 2021, the Dutch Advisory Committee on Migration Affairs (ACVZ) issued a position paper in which it advocated in favour of the creation of a body to advise the Dutch government on labour migration from outside of the EU in a structured manner (ACVZ, 2021). The ACVZ's standpoint was that structural advice on labour migration was necessary 'to ensure that labour migration contributes to well-being'. The ACVZ defined a 'well-being approach to labour migration' as involving 'looking at the opportunities that this migration offers to society in the Netherlands (and other parts of the world) and weighing up all the interests at stake: of employers, employees, migrants and society at large' (ACVZ, 2021, p. 3). Such a policy should lead to a system that is flexible, future-oriented, and proactive, in the sense that it is based on asking which forms of labour migration will be necessary and desirable in the future. It thereby links migration policy with broader societal and political discussions on well-being in relation to housing, labour participation, and economic needs (ACVZ, 2021, pp. 6–7).

Questions around sustainability, migration, and the environment are more evident in relation to certain economic sectors with a considerable impact on the climate, such as agriculture, horticulture, intensive livestock farming, and the meat processing industry where companies are dependent on cheap, low-skilled workers (ACVZ, 2021, pp. 6/10). EU workers are concentrated in some of the sectors which are perceived as problematic from an environmental perspective, which makes them an interesting case study on how the logics of migration and sustainability interact and potentially contradict one another. TCN researchers and students are, on the contrary, seen as a group of migrants who could potentially increase Dutch well-being by aiding the transition to a knowledge-based and environmentally sustainable society. The ACVZ argues that 'a wellbeing approach can help prioritise labour migration in socially valuable sectors and labour migration that makes a contribution to solving societal challenges' (ACVZ, 2021, p. 11).

5.1 EU Migrant Workers in the Meat Sector: Precarity, Vulnerability and a Question of Desirability

The free movement of persons is a fundamental aspect of the establishment, development, and completion of the EU internal market. Its distinctiveness stems from how the right to migrate is framed: rather than a sovereign prerogative of the State to allow entrance into State territory, EU citizens have an individual right to move and reside in other EU states. The free movement of workers entitles EU nationals to a right to work, including seeking work, in a different Member State than that of nationality coupled with a right to enjoy equal treatment with national workers. However, EU mobility is not limited to economically active persons. The TFEU provisions on EU citizenship award every EU citizen a right to move and reside freely in the territory of the other Member States. Free movement is coupled with a right to equal treatment (Article 18 TFEU and Article 45(2) TFEU), although secondary law allows for exceptions.

Looking at primary and secondary law, we can conclude that sustainability was not an issue on the minds of the drafters of the EU regime of free movement of persons. Yet, both politically and legally the right to free movement for EU workers is rooted in the idea that migration is a way for individuals to achieve a better life for themselves and their families (Preamble Regulation 1612/68, now replaced by Regulation 492/2011). This legal reality predates by several decades the discussion on migration as development. The rules on free movement of EU citizens have a strong social dimension that is grounded in primary and secondary law (Articles 18 and 45 TFEU, Directive 2004/38 and Regulations No. 492/2011 and No. 883/2004) as well in the interpretation given to these rights by the European Court of Justice. For EU workers, equal treatment is seen as a tool to ensure their integration in the host State and that of their families (Mantu, 2021). For migrant EU citizens study and labour migration is seen as fostering personal development, while the principle of equal treatment ensures social sustainability by preventing social dumping.

However, the free movement of EU workers has been criticized for its supposed negative and distortive effects on the labour markets, welfare States, and public service provision of the EU States of destination. In the Netherlands the debate follows two tracks. First, there is a focus on the negative impact of free movement on sending EU States in terms of depopulation and brain drain (De Wispelaere et al., 2021; Goldner-Lang & Lang, 2020), borrowing arguments from the debates on migration, development, and brain drain that have been ongoing in relation to Global North–Global South migration flows. The Dutch debate is an example of similar concerns in relation to East–West migration in the EU, since most migrant EU workers move from Central and Eastern European countries to Western European countries in search of work and higher wages. The question raised is what measures are necessary and can be taken to offset the negative consequences of EU-free movement of workers. Proposed solutions include compensatory payments to EU States of origin, the promotion of circular or temporary migration, encouraging return migration, again measures that seem to borrow from the tool kit of migration governance used by EU States in relation to TCN migration. The possibility to apply restrictions to EU-free movement of workers has been played down in the light of its fundamental character under EU primary law, while attempts to promote circular and temporary migration schemes are seen problematic due to the precarity and vulnerability experienced by temporary EU migrants in the Netherlands (Vonk, 2021). Herwig Verschueren (2021) has discussed the possibility that EU States of origin may seek to limit the emigration of their highly skilled workers by conditioning the award of study scholarships on the person working for a number of years in the State of origin. Another possibility explored by Verschueren is the insertion of a provision in EU-free movement law similar to that in the Recast Blue Card (Directive (EU) 2021/1883) that would give a receiving State the power to reject a residence permit on ethical grounds linked to brain drain in specific professions.

Second, the debate on the effects of EU-free movement links with social sustainability concerns for the Netherlands itself and serves as illustration of the divide between low-skilled and skilled migration, more generally. For low-skilled migration the EU already operates a circular and seasonal migration regime in respect of TCN migrants, while for EU citizens circular and temporary migration are the result of labour market transformations and the move towards flexibilization and insecurity. For example, the Dutch meat industry is highly dependent on migrant workers: from the estimated 12,000 workers, about 7,500 are migrants from mainly Poland and Romania; they are overrepresented on the work floor. The COVID-19 pandemic has magnified the vulnerability of these workers linked to the type of work they perform (low-skilled), the conditions of their employment (highly flexible), and their lack of social networks (Berntsen et al., 2022). Most migrant meat workers are not hired directly but via specialized temporary employment agencies that supply workers almost exclusively to the meat industry. The generalized use of zero-hour employment contracts leads to employment and income insecurity. These contracts stipulate payment only for the number of hours worked and allow for dismissal should no work be available, or the worker calls in sick. Especially in the first phase of agency contracts, working hours are not guaranteed, and contracts can be dissolved easily, leading to loss of employment, housing, and healthcare, if the latter are contractually organized by the employer. All this results in a heightened dependency on their employers for, work, housing, healthcare, transportation, and information with limited oversight from the Dutch State via its inspection and enforcement agents.

The experiences of these EU workers during the pandemic and the public and political attention given to their work and living conditions have led to legislative changes that seek to strengthen their legal position vis-a-vis their employers The collective labour agreements for the temporary agency sector include a two months statutory minimum income guarantees for workers who arrive for the first time in the Netherlands and a four-week transition period to vacate employer organized accommodation upon termination of employment contract. These provisions apply also to the CBA in the meat sector (Berntsen et al., 2022). At the same time, there has been increased public and political attention to questions around the desirability of the Dutch meat industry to continue to produce at the high levels of export that it currently does in the light of the social (read in, terms of migrant workers treatment) but also environmental implications of this sector (Geertsema et al., 2023). This is an example of how social sustainability concerns feed into environmental concerns in a migration setting while remaining within the imaginary of sustainability for the Dutch society.

5.2 Third Country National Students and Researchers: Competition and Unequal Exchange

Many States in the EU and the Global North, including the Netherlands, see themselves today as engaged in a 'global race for talent', demonstrated at the EU level by the Commission's communications on 'attracting skills and talent to the EU' (European Commission, 2022) and 'skills and talent mobility' (European Commission, 2023). Highly educated migrants are seen to bring skills and expertise needed for high-value industries and the transition towards a knowledge-based (and possibly green) economy (Nederlands Ministerie van Justitie en Veiligheid, 2022). The recruitment and retention of international students, graduates, and researchers in higher education can thus help countries like the Netherlands to maintain a competitive edge within this 'global race'. This does not mean that TCN students and researchers are unequivocally welcomed in the Netherlands. Their actual position is, in fact, somewhat more ambivalent; the Dutch State weighs their economic contribution against other concerns such as an acute shortage of accommodation and teaching space, and the rise of English at the expense of the Dutch language in higher education (Coello Eertink, 2022). Directive 2016/801, which provides the legal framework for TCN students and researchers in the EU, to some extent reflects this ambivalent attitude. The directive sets out the respective admission requirements for 'students and researchers, as well as other groups' such as au pairs. Students and researchers must have an existing offer from a university or research institution before they enter, and students must also prove that they have sufficient resources (or a scholarship) to cover their tuition fees. The directive also regulates and limits the right of students to work alongside their studies and provides for graduates and researchers to remain and seek employment after they have completed their studies or research activities.

Of course, this raises questions of social and economic sustainability. The idea of a competitive global 'race' implies that other countries must lose for the Netherlands to win. The emigration of skilled and educated people can lead to domestic skills shortages and underdevelopment in countries of origin. Countries of origin may also bear the high costs of training and education, only for graduates to apply their expertise abroad. Skeldon (2020) suggests that poorer countries in particular may 'view the emigration of their skilled to the more developed world with concern'. Conversely, De Haas (2023) argues that emigration is much more a symptom than a cause of disparities between countries, with people tending to pursue opportunities wherever they can find them (De Haas, 2023). Emigration may also enable people to send financial and social remittances, which may (partly) compensate for the loss of people's skills and income. Many students also return to work in their country of origin after studying abroad or migrate again to another third country. In this case, however, the cost of education can still be felt in the form of tuition fees being paid to institutions in the Netherlands—which may also reinforce the dominance of universities in the Global North and limit opportunities for the higher education sector elsewhere. It is difficult to estimate whether the costs outweigh the benefits for the sending country overall. For receiving countries in Europe, Noja et al. (2018) suggest that the social and economic impacts are highly dependent on the flexibility of the domestic labour market and the (dis)parities in skill and education levels between nationals and migrants. Highly educated migrants, therefore, are likely to be welcomed differently within a similarly educated domestic labour force.

In terms of environmental sustainability, it is notable that a significant proportion of TCN students and researchers come to the Netherlands from some of the world's most severely climate-impacted and exposed regions. According to EUROSTAT data on the admission and renewal of permits for the purposes of research or study, citizens of India were consistently the second largest group (after China) each year from 2019 to 2022, with Pakistan and Bangladesh also registering quite highly. South Asia is one of the regions where the largest-scale environmental displacements have been projected (World Bank, 2021). In this regard, the Netherlands is an interesting destination—not only because it contains some of the world's lowest-lying territory. It is a major industrialized State and a former colonial power, which has had a significant impact on today's global economy and ecology. It is also a leader in the green energy transition and in climate litigation cases as *Urgenda* (2019). The Netherlands positions itself as a world leader in environmental expertise and technologies, much like the EU in its Green Deal, and thus cannot escape the serious questions that persist regarding the feasibility of wedding competitive economic growth and environmental sustainability in this kind of 'green growth' strategy. South Asian students and researchers in the Netherlands, therefore, embody many of the intersecting paradoxes and questions of sustainability and ontological security that characterize state-centric migration law and policy in the Anthropocene.

6 Discussion and Conclusions

The effectiveness of regular migration as a means to respond to climate change and sea level rise remains underexplored, as does the development of a sustainable migration law and policy to protect the human rights of migrants in the Anthropocene. In this chapter, we have tried to map some of the interactions between environmental and social sustainability in EU migration law and policy by examining how they play out in the case of two groups of regular migrants in the Netherlands. We have used the examples of EU migrant workers in low-paid and flexible jobs in the meat sector, on the one hand, who are seen as necessary but not per se desirable and, on the other hand, TCN students and researchers who are seen as highly desirable but nonetheless have an ambivalent position in the Netherlands. The desirability and acceptance of these migrants is linked to the perceived social and environmental impact of the industries in which they work or are expected to work in the Netherlands, as well as their implications for countries of origin and the wider world of the Anthropocene. The meat sector is highly polluting and resource-intensive, while students and researchers can be seen as part of the transition to a knowledge-based (and green) economy—yet we have shown that there is much more to consider in terms of the environmental and wider sustainability of both sectors.

Our discussion illustrates that the protection needs of persons affected by environmental change and degradation are a missing link. The role of social and environmental sustainability in the formulation and operationalization of laws and policies is highly obscure, with very little guidance in the laws we have discussed. In the case of free movement, this may be because the laws were written at a time when the notion of sustainability was less ubiquitous in policy circles—although policy and soft law on free movement during the COVID-19 pandemic have forced the EU institutions to engage with questions of social sustainability, particular in relation to migrants' rights.

In the case of TCN students and researchers, sustainability considerations seem to be more connected to the impacts on countries of origin, similar to sustainable development. However, the law we discuss (Directive 2016/801) falls within the domains of EU migration and external relations policy. In the latter policy domain, sustainable development and the protection of the environment are recognized goals of external EU policy, but it is unclear how sustainability affects Directive 2016/801.

Should the directive be interpreted with sustainable development in mind, e.g. with a moratorium on recruitment from countries suffering from 'brain drain' and underdevelopment, or by giving priority to students and researchers from regions where environmental degradation is most severe?

Study and labour migration have to date received very little attention in debates about sustainable migration, which tend to focus on how refugee and asylum law can accommodate displaced people and 'climate migrants'. There is also a continuing lack of clear data on how environmental changes and degradation, including slow-onset events such as sea level rise, impact migrant trajectories, from the decision to migrate to long-term plans for the future.

Linked to this, there is a poor understanding of the ways in which existing legal categories can incorporate the novel concerns and realities of the Anthropocene, including how environmental concerns may figure among the factors that drive migration aspirations. In any case, existing laws already fail to reflect the complexity and mixed nature of people's motivations to move, which subsequently complicates discussions about the creation of new migration or asylum categories. To add another layer of complexity, it is difficult to trace a coherent change of causality between EU policy, global environmental change, particular social, economic and environmental impacts, personal aspirations and capabilities, and actual migration flows. The result is a rather nebulous notion of 'sustainable' migration that incorporates elements of earlier debates around development and the eradication of poverty, environmental issues (which also aggravate the former), and social sustainability concerns.

From states, we see much resistance to the creation of new legal pathways in response to the ontological insecurity of environmental degradation, lest they exacerbate the (perceived) ontological insecurity of immigration. Climate change and migrants are both threatening according to the conventional outlook of States in Europe and the Global North, but how to reconcile this with the realities of the Anthropocene? Migration and the environment are more closely connected than ever before, and neither can be fully controlled or externalized. A genuinely sustainable migration policy for the Anthropocene means engaging with both migration and the environment as complex systems that are beyond the reach of any one State (or group of states), but within which States and other actors can navigate to seek mutual benefits. If a triple win has proved 'elusive' in the past, in the Anthropocene it may also prove inescapable. As we have shown in this paper, however, this requires a much fuller interrogation of sustainability than the one we currently see in migration policy debates.

Competing Interests The authors have no conflicts of interest to declare that are relevant to the content of this chapter.

References

Adviesraad Migratie (ACVZ). (2021, November 1). Towards a Well-Being Approach in Labour Migration Policy: Advisory Models on Labour Migration in the Netherlands and Abroad. https://www.adviesraadmigratie.nl/pub licaties/publicaties/2021/11/01/towards-a-well-being-approach-in-labour-migration-policy

Aleinikoff, T. A. (2019). The Unfinished Work of the Global Compact on Refugees. *International Journal of Refugee Law, 30*(4), 611–617. https://doi.org/10.1093/ijrl/eey057

Baldwin, A. (2013). Racialisation and the Figure of the Climate-Change Migrant. *Environment and Planning A: Economy and Space, 45*(6), 1474–1490. https://doi.org/10.1068/a45388

Baldwin, A., & Fornalé, E. (2017). Adaptive Migration: Pluralising the Debate on Climate Change and Migration. *The Geographical Journal, 183*(4), 322–328. https://doi.org/10.1111/geoj.12242

Baldwin, A., Fröhlich, C., & Rothe, D. (2019). From Climate Migration to Anthropocene Mobilities: Shifting the Debate. *Mobilities, 14*(3), 289–297. https://doi.org/10.1080/17450101.2019.1620510

Baubock, R., & Ruhs, M. (2021). The Elusive Triple Win: Can Temporary Labour Migration Dilemmas Be Settled Through Fair Representation? *Robert Schuman Centre for Advanced Studies Research Paper No. RSC 2021/60.* https://doi.org/10.2139/ssrn.3913992

Berntsen, L., Böcker, A., De Lange, T., Mantu, S., & Skowronek, N. (2022). State of Care for EU Mobile Workers' Rights in the Dutch Meat Sector in Times of, and Beyond, COVID-19. *International Journal of Sociology and Social Policy, 43*(3/4), 356–369. https://doi.org/10.1108/IJSSP-06-2022-0163

Bettini, G. (2019). And Yet It Moves! (Climate) Migration as a Symptom in the Anthropocene. *Mobilities, 14*(3), 336–350. https://doi.org/10.1080/17450101.2019.1612613

Bisong, A. (2024). Legal Pathways to Migration: Labour Migration Arrangements Between West Africa and Europe. In A. O. Akinola & J. Bjarnesen (Eds.), *Worlds Apart?* (pp. 193–218). Routledge.

Boas, I. (2015). *Climate Migration and Security, Securitisation as a Strategy in Climate Change Politics*. Routledge. https://doi.org/10.4324/9781315749228

Boas, I., Farbotko, C., Adams, H., Sterly, H., Bush, S., Van der Geest, K., Wiegel, H., Ashraf, H., Baldwin, A., Bettini, G., Blondin, S., De Bruijn, M., Durand-Delacre, D., Fröhlich, C., Gioli, G., Guaita, L., Hut, E., Jarawura, F. X., & Hulme, M. (2019). Climate Migration Myths. *Nature Climate Change*, 9(12), 901–903. https://doi.org/10.1038/s41558-019-0633-3

Boas, I., Wiegel, H., Farbotko, C., Warner, J., & Sheller, M. (2022). Climate Mobilities: Migration, Im/Mobilities and Mobility Regimes in a Changing Climate. *Journal of Ethnic and Migration Studies*, 48(14), 3365–3379. https://doi.org/10.1080/1369183X.2022.2066264

Boelhouwer, J. (2016, April 20). *Het brede-welvaartsbegriep volgens het SCP*. https://www.scp.nl/publicaties/publicaties/2016/04/20/het-brede-welvaartsbegrip-volgens-het-scp

Browning, C. S. (2017). Security and Migration: A Conceptual Exploration. In P. Bourbeau (Ed.), *Handbook on Migration and Security* (pp. 39–60). Edward Elgar Publishing.

Centraal Bureau voor de Statistiek. (2023). *Monitor Brede Welvaart en de Sustainable Development Goals 2023*. https://www.cbs.nl/nl-nl/dossier/dossier-brede-welvaart-en-de-sustainable-development-goals/monitor-brede-welvaart-en-de-sustainable-development-goals-2023

Chakrabarty, D. (2018). Anthropocene Time. *History and Theory*, 57(1), 5–32. https://doi.org/10.1111/hith.12044

Coello Eertink, L. (2022, November 20). *Netherlands: Challenges for International Students*. European Website on Integration. https://ec.europa.eu/migrant-integration/news/netherlands-challenges-international-students_en

Cole, M. A., Elliott, R. J. R., Okubo, T., & Zhang, L. (2021). Importing, Outsourcing and Pollution Offshoring. *Energy Economics*, 103. https://doi.org/10.1016/j.eneco.2021.105562

Czaika, M., & Münz, R. (2022). *Climate Change, Displacement, Mobility and Migration: The State of Evidence, Future Scenarios, Policy Options*. Delegationen för migrationsstudier (Delmi).

De Haas, H. (2023). *How Migration Really Works: A Factful Guide to the Most Divisive Issue in Politics*. Penguin Books. https://doi.org/10.1080/01419870.2024.2311816

De Lange, T. (2021). A new Narrative for European Migration Policy: Sustainability and the Blue Card Recast. *European Law Journal*, 26(3–4), 274–282. https://doi.org/10.1111/eulj.12381

De Lange, T., Guild, E., Brandl, U., Tsourdi, L., De Kruijff, J., Hardiek, S., & Honuskova, V. (2022). The EU Legal Migration Package: Towards a Rights-Based Approach to Attracting Skills and Talent to the EU (Study for the LIBE Committee of the European Parliament, PE 739.031). European Parliament. https://www.europarl.europa.eu/thinktank/en/document/IPOL_STU(2022)739031

De Wispelaere, F., Pacolet, J., & Gillis, D. (2021). De keerzijde van intra-Europese arbeidsmobiliteit voor de lidstaat van herkomst. Een vergeten perspectief? *SEW: tijdschrift voor Europees en economisch recht*, 5, 178–185.

Dun, O., Klocker, N., Farbotko, C., & McMichael, C. (2023). Climate Change Adaptation in Agriculture: Learning From an International Labour Mobility Programme in Australia and the Pacific Islands Region. *Environmental Science & Policy*, 139, 250–273. https://doi.org/10.1016/j.envsci.2022.10.017

Dussaux, D., Vona, F., & Dechezleprêtre, A. (2023). Imported Carbon Emissions: Evidence From French Manufacturing Companies. *Canadian Journal of Economics/revue Canadienne D'économique*, 56(2), 593–621. https://doi.org/10.1111/caje.12653

European Commission. (2019). *Communication From the Commission to the European Parliament, the European Council, the Council, the European Economic and Social Committee and the Committee of the Regions on the European Green Deal* (COM/2019/640 final). https://eur-lex.europa.eu/legal-content/EN/TXT/?uri=CELEX%3A52019DC0640

European Commission. (2022). *Communication From the Commission to the European Parliament, the European Council, the Council, the European Economic and Social Committee and the Committee of the Regions on Attracting Skills and Talent to the EU* (COM/2022/657 final). https://eur-lex.europa.eu/legal-content/EN/TXT/?uri=CELEX%3A52022DC0657

European Commission. (2023). *Communication From the Commission to the European Parliament, the European Council, the Council, the European Economic and Social Committee and the Committee of the Regions on Skills and Talent Mobility* (COM/2023/715 final). https://eur-lex.europa.eu/legal-content/EN/TXT/?uri=CELEX%3A52023DC0715

Farbotko, C. (2019). Climate Change Displacement: Towards Ontological Security. In C. Klöck & M. Fink (Eds.), *Dealing With Climate Change on Small Islands: Towards Effective and Sustainable Adaptation?* (pp. 251–266). Göttingen University Press. https://doi.org/10.17875/gup2019-1219

Farbotko, C. (2022). Anti-Displacement Mobilities and Re-Emplacements: Alternative Climate Mobilities in Funafala. *Journal of Ethnic and Migration Studies*, 48(14), 3380–3396. https://doi.org/10.1080/1369183X.2022.2066259

Farcy, J.-B., & Sarolea, S. (2022). Labour Migration in the 'New Pact': Modesty or Unease in the Berlaymont? In D. Thym (Ed.), *Reforming the Common European Asylum System* (pp. 277–287). Nomos. https://doi.org/10.5771/9783748931164

Felli, R. (2021). *The Great Adaptation: Climate, Capitalism and Catastrophe.* Verso. https://doi.org/10.14350/rig.60600

Ferris, E. (2020). Research on Climate Change and Migration Where Are We and Where Are We Going? *Migration Studies, 8*(4), 612–625. https://doi.org/10.1093/migration/mnaa028

Filipović, S., Lior, N., & Radovanović, M. (2022). The Green Deal: Just Transition and Sustainable Development Goals Nexus. *Renewable and Sustainable Energy Reviews, 168.* https://doi.org/10.1016/j.rser.2022.112759

Franco Gavonel, M., Adger, W. N., De Campos, R. S., Boyd, E., Carr, E. R., Fabos, A., Fransen, S., Jolivet, D., Zickgraf, C., Codjoe, S. N. A., Abu, M., & Siddiqui, T. (2021). The Migration-Sustainability Paradox: Transformations in Mobile Worlds. *Current Opinion in Environmental Sustainability, 49,* 98–109. https://doi.org/10.1016/j.cosust.2021.03.013

Geddes, A., & Jordan, A. (2012). Migration as Adaptation? Exploring the Scope for Coordinating Environmental and Migration Policies in the European Union. *Environment and Planning c: Government and Policy, 30*(6), 1029–1044. https://doi.org/10.1068/c1208j

Geertsema, K., Grütters, C., De Lange, T., Mantu, S., Minderhoud, P., Van Oers, R., Strik, T., Terlouw, A., & Zwaan, K. (2023). Migratiemaatregelen: Voer voor de formatie. *Nederlands Juristenblad, 98*(40), 2890–3495.

Goldner-Lang, I., & Lang, M. (2020). The Dark Side of Free Movement: When Individual and Social Interest Clash. In S. Mantu, P. Minderhoud, & E. Guild (Eds.), *EU Citizenship and Free Movement Rights: Taking Supranational Citizenship Seriously* (pp. 382–409). Brill.

Guild, E. (2021). Why the Sustainable Development Goals? Examining International Cooperation on Migration. In C. Dauvergne (Ed.), *Research Handbook on the Law and Politics of Migration* (pp. 355–368). Edward Elgar Publishing.

Guild, E. (2020a). Promoting the European Way of Life: Migration and Asylum in the EU. *European Law Journal, 26*(5–6), 355–370. https://doi.org/10.1111/eulj.12410

Guild, E. (2020b). The UN Global Compact for Safe, Orderly and Regular Migration: To What Extent Are Human Rights and Sustainable Development Mutually Compatible in the Field of Migration? *International Journal of Law in Context, 16*(3), 239–252. https://doi.org/10.1017/S1744552320000294

Honarmand Ebrahimi, S., & Ossewaarde, M. (2019). Not a Security Issue: How Policy Experts De-Politicize the Climate Change-Migration Nexus. *Social Sciences, 8*(7), 214. https://doi.org/10.3390/socsci8070214

International Law Association (ILA). (2022, June). *International Law and Sea Level Rise: Report of the Lisbon Conference.* https://www.ila-hq.org/en/documents/interim-report-sea-level-rise-committee-as-presented-at-ila-lisbon-june-2022

International Law Association (ILA). (2024, June). *International Law and Sea Level Rise: Final Report of the Athens Conference*. https://www.ila-hq.org/en/documents/final-report-committee-on-international-law-and-sea-level-rise-22-05-2024

Kälin, W. (2018). The Global Compact on Migration: A Ray of Hope for Disaster-Displaced Persons. *International Journal of Refugee Law, 30*(4), 664–667. https://doi.org/10.1093/ijrl/eey047

Kitara, T., & Farbotko, C. (2023). Picking fruit is not climate justice. *npj Climate Action 2*(1). https://doi.org/10.1038/s44168-023-00057-2

Klöck, C., & Fink, M (2019). *Dealing With Climate Change on Small Islands: Towards Effective and Sustainable Adaptation*. Göttingen University Press.

Loxa, A. (2023). *Sustainability and EU Migration Law: What Place for Migrants' Rights?* Lund University. https://portal.research.lu.se/en/projects/sustainability-and-eu-migration-law-what-place-for-migrants-right

Mantu, S. (2021). *EU Citizenship, Integration and Equal Treatment: The Worker-Citizen Divide* (Nijmegen Migration Law Working Papers Series No. 2021/01). Radboud University Nijmegen. https://repository.ubn.ru.nl/bitstream/handle/2066/230607/230607.pdf

McAdam, J. (2010). 'Disappearing States', Statelessness and the Boundaries of International Law (UNSW Law Research Paper No. 2010–2). https://papers.ssrn.com/abstract=1539766

McAdam, J. (2011). Climate Change Displacement and International Law: Complementary Protection Standards (PPLA/2011/03). UNHCR. https://www.unhcr.org/media/no-19-climate-change-displacement-and-international-law-complementary-protection-standards

Mitzen, J. (2018). Feeling at Home in Europe: Migration, Ontological Security, and the Political Psychology of EU Bordering. *Political Psychology, 39*(6), 1373–1387. https://doi.org/10.1111/pops.12553

Nederlands Ministerie van Justitie en Veiligheid. (2022). De Staat van Migratie. https://open.overheid.nl/documenten/ronl-2cf0251dee3fec7c64207480c2720226feb45_0f/pdf

Noja, G. G., Cristea, S. M., Yüksel, A., Pânzaru, C., & Dracea, R. M. (2018). Migrants' Role in Enhancing the Economic Development of Host Countries: Empirical Evidence from Europe. *Sustainability, 10*(3), 894. https://doi.org/10.3390/su10030894

Oliver, S. (2009). A New Challenge to International Law: The Disappearance of the Entire Territory of a State. *International Journal on Minority and Group Rights, 16*(2), 209–243. https://doi.org/10.1163/157181109X427743

Piguet, E. (2022). Linking Climate Change, Environmental Degradation, and Migration: An Update After 10 Years. *WIREs Climate Change, 13*(1). https://doi.org/10.1002/wcc.746

Pörtner, H.-O., Roberts, D. C., Tignor, M. M. B., Poloczanska, E., Mintenbeck, K., Alegria, A., Craig, M., Langsdorf, S., Löschke, S., Möller, V., Okem, A., & Rama, B. (2022). Climate Change 2022: Impacts, Adaptation and Vulnerability. Contribution of Working Group II to the Sixth Assessment Report of the Intergovernmental Panel on Climate Change. IPCC. https://www.ipcc.ch/report/ar6/wg2/downloads/report/IPCC_AR6_WGII_FullReport.pdf

Rahim, A., Rayp, G., & Ruyssen, I. (2021). Circular Migration: Triple Win or Renewed Interests of Destination Countries (UNU-CRIS Working Paper Serie 3). https://doi.org/10.13140/RG.2.2.11670.06726

Remling, E. (2020). Migration as Climate Adaptation? Exploring Discourses Amongst Development Actors in the Pacific Island Region. *Regional Environmental Change, 20*(3). https://doi.org/10.1007/s10113-020-01583-z

Remling, E. (2023). Exploring the Affective Dimension of Climate Adaptation Discourse: Political Fantasies in German Adaptation Policy. *Environment and Planning C: Politics and Space, 41*(4), 714–734. https://doi.org/10.1177/23996544231154368

Rikani, A., Otto, C., Levermann, A., & Schewe, J. (2023). More People too Poor to Move: Divergent Effects of Climate Change on Global Migration Patterns. *Environmental Research Letters, 18*(2), Article 024006. https://doi.org/10.1088/1748-9326/aca6fe

Salazar, N. B. (2022). Labour Migration and Tourism Mobilities: Time to Bring Sustainability Into the Debate. *Tourism Geographies, 24*(1), 141–151. https://doi.org/10.1080/14616688.2020.1801827

Scott, M. (2016). Applying the Refugee Convention in the Context of Disasters and Climate Change. *Refugee Survey Quarterly, 35*(4), 26–57. https://doi.org/10.1093/rsq/hdw018

Skeldon, R. (2020). Skilled migration. In T. Bastia & R. Skeldon (Eds.), *Routledge Handbook of Migration and Development* (pp. 136–145). Routledge. https://doi.org/10.4324/9781315276908

Stewart, M. (2023). Cascading Consequences of Sinking States. *Stanford Journal of International Law, 59*(2), 131–186.

Stoyanova, V. (2023). Addressing the Legal Quagmire of Complementary Legal Pathways. *European Journal of Migration and Law, 25*(2), 164–199. https://doi.org/10.1163/15718166-12340149

Triandafyllidou, A., & Yeoh, B. S. A. (2023). Sustainability and Resilience in Migration Governance for a Post-Pandemic World. *Journal of Immigrant & Refugee Studies, 21*(1), 1–14. https://doi.org/10.1080/15562948.2022.2122649

UN General Assembly. Global Compact for Safe, Orderly and Regular Migration. A/RES/73/195. 19 December 2019.

UN General Assembly. Transforming Our World: the 2030 Agenda for Sustainable Development. A/RES/70/1. 21 October 2025.

Vankova, Z. (2022). Work-Based Pathways to Refugee Protection Under EU Law: Pie in the Sky? *European Journal of Migration and Law*, 24(1), 86–111. https://doi.org/10.1163/15718166-12340120

Verschueren, H. (2021). De keerzijde van intra-Europese arbeidsmobiliteit: wat kan het Unierecht bijdragen. *SEW: tijdschrift voor Europees en economisch recht*, 5, 189–192.

Vidas, D. (2014). Sea Level Rise and International Law: At the Convergence of Two Epochs. *Climate Law*, 4, 70–84. https://doi.org/10.1163/18786561-00402006

Vidas, D., Fauchald, O. K., Jensen, O., & Walløe Tvedt, M. (2015). International Law for the Anthropocene? Shifting Perspectives in Regulation of the Oceans, Environment and Genetic Resources. *Anthropocene*, 9, 1–13. https://www.fni.no/publications/international-law-for-the-anthropocene-shifting-perspectives-in-regulation-of-the-oceans-environment-and-genetic-resources

Vonk, G. (2021). Vrij verkeer als persoonlijk recht - Enkele juridische reflecties op 'De keerzijde van intra- Europese arbeidsmobiliteit voor de lidstaat van herkomst. Een vergeten perspectief?' door F. De Wispelaere, J. Pacolet en D. Gillis. *SEW: Tijdschrift voor Europees en Economisch Recht*, 5, 186–188.

Witze, A. (2024). Geologists Reject the Anthropocene as Earth's New Epoch: After 15 Years of Debate. *Nature*, 627, 249–250. https://doi.org/10.1038/d41586-024-00675-8

World Bank. (2021). *Groundswell: Preparing for Internal Climate Migration*. https //www.worldbank.org/en/news/infographic/2018/03/19/groundswell---preparing-for-internal-climate-migration

Yamamoto, L., & Esteban, M. (2017). Migration as an Adaptation Strategy for Atoll Island States. *International Migration*, 55(2), 144–158. https://doi.org/10.1111/imig.12318

Open Access This chapter is licensed under the terms of the Creative Commons Attribution 4.0 International License (http://creativecommons.org/licenses/by/4.0/), which permits use, sharing, adaptation, distribution and reproduction in any medium or format, as long as you give appropriate credit to the original author(s) and the source, provide a link to the Creative Commons license and indicate if changes were made.

The images or other third party material in this chapter are included in the chapter's Creative Commons license, unless indicated otherwise in a credit line to the material. If material is not included in the chapter's Creative Commons license and your intended use is not permitted by statutory regulation or exceeds the permitted use, you will need to obtain permission directly from the copyright holder.

CHAPTER 8

Human Rights and Justifications in Climate Litigation: A First Attempt at Conceptualization

Federica Cristani and Elisa Fornalé

1 Introduction: Climate Litigation on the Rise

Climate litigation[1] has been growing exponentially in recent years (Němec et al., 2024).[2] The first remarkable cases in climate litigation were brought before national courts during these years, with the 2019

---

[1] Though a single specific definition of climate litigation does not exist at the international level, we rely on the one reported in Savaresi and Setzer (2022), according to which climate litigation refers to 'lawsuits raising questions of law or fact regarding climate science, mitigation or adaptation, which are brought before international or domestic judicial, quasi-judicial and other investigatory bodies' (p. 8).

[2] This chapter is based on the work carried out within the framework of the Horizon Europe HRJust project States' Practice of Human Rights Justification: A Study in Civil

F. Cristani (✉)
Institute of International Relations Prague, Prague, Czech Republic
e-mail: cristani@iir.cz

E. Fornalé (✉)
World Trade Institute of the University of Bern, Bern, Switzerland
e-mail: elisa.fornale@unibe.ch

© The Author(s) 2026
E. Fornalé (ed.), *Sea Level Rise*,
https://doi.org/10.1007/978-3-031-89171-7_8

decision in the case *Urgenda Foundation v. The Netherlands* being widely considered a milestone in the field.[3]

Within just a few years, a number of similar actions followed in several countries around the world. According to the 2023 Global Climate Litigation Report issued by the UN Environment Programme, as of December 2022, 2,180 climate-related cases have been filed in 65 jurisdictions, including arbitrational tribunal, international and regional courts. 'Children and youth, women's groups, local communities, and indigenous peoples, among others, have been taking a prominent role in bringing these cases to and driving climate change governance reform in more and more countries around the world' (UNEP, 2023).

In 2023 more than 200 cases have been filed against governments and private companies for their alleged climate (in)action, as highlighted in the 2024 report released by the Grantham Research Institute on Climate Change and the Environment, London School of Economics and Political Science (Setzer & Higham, 2024). The report has shed light on several recent key features of climate litigation, including the fact that '[c]limate litigation impacts extend beyond courtroom decisions, influencing policy, governance and public discourse' (Setzer & Higham, 2024, p. 6).[4] Accordingly, national governments are starting to discuss the effect of climate decisions on the relevant national legislative and policymaking processes; also national central banks, financial institutions as

---

Society Engagement and Human Rights through the Lens of Gender and Intersectionality, GA No. 101094346 and the SERI (Swiss State Secretariat for Education, Research and Innovation - SERI) under grant agreement n. 23.00131/ 1010943546. Federica Cristani has written Sects. 2.1 and 2.2, the Tables, the Figures and the Annex; Sects. 2 and 2.3 are by Elisa Fornalé. The introduction and conclusions are the outcome of a common reflection.

[3] The Urgenda Foundation argued that the State had an obligation to reduce greenhouse gas emissions at a higher rate than the government planned. The court agreed and ordered the government to reduce the emissions by at least 25% below 1990 levels by 2020. The Dutch Supreme Court, which, in its reasoning, argued by referring to Article 2 (the right to life) and Article 8 (the right to private and family life) of the European Convention on Human Rights, upheld the decision in 2019. The decision in Urgenda Foundation v. The Netherlands is available in English at: https://www.urgenda.nl/wp-content/uploads/ENG-Dutch-Supreme-Court-Urgenda-v-Netherlands-20-12-2019.pdf. For more information, see Urgenda, Landmark Decision by Dutch Supreme Court, https://www.urgenda.nl/en/themas/climate-case/.

[4] This has also raised the question of what is the role of national courts in climate-related disputes and issues. In this respect, see Burgers (2020) and Carnwath (2022).

well as the insurance sector are increasingly including climate litigation risk considerations in their policies.

Not only national courts have been involved in climate disputes; the years 2023–2024 indeed marked an important milestone in international and regional climate litigation, with requests for advisory opinions on climate change, international law and human rights issues that have been filed before the major international courts and tribunals (i.e. the Inter-American Court of Human Rights (IACtHR, 2023), the International Court of Justice (ICJ, 2023), and the International Tribunal on the Law of the Sea [ITLOS]).

In the last years, an extensive and growing bulk of academic articles, reports and policy publications on the analysis of climate litigation cases has been produced, also engaging in mapping and classification exercises (e.g. Network for Greening the Financial System, 2023; Setzer & Higham, 2024; UNEP, 2023).[5] In parallel, publicly accessible databases on climate litigation have been developed (Urgenda, 2024; UZH, 2024; Sabin Center for Climate Change Law, 2024; NYU, 2024).

While there has generally been a focus on the analysis of the requests and arguments put forward by the claimants in climate litigation cases, a comprehensive and systematic review of the arguments put forward by States as the defendants has been generally missing in the relevant literature and reporting. The research carried out by the HRJust project took a first step towards filling these gaps. The project's comprehensive data collection of climate litigation cases has been devoted to a scrutiny of the arguments put forward by states[6]. In particular, the research question has

---

[5] See, among others, the reports of the UN Special Rapporteur on the Human Right to a Clean, Healthy and Sustainable Environment, all available at the official OHCHR website, retrieved October 10, 2024, from: https://www.ohchr.org/en/special-procedures/sr-environment, as well as the reports produced in the framework of the United Nations Environment Programme, retrieved October 10, 2024, from: https://www.unep.org/resources/report/global-climate-litigation-report-2023-status-review, and the reports drafted by the Sabin Center for Climate Change Law of Columbia University, retrieved October 10, 2024, from: https://climate.law.columbia.edu/content/searchable-library#!#%2Ffilter%2Fcategories%2FClimate%20Litigation.

[6] This chapter makes reference to data collected by the research teams of the Institute of International Relations in Prague (Dr. Federica Cristani, Dr. Petra Ditrichová, Dr. Jan Lhotský and Ms. Elif Naz Nemec), the World Trade Institute of the University of Bern, and the University of Trento. As detailed in the Annex to this chapter, the HRJust Climate Claims Dataset includes 365 national climate litigation cases (at the time of writing—June 2025; www.hrjust-climate-claims.eu).

focused on how States develop different justifications in connection with human rights, what it means for human rights claims, and how they can be framed within the current international legal framework in terms of implications, as detailed in the next sections.

2 Human Rights and State Justifications in Climate Litigation

Regarding climate litigation, the first step of the present research has been to identify and analyse which kinds of justifications have been raised by States to explain their actions or omissions in implementing climate change obligations and how this could (not) have impaired individual rights.[7] In particular, what is becoming a larger issue of concern is whether the justifications invoked represent a positive enforcement of existing rights or whether, on the contrary, this way of proceeding could result in undermining the overall human rights systems.

As described by Besson, there is a tendency to think that 'human rights are self-justificatory' but in reality there is a growing need to clarify the terms of the debate by conciliating the human rights theory with the human rights practice (Besson, 2022, p. 23). In particular, sea level rise and other slow-onset events are raising new protection claims by requiring States and the international community to engage with positive obligations to prevent and adapt to this changing situation. This has been very well delineated in, among others, the reports drafted by the International Law Association (ILA) Committee on International Law and Sea Level Rise. The Final Committee Report adopted in Athens in May 2024 makes it clear that 'States have duties to (i) respect, (ii) protect, and (iii) fulfil human rights, and (iv) to do so on a non-discriminatory basis'; such duties go along with the 'duty to cooperate for the realization of human rights [...which] is becoming increasingly significant, particularly where climate change impacts undermine the capacities of affected States to discharge their enduring obligation to respect, protect and fulfil rights' (ILA, 2024, p. 28–31).

---

[7] The research path will then develop and integrate with a deeper analysis and understanding of the framework for human rights justification—also thanks to the data that is being collecting through the civil society engagement exercise (discussed below in the text).

Climate litigation offers the ideal situation for testing how human rights theory and human rights practice operate. As illustrated below, climate claims are generally brought by civil society representatives and/or individuals, and are mostly grounded on human rights—i.e. they claim alleged violations of human rights-related obligations by States in the context of climate change; in turn, the States tend to present justifications for their conduct—some of them also based on human rights arguments, or at least affirming the State's compliance with international climate change-related obligations. At this point, the question arises whether such (human rights) justifications find their place within the relevant international legal framework; and if not, the following-up question would be 'Which consequences arise for the State and which (legal) instruments can be used by the affected individual(s)?'

The mapping exercise has so far covered climate litigation cases brought before national and European courts. The following sections give an account of the preliminary findings of the data collection, with a special focus on the Swiss *Verein KlimaSeniorinnen* case.

2.1 Introducing the HRJust Climate Change Claims Dataset: Preliminary Empirical Insights

As mentioned, there have been several efforts to map climate litigation cases, especially national cases. In such exercises, case law has been categorized according to different variables, the most frequent ones being: (1) those based on the types of applicants (e.g. NGOS, individuals, governments) and defendants (government and corporations) (e.g. Markell & Ruhl, 2012; Savaresi & Setzer, 2022); and (2) those based on the type of climate action, i.e. whether it concerns measures of climate change mitigation, adaptation or both (e.g. Markell & Ruhl, 2012; Setzer & Higham, 2024).

Additionally, climate cases can in turn be classified according to the strategy adopted: indeed, some claimants have aimed to achieve broader societal impacts, while others have targeted (the lack of) climate-related policies of companies or the (mis)management of transition policies (Setzer & Higham, 2024). According to the 2024 report of the Grantham Research Institute on Climate Change and the Environment of the London School of Economics and Political Science, the most frequent type of climate case so far has been those brought against national governments 'that challenge the ambition or implementation of climate targets and policies affecting the whole of a country's economy and society' (Setzer & Higham, 2024, p. 24).

These kinds of cases usually develop human rights arguments that rely on relevant international and regional treaties, as well as national and constitutional regulations. In turn, climate litigation cases can be classified (3) depending on the types of rights invoked. In this respect, any classification has been challenged by the fact that usually human rights complaints are not limited to the invocation of one human right, and/or one type of obligation (e.g. Anderson, 1998; Anton & Shelton, 2011; Anton & Shelton, 2011; Boyle, 2006; Iyengar, 2023; Savaresi & Setzer, 2022; Setzer & Higham, 2024; Shelton, 2006, 2010). Usually, reports tend to distinguish between substantive obligations (usually linked to 'the right to life, adequate housing, food, and the highest attainable standard of health) and procedural obligations [(including access] to remedies') (Savaresi & Setzer, 2022).

From the data collected so far, a few preliminary considerations can be elaborated. Firstly, as for the argumentations put forward by the applicants, though they vary from case to case, usually they are grounded on constitutional provisions, domestic environmental laws, as well as international instruments, in particular the Paris Agreement and the European Convention on Human Rights (ECHR). The most common human rights violations that have been claimed include those of the right to life, the right to the protection of private and family life, and the right to a healthy environment. Other alleged human rights infringements encompass those related to non-discrimination, as well as the right to property, and the right to health. Some cases also include alleged breach of the prohibition of torture, or the right to a fair trial, the right to an effective remedy, or again freedom of expression, assembly and association, and freedom of information.

When it comes to the defences of States, human rights justifications have been used in their arguments, as in the Indian case of *Hindustan Zinc Ltd. v. Rajasthan Electricity Regulatory Commission* (2015), where the Renewable Energy Regulatory Commission defended the imposition of regulations mandating captive power plants and open-access consumers to procure a minimum amount of their energy from renewable sources. In this case the defence is grounded on human rights justifications, specifically the right to life and the duty to protect the environment according to Articles 21 and 51 A(g) of the Constitution of India, respectively, which the court has interpreted as guaranteeing the right to live in a healthy and pollution-free environment. In this case, the court sided with the national commission's arguments, stating that 'the said Regulations are consistent with the International obligations of India, as India has ratified to the Kyoto Protocol' (Hindustan Zinc Ltd. v. Rajasthan Electricity Regulatory Commission, 2015, p. 32).

However, most of the time States do not rely expressly on specific human rights—rather, they rely more generally on the fact that they are complying with their obligations and duties, e.g. those towards the environment or climate change commitments.

In *Future Generations v. Ministry of the Environment and Others*, where the Columbian Supreme Court declared that 'the Colombian Amazon is recognized as a 'subject of rights', entitled to protection, conservation, maintenance and restoration led by the State and the territorial agencies' (Future Generations v. Ministry of the Environment and Others, 2018, p. 45), the governmental actors had (unsuccessfully) argued that they had complied with their rights and obligations towards the environment and the protection of the Amazon (Future Generations v. Ministry of the Environment and Others, 2018, p. 6; Tigre, 2024)

States also tend to rely on climate action-based justifications where the argument is that democratically legitimized government is the appropriate body to adopt policy choices. This emerged very clearly in the case *Urgenda Foundation v. State of the Netherlands*, where the 'State has asserted that it is not for the courts to undertake the political considerations necessary for a decision on the reduction of greenhouse gas emissions. In the Dutch system of government, the decision-making on greenhouse gas emissions belongs to the government and parliament' (Urgenda Foundation v. State of the Netherlands, 2019, paras. 8.1–8.3.5).

Also in *Sharma and others v. Minister for the Environment*, the Australian Ministry of the Environment emphasized the limited role of the courts in decision-making by adopting a separation of powers argumentation. Moreover, the Minister's response seemed to rely on the need to balance competing interests and potential benefits and harms, suggesting that it is up to the government to engage in such an assessment. As reported in the judgement, 'the Minister contended that how to manage the competing demands of society, the economy and the environment over the short, medium and long term, is a multifaceted political challenge. In the context of climate change, measures to manage those competing demands occur within the context of evolving national and international strategies. It was said that reducing greenhouse gas emissions while simultaneously managing the demands of society and the economy is a complex and nuanced task' (Sharma & others v. Minister for the Environment, 2021, para. 478).

Alternatively, States tend to rely on more procedural-based defences, arguing by pointing to a lack of a sufficient substantiation justification, as in the *In re Federal Climate Protection Act Austria* case, where the Austrian government held that the concerns about the proportionality of the regulation had not been precisely described or conclusively and verifiably explained in the application (In re Federal Climate Protection Act Austria, 2023). Similarly, in the *Children of Austria v. Austria* case, the government put forward arguments based on the alleged inadmissibility of the application (Children of Austria v. Austria, 2023).

While States have relied on diverse types of justifications, most of them can be included in some overarching categories—in this respect, we have attempted to delineate the first typologies of justification, as they are listed in the sections below. Before delving into the methodology of our classification, it is worth it to zoom in on the first climate case decided by the ECtHR, namely the *KlimaSeniorinnen* case.

2.2 A Zoom in on the ECtHR Level: The Verein KlimaSeniorinnen Schweiz Case

On 9 April 2024, the ECtHR decided its first three climate cases. *Duarte Agostinho and Others v. 32 Member States* (ECtHR, 2024a) and *Carême v. France* (ECtHR, 2024b) were declared inadmissible. In the *Verein KlimaSeniorinnen and Others v. Switzerland* (ECtHR, 2024c) the ECtHR recognized that Switzerland had violated human rights through its climate inaction. Given the potential implications of such a decision for ongoing and future domestic climate litigation cases, it is worth deeper attention.

On 25 November 2018, the non-governmental organization *KlimaSeniorinnen* (representing 'senior women') requested the issuing of a ruling by the Swiss Federal Council, the Federal Department of the Environment, Transport, Energy, and Communications, the Federal Office for the Environment, and the Federal Office of Energy to address a series of omissions regarding climate change protection. They also requested to adopt the necessary measures to ensure that Switzerland contributes to the Paris Agreement goal of reducing global warming by 2030. All the above offices declared the requests inadmissible. On 27 November 2018, the Federal Administrative Court rejected the appeal by KlimaSeniorinnen against the governmental offices' decisions as well. In turn, on 21 January 2019, KlimaSeniorinnen appealed to the Federal Supreme Court, claiming in particular that given the fact that as senior women, 'they were particularly affected by and vulnerable to climate change, the State had a duty to protect them'. By failing to take the necessary measures in line with the Paris Agreement, the governmental authorities had violated their rights. The Federal Supreme Court rejected all the claims and the case was finally brought before the ECtHR (Ammann & Marti, 2020).

The Court found that Switzerland breached its positive obligations related to climate change: in particular, its domestic regulatory framework presented 'critical gaps', especially linked to the topics of carbon budget and reduction of national GHG emission. Additionally, Switzerland did not manage to meet its past GHG emission reduction targets. Accordingly, the Court found a violation of the right to respect for private and family life (Article 8 ECHR) and the right to access to court (Article 6 ECHR). As the Court highlighted, 'Article 8 must be seen as encompassing a right for individuals to effective protection by the State authorities from serious adverse effects of climate change on their life, health, well-being and quality of life' (ECtHR, 2024c, para. 519).

Also, the Court found a violation of Article 6 ECHR, affirming that the 'domestic courts did not engage seriously or at all with the action brought by the applicant association' (ECtHR, 2024c, para. 636). At the same time, the Court recalled the important role of national courts in climate litigation as the first places where the implementation of the ECHR is to be ensured (ECtHR, 2024c, para. 639).

The *Verein KlimaSeniorinnen Schweiz* case raises a number of relevant issues and questions. What is relevant here, however, is the analysis of the defence put forward by the Swiss government.

At the forefront, Switzerland underlined its compliance with international commitments in the field of climate change: '[b]y ratifying the Paris Agreement, Switzerland had made a definite commitment to halve its GHG emissions by 2030' (ECtHR, 2024c, para. 92) and illustrated how its national regulations had been drafted and implemented accordingly (ECtHR, 2024c, para. 100).

At the same time, the Swiss government revindicated that the right place to address climate change issues—including demands from civil society—is the policy domain, and not the European courtroom. The government recalled in this respect the importance of respecting the principle of subsidiarity and separation of powers; accordingly, it contended that the democratic system of Switzerland already provides opportunities for citizens to address climate change concerns without the need to eventually bring the case to the ECtHR, which should not become 'a supreme court for the environment'.[8]

The government thus relied on a separation of power argument (as other States have been also doing in domestic climate litigation, as evidenced above). This argumentation is somehow linked to the interpretation of the margin of appreciation concept.[9] As recalled by the Court, national governments have the 'primary responsibility' to ensure that the ECHR is adequately implemented, while also exercising a margin of appreciation (ECtHR, 2024c, para. 542).

And with regard to the margin of appreciation, the ECtHR introduced a distinction:

> 'between the scope of the margin as regards, on the one hand, the State's commitment to the necessity of combating climate change and its adverse effects, and the setting of the requisite aims and objectives in this respect, and, on the other hand, the choice of means designed to achieve those objectives. As regards the former aspect, the nature and gravity of the threat and the general consensus as to the stakes involved in ensuring the overarching goal of effective climate protection through overall GHG reduction targets in accordance with the Contracting Parties' accepted commitments to achieve carbon neutrality, call for a reduced margin of appreciation for the States. As regards the latter aspect, namely their choice of means, including operational choices and policies adopted in order to meet internationally anchored targets and commitments in the light of

---

[8] As recalled by the Swiss government, '[w]hile the Government accepted that in democratic societies the public legitimately sought to put pressure on the authorities to address climate change, they were of the view that the system of individual application under the Convention was not the appropriate means to do that given, in particular, the principle of subsidiarity. The democratic institutions in the political system of Switzerland provided sufficient and appropriate means to address concerns relating to climate change, and a 'judicialisation' of the matter at the international level would only create tension from the perspective of the principle of subsidiarity and the separation of powers. In any event, the Court could not act as a supreme court for the environment, given, in particular, the evidentiary and scientific complexity of the matter". (ECtHR, 2024c, para 338).

[9] The margin of appreciation has been defined as 'one of the main sources for the ECtHR's exercise of deference' (Chagas, 2022) and an important argumentative framework used by the ECtHR (among others, Gerards, 2018; Legg, 2012).

priorities and resources, the States should be accorded a wide margin of appreciation' (ECtHR, 2024c, para. 543).

Thus, when it comes to international obligations which States have already voluntarily agreed on, and for which there are already goals that have been commonly determined by the States themselves, the Court grants 'a reduced margin of appreciation', which is instead interpreted more 'wide[ly]' when it comes to national choices and policies for meeting the international goals. It seems that the Court is thus introducing two modalities of the margin of appreciation assessment. And on the first one—linked to the international commitments—it also develops a 'check-list' that would help the Court in its assessment of the measures that States adopt. In this respect, the Court contends that 'in order for the measures to be effective, it is incumbent on the public authorities to act in good time, in an appropriate and consistent manner' (ECtHR, 2024c, para. 548).

And after having engaged in an in-depth analysis according to the proposed check-list, the Court found that the Swiss government had 'failed to meet its past GHG emission reduction targets [...]. By failing to act in good time and in an appropriate and consistent manner regarding the devising, development and implementation of the relevant legislative and administrative framework, the respondent State exceeded its margin of appreciation and failed to comply with its positive obligations in the present context' (ECtHR, 2024c, para. 573), which in turn led to a 'violation of Article 8 of the Convention' (ECtHR, 2024c, para. 574).

The above-mentioned arguments developed by Switzerland, basically based on the separation of power arguments and the need to respect the margin of appreciation, were also sided with by some other European States, which submitted third party interventions (namely Austria, Ireland, Italy, Latvia, Norway, Portugal, Romania and Slovakia).[10] The third party intervention of Norway is particularly telling in this respect. Norway made it clear that '[p]olitical decisions regarding the choice of climate targets and policies are within a state's sovereignty to decide' (Norway. Attorney General for Civil Affairs, 2022, para. 2), adding that 'every country has its own national system [that] it needs to transform to limit greenhouse gas emissions. The choice of regulations, solutions,

---

[10] For the full list of third party intervention, see the official website of the association Verein KlimaSeniorinnen Schweiz. Retrieved October 14, 2024, from https://www.klimaseniorinnen.ch/drittparteien-interventionen/.

when to implement them and how should be up to sovereign States to decide' (Norway. Attorney General for Civil Affairs, 2022, para. 3).

And when it came to the ECHR, it claimed that it 'is not an instrument for the protection of collective interests, and the Court is not a supervisor of society-wide policy decisions' (Norway. Attorney General for Civil Affairs, 2022, para. 4), given also the fact that already '[w]ithin the context of the Council of Europe the Member States may choose to introduce enforceable rights pertaining directly to the environment and climate [...]. Such developments belong, however, within the competence of the legislators' (Norway. Attorney General for Civil Affairs, 2022, para. 26).

It is quite interesting how Norway has highlighted that it is not only within the domain of States to decide how to implement climate goals, but that also new 'enforceable rights' in the area of climate change can be 'introduced' only by states within international political platforms like the Council of Europe. This way, all climate change-related debates and regulatory developments seem to be brought under the 'political' umbrella of State sovereignty.

The above arguments in turn raise other questions, which turn back to the HRJ research questions: to what extent can States claim 'sovereignty' over climate change policy decisions, where and when can/should the judicial scrutiny intervene, and with which instruments? Would the twofold path of the margin of appreciation illustrated by the Court in the *KlimaSeniorinnen* case be adequate or should the courts rely on different arguments?

2.3   Climate Litigation and the Emergence of Different Typologies of Justifications

This analysis offered the first insights into the major arguments formulated by States in this respect and their practices of adapting to the rapid changes and new protection needs across a range of jurisdictions. Climate change offers an increasingly complex situation that could require States to have obligations to face multiple risks, including solving challenges raised by competing or conflicting values and principles (e.g. security/ development). As illustrated by Kumm, despite the claim that 'human rights are universal', their interpretation may vary between jurisdictions and across different contexts, giving rise to different justifications invoked by States (Kumm, 2018, p. 238). For this, it is increasingly relevant to explore how a systemic framework apt for anticipating and tackling current and future challenges could be developed by understanding the motivations and causes of the actions of various actors.

Defining the typologies of the justifications used by States could help to reveal major trends in climate change litigation, explore the legal challenges faced in developing climate norms, and advance the conceptualization of the overarching concept of human rights justification (Besson, 2022; Collier et al., 2008). Drawing, as mentioned above, from the dataset built by the HRJust project, we suggest with our analysis that one should start with three main typologies that build on the work of Ghaleigh (2010) and we consider as promising approaches.

First, in some cases States have addressed climate change—or the human rights concerns raised by the applicants—by adopting a 'defensive approach' that, in the end, risks resulting in defending the *status quo* of the alleged violations (Ghaleigh, 2010). This type of litigation seeks to dismiss the given cases due to a 'lack of causality', 'procedural obstacles', or even 'confidentiality reasons'.

Second, in a distinct type of litigation, States embrace a more 'promotive approach' by illustrating how duties to protect, respect, and fulfil have been implemented. In their observations, they provide accurate descriptions of the measures adopted. For instance, several cases enumerated climate change mitigation and adaptation strategies. In this context, States particularly reveal their responsibility to assess a range of threats and challenges that have an impact on the enjoyment of human rights. They assume a major role in taking decisions on distributing limited domestic resources to address these issues. To this end, States could be required to adopt compromises and face trade-offs (HRC, 2022, paras. 6.10–6.11).

In fact, climate policies are adopted by making use of the discretion that States have, a discretion that needs to be exercised in good faith. Overall, States in their observations acknowledge the need for significant investment at domestic level to tackle these challenges by not refraining from 'regulatory activism'—see cooperative justification—and their engagement with climate concerns is increasingly shaped by the domestic and international contexts.

A further articulation of the motivations adopted to take action is the 'boundary-testing' approach, our final category (Ghaleigh, 2010). It occurs at the intersections of competing or conflicting interests faced by States at domestic level that could undermine the scope of human rights. The related discussion focuses, for instance, on invoking 'public interest' justifications to 'limit' human rights claims. This type could be particularly problematic if left without a conceptual accuracy that would clarify the legitimacy of its use and the conditions under which infringements could be authorized. For instance, they could be authorized by law. At the same time, this type could also include positive achievements by mobilizing the 'sustainable development' principle to shape the present and future development of States' climate policies. In shifting up to sustainability, this justification offers an opportunity for adopting counter-mobilization as an alternative strategy to those for dealing with the legal challenges that could arise to challenge economic development prerogatives.

The use of justifications in climate change litigation provides new venues for exploring the possibilities and limitations of human rights-based climate litigation. This preliminary analysis offers an opportunity to reflect on the values needed in the long term and how to act when facing global challenges at both theoretical and operational levels (Table 1, Fig. 1).

3 The Way Forward

Current and future domestic climate litigation will most likely engage with the arguments developed by the ECtHR in the *Verein KlimaSeniorinnen Schweiz* case (Abel, 2024; Hamilton, 2024; Higham et al., 2024). Moreover, thanks to the already released advisory opinion of ITLOS of 21 May 2024, and the ones from the IACtHR of 29 May 2025 (IACtHR, 2025) and the ICJ of 23 July 2025 (ICJ, 2025). Discussions on the development of climate litigation will be further enriched, given also the significant volume of State and general public engagement in the relevant proceedings (Setzer & Higham, 2024).[11]

In particular, the advisory opinion of the IACtHR seems to provide some guidance on the use of (human rights) justifications by states in climate litigation cases and the relevant courts' assessment. Indeed, the IACtHR observes that 'States must refrain from adopting retrogressive measures. This obligation derives from the principle of progressivity and non-retrogression which is applicable to all the rights that are under threat

---

[11] Suffice to recall that ITLOS's advisory process attracted statements from around 50 States, international organizations, and NGOs; the IACtHR received more than 260 submissions from States, international organisations, NGOs and individuals; also the ICJ received more than 80 written statements, which is the highest number of submissions ever received in an advisory proceedings.

Table 1 Building different typologies of justifications

Typologies of justifications		Examples of cases
Defensive	**Lack of Causality Justification** When States deny a causal link between an alleged act or policy and a harmful violation that is generally a result of GHG emissions, or other adverse climate change impacts.	The case *Urgenda Foundation v. State of the Netherlands*
	Procedural Justification When States disregard the allegations, asserting that the claimant lacks the legal standing or right to bring the case before the court.	The case *Tsama William and Others v. Uganda's Attorney General and Others*
	Confidentiality Justification When States challenge or limit freedom of information by relying on confidentiality concerns.	The case *Millar v. Department of Premier and Cabinet (General)*
Promotive	**Climate Action-Based Justification** When States highlight the role of policy decisions in the context of the allegations. States defend their actions as being rooted in legitimate, considered policy choices.	The case *Urgenda Foundation v. State of the Netherlands*
	Human Rights Justification When States rely on HRJs in their defences.	The case *Hindustan Zinc Ltd. v. Rajasthan Electricity Regulatory Commission*
	Cooperative Response When States admit to certain allegations and highlight their efforts to address the issues raised. It signifies a willingness to acknowledge shortcomings and engage in remedial actions or dialogue to improve.	The case *In re Court on its own motion v. State of Himachal Pradesh and others*
Boundaries-testing	**Public Interest Justification** When States attempt to justify an act by claiming it serves the public interest.	The case *Sukhdev Vihar Welfare Residents Association v. Union of India*
	Sustainable Development Justification When States present arguments based on intergenerational equity or sustainable development principles.	The case *Dual Gas Pty. Ltd. and Others v. Environment Protection Authority*

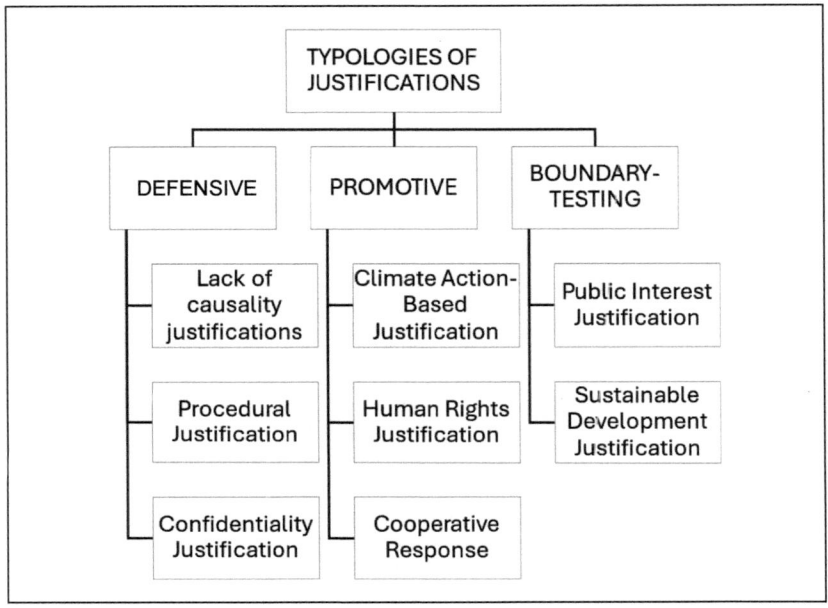

Fig. 1 Typologies of States' justification

in this context, and requires than any setback in climate and environmental policies that harm human rights must be exceptional, duly justified based on objective criteria, and comply with the standards of necessity and proportionality' (IACtHR, 2025, para. 222). Additionally, the Court recalled 'the obligation of States to justify their climate-related decisions by reference to the best available science' (IACtHR, 2025, para. 503), adding that 'States must refrain from acts or omissions that, directly or indirectly, obstruct, restrict or harm effective access, in equal conditions, to the enjoyment of human rights of individuals affected by the climate emergency' (IACtHR, 2025, para. 223). This seems to suggest that States may not justify their inaction on climate change by simply invoking human rights or other competing interests, without any sound references to scientific justifications. According to the Advisory Opinion, any rollback—or even delay—of climate and environmental protection is, in principle, incompatible with core human rights obligations under the American Convention on Human Rights. The IACtHR adopts a strict framing of the obligation of non-retrogression, meaning that States must not reverse established safeguards unless such reversals are objectively justified, strictly necessary and proportionate (IACtHR, 2025, para. 240). Also, when it comes to climate litigation proceedings, the Advisory Opinion can assist in overcoming relevant procedural obstacles. Indeed, the Court affirms that 'the conduct of judicial proceedings […] should be guided by the application of the pro actione principle. According to this principle, the interpretation or application of procedural rules must not unjustifiably prevent or hinder a court from hearing and ruling on the claims submitted to it in accordance with the law, and the interpretation most favorable to access to justice must always prevail. Consequently, judicial bodies must interpret and apply the relevant rules in such a way as to effectively guarantee access to substantive justice for those who require it in the context of the climate emergency' (IACtHR, 2025, para. 543). This might again serve as guidance for national and regional courts when facing arguments of states based on procedural grounds. This will certainly enrich the ongoing discussion on human rights justifications and in general, on arguments employed by states in climate change proceedings, as delineated in the previous paragraphs. This makes the mapping exercise of the HRJust project a very timely one. The data collected so far is already revealing different typologies of justifications put forward by States in the context of domestic climate litigation.

Conceptualizing each typology would not only be a theoretical exercise but would also have practical implications as it would serve to help us understand whether each justification has its 'place' in the relevant international legal framework. This would, in turn, be the first step in understanding whether States face some consequences in cases where we deal with justifications which impair individual human rights, and/or which (legal) instrument(s) individuals can use in this respect (Fig. 2).

The aim of the new database is to offer a comprehensive picture of justifications elaborated by States: this can help us to understand the climate litigation phenomenon, while also identifying the pros and cons of grounding climate disputes on human rights arguments and on the relevant international instruments. In turn, it will support the analysis of the developing roles of the main actors involved, namely courts and tribunals, civil society (as the applicants in the majority of cases), and national governments.

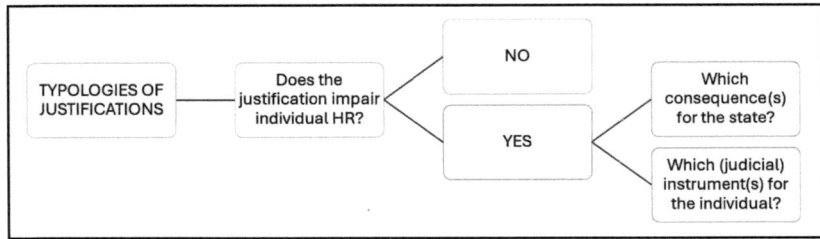

Fig. 2 Conceptualization of States' justifications

Annex

Introduction to the Methodology: The HRJust Climate Claims Dataset

The HRJust Climate Claims Dataset includes domestic climate litigation cases. The scrutinized national jurisdictions comprise so far the HRJust core case studies (i.e. Switzerland, the Czech Republic, Finland, Sweden, Taiwan, Ukraine, and India), plus other national jurisdictions from Europe (Austria, Belgium, Bulgaria, Denmark, Cyprus, Estonia, the Netherlands, Norway, Portugal), Latin America (Argentina, Brazil, Chile, Colombia, Peru), as well as Australia, Canada, Papua New Guinea, Russia, Thailand, Turkey, and Uganda (324 cases overall at the time of writing—October 2024). The cases have been retrieved from the relevant academic literature and the above-mentioned reports, as well as from the open-access databases developed by the Sabin Center for Climate Change Law, the Climate Rights and Remedies Project of the University of Zurich, and the Climate Law Accelerator CLX of New York University.

Each case has been analysed under several variables, according to the following codebook (Table 2).

Table 2 Structure of the HRJust Climate Claims Dataset[12]

	Variable	Meaning	Examples
General information	Country	The country in which the case was filed or originated.	In the case *Klimatická žaloba ČR v. Czech Republic*: Czech Republic.
	Case name	The name of the case.	*Klimatická žaloba ČR v. Czech Republic*.
Information on parties	Claimant	The party (or parties) who initiate(s) the lawsuit.	In the case *Klimatická žaloba ČR v. Czech Republic*: Klimatická žaloba ČR, z. s.
	Respondent	The party (or parties) against whom the legal action is initiated.	In the case *Klimatická žaloba ČR v. Czech Republic*: the Government of the Czech Republic, Ministry of the Environment, Ministry of Industry and Trade, Ministry of Agriculture, Ministry of Transport.
Type of claimant	Public	The claimant is a government entity or public institution at any level (local, regional, national, or international).	The case *Saonu and Morobe Provincial Government v. Minister for Environment and Conservation and Climate Change and Others*.
	Civil Society	The claimants are non-governmental organizations (NGOs), advocacy groups, or any non-profit entities that are part of civil society.	The case *Klimatická žaloba ČR v. Czech Republic*.
	Individual	The claimants are natural persons acting in their personal capacity.	The case *Álvarez et al. v. Peru*.

(continued)

[12] The research has been carried out as part of the tasks and activities of WP6 (Climate) for developing the digital Climate Claims Visual Maps (www.hrjust-climate-claims.eu). Data here presented have been collected by the research teams at the World Trade Institute of the University of Bern, Switzerland (Prof. Elisa Fornalè and Ms. Riccarda Heepen) and at the Institute of International Relations in Prague, Czech Republic (Ms. Elif Naz Němec, Dr. Petra Ditrichová, Dr. Jan Lhotský and Dr. Federica Cristani).

Table 2 (continued)

Variable	Meaning	Examples
Corporation	The claimant that initiated the case is a private company, business, or corporate entity.	The case *KEPCO Bylong Australia v. Independent Planning Commission and Bylong Valley Protection Alliance*.
Type of respondent		
Public	Respondents acting as a government entity or public institution at any level (local, regional, national, or international).	The case *Klimatická žaloba ČR v. Czech Republic*.
Civil	The respondent is a private individual, corporation, or non-governmental organization.	The case *Joe Davidson Town Planning v Byron Shire Council*.
Both	The respondents in the case include both public (government or public institutions) and civil entities (private individuals, corporations, NGOs).	The case *KEPCO Bylong Australia v. Independent Planning Commission and Bylong Valley Protection Alliance*.
Content of the case		
What is at stake	A brief description of the core dispute or legal question at the heart of the case, typically in one sentence.	In the case *Klimatická žaloba ČR v. Czech Republic*: assessing alleged breach of human rights as a result of the government's climate inaction.

Variable	Meaning	Examples
Summary	An overview of the main arguments presented by both the claimant and the respondent.	In the case *Klimatická žaloba ČR v. Czech Republic*: In this case, an NGO prepared a climate lawsuit and filed it at the Municipal Court in Prague in 2021, claiming the relevance of a number of international law or human rights instruments, such as the Paris Agreement, the European Convention on Human Rights, the EU Charter on Fundamental Rights, and the Czech Constitution/Charter of Fundamental Rights and Freedoms, to the case. By pointing to the insufficient activity of the State in GHG emission reduction and its plans, a number of human rights violations were claimed in the lawsuit.

(continued)

222 F. CRISTANI AND E. FORNALÉ

Table 2 (continued)

Variable	Meaning	Examples
Outcome	It details the court's decision and the reasoning behind it. If the case is still pending, this section will simply note that status.	In the case *Klimatická žaloba ČR v. Czech Republic*: The Municipal Court (in its first decision) partially upheld the action, explaining, among other things, that the right to a favourable environment according to the Czech Constitution (para 328 of the judgement) was violated by the insufficient activity of the state. According to the Court, the government inaction breached the rights of the complainants. The Municipal Court (in its first decision) also cited the Urgenda judgement. Following an appeal, the Supreme Administrative Court (SAC) overturned the decision of the first instance court and annulled the ruling, emphasizing the restraint of the judiciary. The SAC explained that in 2020 the European Union had made a collective commitment under the Paris Agreement to reduce emissions by 2030. Nevertheless, this collective commitment had not yet been transformed into EU law. It was therefore not possible to determine a specific commitment for the Czech Republic, as the specification of the national commitments was still subject to legislative and political negotiations. The SAC returned the lawsuit to the Municipal Court in Prague for further consideration. The Municipal Court subsequently issued a new verdict, in which it sided with the arguments of the Supreme Administrative Court and completely dismissed the lawsuit.

Variable	Meaning	Examples
Date of Submission	The date when the case was officially filed or brought before the court.	In the case *Klimatická žaloba ČR v. Czech Republic*: 2021
Date of the Decision	The date on which the court made its decision regarding the case. In cases where multiple decisions were made at different levels of the judicial process, this category includes the dates of all significant decisions.	In the case *Klimatická žaloba ČR v. Czech Republic*: 15/06/2022 (Municipal Court); 20/02/2023 (SAC); 25/10/2023 (Municipal Court).
Type of Court	The specific court that issued the decision on the case. If a case was adjudicated by multiple courts, this category lists all the relevant courts that issued significant decisions on it.	In the case *Klimatická žaloba ČR v. Czech Republic*: Municipal Court; Supreme Administrative Court.
Status of the Proceedings	This variable indicates the current stage of the legal proceedings.	In the case *Klimatická žaloba ČR v. Czech Republic*: Pending (an appeal to the Supreme Administrative Court has been filed).
National Law Invoked (National Law invoked in the case)	Any national legislation, other than the constitution, cited or applied in the case. This includes the full title(s) of the law(s) invoked.	In the case *Klimatická žaloba ČR v. Czech Republic*: National Action Plan for Adaptation to Climate Change.
National Constitution Invoked	A binary (Yes/No) indicator indicating whether any national constitutions were cited or relied upon in the arguments or the court's decision.	In the case *Finnish Association for Nature Conservation and Greenpeace v. Finland*: Yes.
Invoked Articles of the National Constitution	If the national constitution was invoked, this field lists the specific articles or sections cited in the proceedings.	In the case *Finnish Association for Nature Conservation and Greenpeace v. Finland*: Art. 20, Art. 21(1), Art. 22, Art. 46(1).
Which international law instrument was invoked? (International law instrument invoked)	This variable lists all the international legal instruments, such as treaties, conventions, protocols, or agreements, that were cited or relied upon in the case.	In the case *Klimatická žaloba ČR v. Czech Republic*: Paris Agreement, European Court of Human Rights, EU Charter.
Human Rights Instrument	A binary (Yes/No) indicator signifying whether any human rights treaties, conventions, or declarations were cited or relied upon in the legal arguments or the court's decision.	In the case in the case *Klimatická žaloba ČR v. Czech Republic*: Yes.

(continued)

Table 2 (continued)

Variable	Meaning	Examples
Climate Change Instrument	A binary (Yes/No) indicator indicating whether any international legal instruments specifically related to climate change were cited or relied upon in the legal arguments or the court's decision.	In the case in the case *Klimatická žaloba ČR v. Czech Republic*: Yes.
Human rights involvement		
Direct or Indirect Classification	This category identifies the nature of the human rights application within the case as either 'Direct' or 'Indirect'. 'Direct' involvement signifies cases where human rights are explicitly used in the arguments or cited by the parties or the court. 'Indirect' involvement is recognized when the case does not explicitly mention human rights but involves principles or reasoning that can be linked to human rights concepts. In case neither type of application is found in the case, then this section is classified as 'None'.	In the case *Klimatická žaloba ČR v. Czech Republic*: Direct. In the case *KEPCO Bylong Australia v. Independent Planning Commission and Bylong Valley Protection Alliance*: Indirect.
Indirect involvement	For cases marked as having 'Indirect' human rights involvement, this field specifies which human rights are implied in the case's context or reasoning without any direct citation.	In the case *KEPCO Bylong Australia v. Independent Planning Commission and Bylong Valley Protection Alliance*: Inter-generational equity.
Direct involvement	In cases identified as having 'Direct' human rights involvement, this field lists the specific human rights that are explicitly cited or used in arguments within the case.	In the case *Klimatická žaloba ČR v. Czech Republic*: The right to life, the right to health, the right to a favourable environment, the right to property, the right to own a business, the right to protection of private and family life, and the right to self-government.

Variable	Meaning	Examples
Human rights found to be violated	This category is used to document instances where the court or tribunal finds a violation of one or more human rights. The specific rights found to be violated are listed here.	In the case *Residents of Omkoi v. Expert Committee on EIA Consideration and the Office of Natural Resources and Environmental Policy and Planning*: the right to live in a good environment, the right to meaningful participation of the community.
State justifications	Notes Concerning State Arguments	This category is used to provide an overview of State argumentation in cases that directly or indirectly involve human rights arguments. If no documents related to the case are available in the relevant database, or if there is no content related to the case in it, the researcher makes a note of it.	In the case *A.S. & S.A. & E.N.B v. Presidency of Türkiye & The Ministry of Environment, Urbanization and Climate Change*: No decision text available. In the case *Gray v. Minister for Planning and Ors*: The Minister argued that the Environmental Assessment Report adequately addressed intergenerational equity.

References

Abel, P. (2024, April 17). *Mixed Signals for Domestic Climate Law: the Climate Rulings of the European Court of Human Rights*. Verfassungsblog. https://verfassungsblog.de/mixed-signals-for-domestic-climate-law

Ammann, O., & Marti, C. (2020, May 5). *Verein KlimaSeniorinnen Schweiz and others v Federal Department of the Environment, Transport, Energy and Communications, Final Appeal Judgment, 1C_37/2019, BGE 146 I 145, ILDC 3193 (CH 2020), 5th May 2020, Switzerland; Federal Supreme Court [BGer]*. Oxford Reports on International Law. https://opil.ouplaw.com/display/https://doi.org/10.1093/law-ildc/3193ch20.case.1/law-ildc-3193ch20

Anderson, M. R. (1998). Human Rights Approaches to Environmental Protection: An Overview. In A. Boyle & M. R. Anderson (Eds.), *Human Rights Approaches to Environmental Protection* (pp. 1–24). Oxford University Press.

Anton, D. K., & Shelton, D. (2011). Procedural Human Rights and the Environment. In D. K. Anton & D. Shelton (Eds.), *Environmental Protection and Human Rights* (pp. 356–435). Cambridge University Press.

Besson, B. (2022). Justifications. In D. Moeckli, S. Shah, S. Sivakumaran & D. Harris (Eds.), *International Human Rights Law* (pp. 23–42). 4th edn. Oxford University Press.

Boyle, A. (2006). Human Rights or Environmental Rights? A Reassessment. *Fordham Environmental Law Review, 18*(3), 471–511.

Brems, E. (1996). The Margin of Appreciation Doctrine in the Case-Law of the European Court of Human Rights. *Zeitschrift Für Ausländisches Öffentliches Recht und Völkerrecht, 56*, 240–314.

Burgers, L. (2020, March). Should Judges Make Climate Change Law?. *Transnational Environmental Law, 9*(1), 55–75. https://doi.org/10.1017/S2047102519000360

Carnwath, R. (2022, January 19). *Climate-Conscious Courts: Reflections on the Role of the Judge in Addressing Climate Change*. LSE Commentary. https://www.lse.ac.uk/granthaminstitute/news/climate-conscious-courts-reflections-on-the-role-of-the-judge-in-addressing-climate-change

Chagas, C. A. (2022). Balancing Competences and the Margin of Appreciation: Structuring Deference at the ECtHR. *ICL Journal, 16*(1), 1–26. https://doi.org/10.1515/icl-2021-0009

Children of Austria v. Austria. (2023, February 21). Constitutional Court. Decision. https://climatecasechart.com/non-us-case/children-of-austria-v-austria

Collier, D., Laporte, J., & Seawright, J. (2008). Typologies: Forming Concepts and Creating Categorical Variables. In J. M. Box-Steffensmeier (Ed.), *The Oxford Handbook of Political Methodology* (pp. 152–173). Oxford University Press.

European Court of Human Rights (ECtHR) (2024a, April 9). *Duarte Agostinho and Others v. Portugal and Others.* Application no. 39371/20. Decision. https://hudoc.echr.coe.int/eng#{%22itemic%22:[%22002-13724%22]}

ECtHR (2024b, April 9). *Carême v. France.* Application no. 7189/21. Decision. https://hudoc.echr.coe.int/eng/#{%22itemid%22:[%22001-233174%22]}

ECtHR (2024c, April 9). *Verein KlimaSeniorinnen Schweiz and Others v. Switzerland.* Application no. 53600/20. Judgment. https://hudoc.echr.coe.int/eng#{%22itemid%22:[%22001-233206%22]}

Fornalé, E., Bilkova, V., Burgorgue-Larsen, L., Cristani, F., De Vido, S., Doebbler, C., & Hertogen, A. (2023). *Amicus Curiae - Request for an Advisory Opinion on the Climate Emergency and Human Rights Submitted to the Inter-American Court of Human Rights by the Republic of Colombia and the Republic of Chile of January 9, 2023.* World Trade Institute, University of Bern. https://doi.org/10.48350/190814

Future Generations v. Ministry of the Environment and Others. (2018, April 5). Supreme Court. Case STC4360-2018. Radicacion No. 1101-22-03-000-2018-00319-01. Decision. https://climatecasechart.com/non-us-case/future-generation-v-ministry-environment-others

Gerards, J. (2018). Margin of Appreciation and Incrementalism in the Case Law of the European Court of Human Rights. *Human Rights Law Review, 18*(3), 495–515. https://doi.org/10.1093/hrlr/ngy017

Ghaleigh, N. S. (2010). 'Six Honest Serving-Men': Climate Change Litigation as Legal Mobilization and the Utility of Typologies. *Climate Law, 1*(1), 31–61.

Hamilton, R. (2024, April 2). *The 'Year of Climate' in International Courts.* LegalPlanet. https://legal-planet.org/2024/04/02/the-year-of-climate-in-international-courts

Higham, C., Keuschnigg, I., Chan, T., & Setzer, J. (2024). *What Does the European Court of Human Rights' First Climate Change Decision Mean for Climate Policy?* Grantham Research Institute on Climate Change and the Environment and Centre for Climate Change Economics and Policy, London School of Economics and Political Science. www.lse.ac.uk/granthaminstitute/news/what-does-the-european-court-of-human-rights-first-climate-change-decision-mean-for-climate-policy

Hindustan Zinc Ltd. v. Rajasthan Electricity Regulatory Commission. (2015, May 13). Civil Appeal No. 4417 of 2015; (2015) 12 SCC 611. https://climatecasechart.com/non-us-case/hindustan-zinc-ltd-v-rajasthan-electricity-regulatory-commission/

Human Rights Committee (HRC). (2022, September 22). *Views Adopted by the Committee Under Article 5 (4) of the Optional Protocol, Concerning Communication No. 3624/2019.* https://ccprcentre.org/files/decisions/CCPR_C_135_D_3624_2019_34335_E.pdf

IACtHR. (2025, May 29). *Advisory Opinion AO-32/25 Requested by the Republic of Chile and the Republic of Colombia. Climate Emergency and Human Rights.* https://jurisprudencia.corteidh.or.cr/en/vid/1084981967

ICJ. (2025, July 23). *Advisory Opinion. Obligations of States in Respect of Climate Change.* https://www.icj-cij.org/case/187

Kumm, M. (2018). The Turn to Justification: On the Structure and Domain of Human Rights Practice. In A. Etinson (Ed.), *Human Rights: Moral or Political?* (pp. 238–261). Oxford University Press.

Iyengar, S. (2023). Human Rights and Climate Wrongs: Mapping the Landscape of Rights-Based Climate Litigation. *Review of European, Comparative & International Environmental Law, 32*(2), 299–309. https://doi.org/10.1111/reel.12498

In re Federal Climate Protection Act Austria. (2023, June 27). Constitutional Court. Decision. https://climatecasechart.com/non-us-case/in-re-federal-climate-protection-act-austria

Inter-American Court of Human Rights (IACtHR). (2023, January 9). *Request for an Advisory Opinion on the Climate Emergency and Human Rights Submitted by the Republic of Colombia and the Republic of Chile.* https://www.corteidh.or.cr/docs/opiniones/soc_1_2023_en.pdf

International Court of Justice (ICJ). (2023, April 12). *Request for Advisory Opinion Transmitted to the Court Pursuant to General Assembly Resolution 77/276 of 29 March 2023: Obligations of States in Respect of Climate Change.* https://www.icj-cij.org/case/187

International Law Association (ILA). (2024). *International Law and Sea Level Rise Committee: Final Committee Report Athens May 2024.* https://www.ila-hq.org/en_GB/committees/international-law-and-sea-level-rise

International Tribunal for the Law of the Sea (ITLOS). (2022, December 12). *Request for an Advisory Opinion Submitted by the Commission of Small Island States on Climate Change and International Law (Request for Advisory Opinion submitted to the Tribunal).* https://www.itlos.org/en/main/cases/list-of-cases/request-for-an-advisory-opinion-submitted-by-the-commission-of-small-island-states-on-climate-change-and-international-law-request-for-advisory-opinion-submitted-to-the-tribunal

ITLOS. (2024, May 21). *Advisory Opinion.* https://www.itlos.org/en/main/cases/list-of-cases/request-for-an-advisory-opinion-submitted-by-the-commission-of-small-island-states-on-climate-change-and-international-law-request-for-advisory-opinion-submitted-to-the-tribunal/

Legg, A. (2012). *The Margin of Appreciation in International Human Rights Law: Deference and Proportionality.* Oxford University Press.

Němec, E. N., Lhotský, J., Cristani, F., & Fornalé, E. (2024). *WP6 Preliminary Working Paper on the National Climate Litigation Case Reviews: State*

Accountability for Its Use of Human Rights Justifications. Zenodo. https://doi.org/10.5281/zenodo.10997255

New York University (NYU). (2024). *Climate Law Accelerator CLX: Rights-Based Climate Litigation Cases*. Retrieved October 10, 2024, from https://clxtoolkit.com/map

Markell, D., & Ruhl, J. B. (2012). An Empirical Assessment of Climate Change in the Courts: A New Jurisprudence or Business as Usual? *Florida Law Review, 64*(1), 15–72.

Network for Greening the Financial System. (2023, September). *Climate-Related Litigation. Recent Trends and Developments. Technical Document*. https://www.ngfs.net/sites/default/files/medias/documents/ngfs_report-on-climate-related-litigation-recent-trends-and-developments.pdf

Norway. Attorney General for Civil Affairs. (2022, December 5). *Third Party Intervention by the Kingdom of Norway. App. no. 53600/20 Verein Klimaseniorinnen Schweiz and Others v. Switzerland*. https://www.klimaseniorinnen.ch/wp-content/uploads/2023/01/53600_20_GC_OBS_G3_Norway_05_12_22.pdf

Sabin Center for Climate Change Law. (2024). *Climate Change Litigation Databases*. Retrieved September 26, 2024, from https://climatecasechart.com

Savaresi, A., & Setzer, J. (2022). Rights-Base Litigation in the Climate Emergency: Mapping the Landscape and New Knowledge Frontiers. *Journal of Human Rights and the Environment, 13*(1), 7–34. https://doi.org/10.4337/jhre.2022.01.01

Setzer, J., & Higham, C. (2024, July). *Global Trends in Climate Change Litigation: 2024 Snapshot*. LSE Policy Report. https://www.lse.ac.uk/granthaminstitute/publication/global-trends-in-climate-change-litigation-2024-snapshot

Sharma and others v. Minister for the Environment. (2021, May 27). Federal Court of Australia. VID 389 of 2021; [2021] FCA 560; [2021] FCA 774; [2022] FCAFC 35; [2022] FCAFC 65. Judgment Establishing Duty of Care but Rejecting Injunction. https://climatecasechart.com/non-us-case/raj-seppings-v-ley/

Shelton, D. (2006). Human Rights and the Environment: What Specific Environmental Rights Have Been Recognized? *Denver Journal of International Law & Policy, 35*(1), 129–71.

Shelton, D. (2010). Human Rights and the Environment: Substantive Rights. In M. Fitzmaurice, D. M. Ong, & P. Merkouris (Eds.), *Research Handbook on International Environmental Law* (pp. 265–284). Edward Elgar Publishing.

Tigre, M. A. (2024, June). *Climate Litigation in the Global South: Mapping Report*. Sabin Center for Climate Change Law. https://scholarship.law.columbia.edu/cgi/viewcontent.cgi?article=1231&context=sabin_climate_change

Tigre, M. A., & Barry, M. (2023, December). *Climate Change in the Courts: A 2023 Retrospective*. Sabin Center for Climate Change Law. https://scholarship.law.columbia.edu/cgi/viewcontent.cgi?article=1213&context=sabin_climate_change

United Nations Environment Programme (UNEP). (2023, July 27). *Global Climate Litigation Report: 2023 Status Review*. https://doi.org/10.59117/20.500.11822/43008

University of Zurich (UZH). (2024). *Climate Rights and Remedies (CRRP) Project: Climate Litigation Database*. Retrieved September 26, 2024, from https://climaterightsdatabase.com

Urgenda. 2024. *Global Climate Litigation*. Retrieved September 26, 2024, from https://www.urgenda.nl/en/themas/climate-case/global-climate-litigation

Urgenda Foundation v. State of the Netherlands. (2019, December 20). [2015] HAZA C/09/00456689. Supreme Court. Judgment. https://climatecasechart.com/non-us-case/urgenda-foundation-v-kingdom-of-the-netherlands

Open Access This chapter is licensed under the terms of the Creative Commons Attribution 4.0 International License (http://creativecommons.org/licenses/by/4.0/), which permits use, sharing, adaptation, distribution and reproduction in any medium or format, as long as you give appropriate credit to the original author(s) and the source, provide a link to the Creative Commons license and indicate if changes were made.

The images or other third party material in this chapter are included in the chapter's Creative Commons license, unless indicated otherwise in a credit line to the material. If material is not included in the chapter's Creative Commons license and your intended use is not permitted by statutory regulation or exceeds the permitted use, you will need to obtain permission directly from the copyright holder.

PART III

Sea Level Rise: New Developments at International Level

Chapter 9

Sovereignty, State Cooperation, and Sea Level Rise

Tamás Vince Ádány

The structure and content of international law have been constantly changing from the early modern times to our days. In the course of the last hundred years, the number of sovereign actors quadrupled, and a far-reaching normative expansion increased the number of relevant subject matters and resulted in more thorough international regulations and several enforcement systems of varying efficiency. Since a major paradigm shift after World War II, the past few decades saw the waning exclusivity of the role of sovereigns in the international system, yet their dominance remains unquestionable. Modern statehood and sovereignty are deeply rooted in territoriality, although globalization has marred the previous exclusivity of these concepts (De Feyter, 2017, Introduction). Arguably, the most important factor in the erosion of the role of States has been the emergence of a number of issues that cannot be addressed from a territorially divided, States-only perspective. Serving the "common good" with regard to these problems is only possible if 'humanity as a whole [is understood as] the publicum, or beneficiary of global public goods' (Kaul et al.,

T. V. Ádány (✉)
Pazmany Peter Catholic University, Budapest, Hungary
e-mail: adany.tamas@jak.ppke.hu

© The Author(s) 2026
E. Fornalé (ed.), *Sea Level Rise*,
https://doi.org/10.1007/978-3-031-89171-7_9

1999, p. 3). The realization that some of these global issues are common concerns for humanity (Kaul et al., 1999, p. 586) demands a coordinated response from States, which raises the need for State cooperation to a whole new level.

Even in this historical perspective on an ever-changing international legal system, it is a preliminary assumption of this chapter that the challenges created by global warming through climate change and the sea level rise are unique and previously never encountered in the modern history of the international community as a whole. There has never been another event in history (at least in modern history) where a human-induced natural phenomenon had major, direct, and adverse impacts on the territory of every single State at the same time, including a simultaneous territorial reduction and a probable destruction of critical coastal infrastructures in more than three quarters of all states, i.e. in every coastal State—to a varying degree. This remains true even if sea level rise will have different impact locally, depending on a number of factors, like regional vertical land movements, and considering a certain level of previous scientific uncertainties deriving from a number of possible scenarios determined by human behaviour (IPCC, 1990, Chapter 9, pp. 278–279; 2019). Since 1990 an overwhelming scientific agreement developed, however, and today it is virtually certain that the global mean sea level is rising, and it can also be stated with high confidence that the rate of this rise is accelerating (IPCC, 2019). Adaptation mechanisms raise a number of legal and administrative issues, both domestically and internationally. Regulating coastal land use, with potentially serious implications for property rights and the related compensations, is mostly an internal issue in developed nations, but issues of maritime borders, jurisdiction problems, and other transboundary matters are also likely to emerge. One of the early reports of the IPCC from 1990 predicted several such examples: 'if a nation loses maritime boundary base points and therefore a legal claim to sea territory, or if beach nourishment measures are required in the vicinity of national borders. An example of the second issue would be if protective measures interrupt or impede the longshore sediment transport benefitting an adjoining coastal state. In the worst case, sea level rise may result in the total land loss of an island nation' (Dronkers et al., 1990, p. 19). It is therefore assumed here that sea level rise is a common issue for humanity that requires collective legal actions for jointly effectuating the necessary policies.

It seems somewhat paradoxical that in the shadow of this truly global threat a number of State leaders instinctually focus on "their own problems" and become somewhat distrustful towards State cooperation. Despite this reluctance, a system of cooperation already exists in this regard, and there is a strong case for its enhancement, yet some legal and political considerations have the potential to undermine such global efforts.

1 State Cooperation: From a Cold War Principle to a Global 'Common Good'

The success of both the League of Nations and the United Nations is often measured against their results—or lack thereof—in the preservation of global peace. From that perspective it is easy to disregard the actual results of State cooperation reached under the auspices of these two organizations (Tung, 1968, pp. 33–34). Even during the late 1950s, at the height of Cold War tensions, UN members from the two blocks could reach an agreement on recommending a frank, friendly, and free cooperation in the fields of economy, culture, science, and communication technology (Service de l'information des Nations Unies, 1960, p. 89). But true enough, it was a mere recommendation rather than an actual legally prescribed obligation.

The movement towards a "duty of States to co-operate" went on a rocky road during the early years of the *détente*. Forging an agreement between the two Cold War blocks on the principles of peaceful coexistence in order to replace the constant threat of mutually assured destruction formed the background to the new principal framework regulation on State cooperation. Defining the contents of the duty to cooperate therefore inevitably caused some difficulties in the 1960s, partly because unlike other principles of the UN (non-intervention, prohibition of the use force, sovereign equality), which 'constitute negative injunctions which serve to delimit the sphere within which each State has inviolable and exclusive competence' (Houben, 1967, p. 721), this principle has always been meant to 'remove barriers'. During the preparatory discussions, major divides unfolded between the Western and Eastern blocks and the non-aligned countries, and the compromise finally included certain limits to the duty to cooperate (Houben, 1967, pp. 722–723). The compromise was necessary for accepting the final text of the

'Declaration on Principles of International Law concerning Friendly Relations and Cooperation among States in accordance with the Charter of the United Nations' (UNGA, 1970a). The final text of the principles, adopted by the General Assembly upon consensus in the Drafting Committee (UN, 1970, p. 40), reads as follows:

> States have the duty to co-operate with one another, irrespective of the differences in their political, economic and social systems, in the various spheres of international relations, in order to maintain international peace and security and to promote international economic stability and progress, the general welfare of nations and international co-operation free from discrimination based on such differences. (UNGA, 1970b, p. 102)

The wording is remarkable, since it actually refers to a *duty* to cooperate, which promises actual normative content. Yet considering that the scope and content of this "duty" was subject to some Cold War fuelled debate, it is hardly surprising that the text is repetitious and fails to meet certain normative standards (Arangio-Ruiz, 1979, p. 102).

The other paragraphs on this principle react to the prior disputes by clarifying some points raised during those discussions. First, special areas for cooperation are designated: the maintenance of international peace and security; and the promotion of universal observance of human rights. In other areas, the duty to cooperate seems much weaker, not only because the operative word was changed from 'shall' to 'should', but also because the principles of sovereign equality and non-intervention are emphasized as the apparent antonyms for cooperation in economic, social, cultural, technical and trade affairs. There is also a distinction in the duties of UN member States and others in the text, but today, with the UN's near-universal membership, this distinction has lost its political edge, and became a simple application of the *pacta tertiis* rule from the 1969 Vienna Convention.

Last, but not least, the special needs of developing States are recognized in the text as a special standard of cooperation to promote economic growth.

According to Gaetano Arango-Ruiz, who served as the rapporteur of the UN Special Committee on Principles of International Law concerning Friendly Relations and Co-operation among States (UN, 1970, p. 24), there was an understanding during the preparatory works of the Friendly Relations Declaration that the principle of 'co-operation should be a sort of procedural super-principle that would be meant to operate, in practice, as a set of as many procedural sub-principles as were the other (six) principles of Friendly Relations. [...] That the principle of co-operation should assume such an ancillary function—but the vital function, to be sure, of ensuring the effective translation of each of the other principles into reality—was proved also by the fact that to 'co-operate'—without any further qualification—does not mean much' (Arangio-Ruiz, 1979, p. 102).

However, he also sounded disappointment in the strength of this duty. The text does not specify the actual obligations of States to be carried out in order for them to observe their duty under this principle, and what remains is a simple reiteration of the message to UN Member States saying that they should cooperate. In the words of the former rapporteur: 'We are frankly unable to find anything in the formulation of this principle, but the repetition of that statement, except perhaps a certain general emphasis on economic co-operation as distinguished from peaceful settlement, peacekeeping and human rights' (Arangio-Ruiz, 1979, p. 102).

Even in this weakened form, the principle has been a counterpoint to the more concrete non-intervention principle. As non-intervention refers to domestic affairs only, it is at least a plausible interpretation that certain affairs are, by nature, not domestic any longer, despite the facts of a case falling within the exclusive jurisdiction of a state. Particularly the questions of peace and human rights are considered to be subject to the duty to cooperate, and consequently, these two aspects have arguably been removed from the field of exclusive domestic affairs.

Gaetano Arango-Ruiz also found a silver lining in his above-cited academic course on the Friendly Relations Declaration. In hindsight, we can approve his estimate that the principle paved the way for subsequent legal developments by opening an angle to include the special needs of developing countries. The related discussion between the global North and South continued since then, and gained a new dimension with the realization of the growing deterioration of the environment.

The rise of international environmental law from the 1970s gained momentum from the post-war legal developments that opened the way for a new international law which is not based on the exclusive representation of States. Multilateralism led to the emergence of new concepts expressing a demand to address the new, inherently global problems for the international community, and indeed for the entire human race. This tendency was closely linked to the realization that the ecosystem of our planet is a complex set of interdependent sub-systems, and that environmental problems, as natural phenomena, tend to bypass human-made social constructs, most prominently State borders. For this reason, environmental problems can be better addressed via a prevention-oriented cooperation than by attempting to remedy damages independently in the traditional, sovereign way. The need for this enhanced cooperation demanded new legal concepts to express that the underlying problems affect each and every member of the international community, and, more importantly, that these new global environmental problems adversely affect every human being all over the world.

Technological advances made it possible for human enterprises to boldly seek settlement in previously marginal areas. Instead of allowing for a new colonial race to gain control of these areas, a new concept of international law was born: 'common areas'. These areas 'lie outside national boundaries, are not reducible to national or private appropriation, [... and in them,] coherent and comprehensive regulation must be international' (Shelton, 2009, p. 35). The two most notable examples of such common areas are Antarctica and outer space, but the legal status of the high seas also changed somewhat in this period. The high seas and outer space were also recognized as the 'common heritage of mankind', which expressed that the right to access these areas is not reserved for a few selected technologically advanced nations, or even for states only (Shelton, 2009, p. 33), but, indeed, for all of humankind. Both of these approaches are based on the common understanding that no State can impose its particular approach to the use of the given area, and due to the strong and direct connection to the interests of mankind, States can be understood as merely the facilitators of solutions (Brunnée, 2008, p. 554).

Shortly after the positive law reflections of the above two concepts were codified, a discussion on the common concern of humanity started in the literature and a number of international instruments. Behind every common concern there is an important value of the international community (Bowling et al., 2016, pp. 5–6), but in itself the importance of such a value is insufficient for establishing a call for an enhanced cooperation. In the words of Dinah Shelton, it is also required that these issues 'inevitably transcend the boundaries of a single State and require collective action in response; no single State can resolve the problems they pose or receive all the benefits they provide' (Shelton, 2009, p. 34). In other terms: an active and meaningful cooperation on these matters is essential if any hope of tackling the problem is to be created, yet there is a certain reluctance on the part of States to undertake actual obligations to fulfil their duty to cooperate.

Such an understanding of the common concern of humankind may *prima facie* add new dimensions to the duty of States to cooperate. Multilateral 'environmental agreements [...] have come to facilitate both the pragmatic coordination of individual states' efforts to address common concerns, and the cultivation and institutionalization of normative communities' (Brunnée, 2008, p. 555).

The treaty-based regulations of common concerns cover a wide range of international environmental regulations. Unlike other legal formulas used to express commonality, like 'common heritage of mankind', treaty regulations on common concerns typically do not refer to a portion of the geosphere, but rather to a specific process in that sub-system or a designated action (Brunnée, 2008, pp. 564–565). The preamble to the 2015 Paris Agreement seemed to differ from this trend when acknowledging "that climate change is a common concern of humankind". However, the context of this reference does not introduce new considerations, but essentially reiterates the *pacta sunt servanda* principle. The preamble refers to the need to observe other obligations in the field of 'human rights, [namely] the right to health, the rights of indigenous peoples, local communities, migrants, children, persons with disabilities and people in vulnerable situations and the right to development, as well as gender equality, empowerment of women and intergenerational equity'. Considering the variety of the obligations in these areas of international law and their nature, the actual legal consequences of the preambular reference remain somewhat ambiguous. Does the obligation to cooperate regarding this common concern amount to an intersection of the State obligations in human rights law, rights of indigenous peoples and intergenerational equity, or does every single area listed in the preamble have its own régime that is applicable to it? This question cannot be reassuringly answered by means of an interpretation of the treaty itself, and will probably require an analysis of future State conduct.

From the perspective of the global sea level rise, the obligations emerging from a perceived common concern that is not yet acknowledged in a dedicated treaty may also lead to some ambiguity. First and foremost the question is whether we need a specific treaty at all. In the past years, the interconnectedness of global warming, atmospheric changes, sea level rise, and degradation of biodiversity became a generally accepted scientific fact echoed in many reports issued by numerous independent and intergovernmental agencies. However, the law of treaties does not make the interpolation of the obligations listed in a treaty into other relations beyond the scope of the treaty self-evident, even if the subject matters are as closely related as the elements of the global ecosystem. In simpler terms: the performance of the obligations prescribed by a climate change treaty will likely have beneficial side effects for the situation of rising sea levels, yet it is not evident that those obligations would be readily applicable in that other area (ILC, 2021, Guideline 9). The reason for this lies in the inherently consent-based nature of international treaty obligations. Until the second half of the twentieth century, voluntarism was the exclusive driving force behind international obligations: no such obligation could have existed if it was contrary to the actual consenting will of the state. In contemporary international law, there are at least two types of legal sources that disrupted the exclusivity of prior State consent. First, although customary international law has been the organic form of international law making for many a millennia, its codification and progressive development techniques have become more institutionalized under the auspices of the United Nations[1]; and second, peremptory norms of international law can be binding on states regardless of their consent or lack thereof.

Peremptory norms (*jus cogens*) entail obligations *erga omnes*: by their very nature, these norms represent and protect possibly the most fundamental common values of the international community (ILC, 2022a). The list of these norms is, however, short, as evidenced by the most recent non-exhaustive list offered by the ILC (2022a, Annex). When examining this list from the perspective of sea level rise, it must be noted that most of the peremptory norms relate to the situation of human beings and not the rights of States. The 2022 ILC list seems inapplicable to sea level rise, except for the most serious cases of disappearing small island nations, which possibly affect the right to self-determination of peoples. As is observed below, current trends in the international discussion seem to move towards the persistence of statehood in these cases, and therefore, it further mitigates the applicability of a *jus cogens* argument to enforcing a cooperation against sea level rise.

---

[1] See below.

A different outcome also seems possible at this time, at least if earlier understandings of peremptory norms are to be revived: the dictum of the ICJ in the Barcelona Traction case, which referred to 'the principles and rules concerning the basic rights of the human person', could be revived. New tendencies in *erga omnes partes* cases (Carli, 2023) in ICJ litigation regarding peremptory norms at least show a jurisdictional opening for the admission of such future cases; however, the evaluation of this potential future development is beyond the scope of the present chapter.

2 Obligations to Cooperate in Sources of International Law

Treaties, if widely ratified, can still be the best form to list the obligations that give meaning and content to the enhanced cooperation duty emerging from a common concern of humanity. Multilateral environmental treaties that consist of an explicit reference to a common concern tend to be widely ratified. The UN Framework Convention on Climate Change, the UN Biodiversity Convention, and the 2015 Paris Agreement are all almost universally ratified. Before the Paris Agreement, there were also other treaties with provisions on concrete forms of cooperation in certain specific matters; however, their ratification status is far from their being universal. Examples of such treaties include the 1989 Basel Convention on the Control of Transboundary Movements of Hazardous Wastes and their Disposal; the 1991 Convention on Environmental Impact Assessment in a Transboundary Context; the 1992 Helsinki Convention on the Protection and Use of Transboundary Watercourses and International Lakes; and the 1992 Helsinki Convention on the Transboundary Effects of Industrial Accidents. Typically, those treaties that provide for a duty to cooperate based upon more precisely formulated obligations are ratified by around 20–30% of states globally.

From the three aforementioned agreements, which are explicitly based upon the realization of a common concern of humankind, the Paris Agreement seems to rely on the broadest concept of this sort; however, in this text the actual duty to cooperate is still subject to limitations. In spite of the many undertakings to cooperate in various matters to be discussed below, the overall concept of the intended global response is still based on a State-driven procedure. Famously, the global efforts to hold the increase in the global average temperature at well below 2 °C above pre-industrial levels in this Agreement are centred around nationally determined contributions (Paris Agreement, 2015, Article 2) that are subject to different thresholds in an apparent representation of the common, but differentiated responsibility. Also, assistance to developing countries is encouraged (Paris Agreement, 2015, Article 7(6)), but the actual cooperation in the realization of the nationally determined contributions remains voluntary (Paris Agreement, 2015, Article 6(1)).

Cooperation seems more important in the realization of adaptation measures based upon the Cancun Adaptation Framework, but the obligation also remains somewhat generic ('should strengthen' being the operative phrase) in that regard as well. Even in this aspect the cooperation is mostly related to academic matters instead of potentially more expensive definitive actions (Paris Agreement, 2015, Article 7(7)). Similarly, the general provision on cooperation in Article 12 also refers mostly to education and access to information regarding climate change.

The operation of the duty to cooperate is most obvious with regard to the rules pertaining to the assistance to developing countries. An obligation to provide continuous and enhanced international support is part of the academic cooperation prescribed in Article 7 (Paris Agreement, 2015, Article 7(13)). Article 9 stipulates the allocation of financial resources to assist developing country parties, and Article 11 imposes obligations for cooperative action in a similarly clear formulation. Article 10 also envisages a cooperation in technology transfer, yet this article is much more carefully worded than the rules on financial assistance and capacity building.

A far more cautious approach of States seems to be present in international customary law. Customary law, by definition, evolves over a prolonged time period; therefore, it is typically, but not necessarily, less quick to respond to pressing global changes than treaty law. Sea level rise is currently on the agenda of the International Law Commission; at this time it is impossible to write about any 'findings' of the ILC as a certainty. The facilitators of the discussions have prepared "issue papers" to be used as the grounds for the discussions. The discussion on the contents of the duty to cooperate at the time of the closing of the present text still shows the conceptual differences among the members of the ILC, with no apparent consensus in sight (ILC, 2024, pp. 18–19, para. 82).

However, it is also important to give some preliminary considerations to recent and related ILC outcomes, such as the 2021 Draft Guidelines on the Protection of the Atmosphere. The connection between global warming, the degradation of the atmosphere and sea level rise is palpable from a scientific point of view, yet the content of the duty to cooperate in these fields may show some variance. Although there was some disagreement about the inclusion of a reference to common concern (Sanchez Castillo-Winckels, 2017), the final text of the Draft Guidelines on the Protection of the Atmosphere has a preambular paragraph stating, 'Considering that atmospheric pollution and atmospheric degradation are a common concern of humankind'. According to the ILC Commentary to the text, common concern 'identifies a problem that requires cooperation from the entire international community, while at the same time that its inclusion does not create, as such, rights and obligations, and, in particular, that it does not entail *erga omnes* obligations in the context of the draft guidelines'. This marks a highly cautious approach from the Commission, and it seems that there is no large impetus to pursue progressive development in the duty to cooperate. Guideline 8 supports this presumption, as it seems to encourage a limited cooperation in scientific matters without a reference to people's rights, or assistance for vulnerable groups.

The ILC work on sea level rise started in 2019 and was immediately slowed down by the COVID-19 pandemic, and the first meeting of the ILC Study Group was held only in June 2021 (ILC, 2021, p. 165). The first set of issues—and actually a precursor to the subsequent cooperation—was identified as those related to changes in maritime borders. In that regard it is noteworthy that Yacouba Cissé, the vice-president of the Study Group, argued for the fixed borders concept being more beneficial for the African states, partly because of an obligation to cooperate based upon UNCLOS Articles 83(3) and 74(3) (ILC, 2021, p. 168). These two articles call for States to make 'every effort to enter into provisional arrangements of a practical nature' regarding the delimitation of the continental shelf and the exclusive economic zones, respectively, if no agreement can be reached by the States concerned.

The first issue paper (ILC, 2020) examined whether the principle of international cooperation may be applied to help States cope with the adverse effects of sea level rise on their populations (ILC, 2020, p. 20, para. 50). The Commission identified a number of regional treaties that already include relevant rules in that regard (ILC, 2020, pp. 50–52), and the Israeli delegation specifically drew attention to 'the delicate balance achieved by existing maritime border agreements, which meaningfully and significantly contributed to increased regional and international stability and positive cooperation' (ILC, 2020, p. 20).

The second issue paper consists of far more statements on cooperation regarding sea level rise. The Maldives has persistently upheld a position that 'sea level rise was not a distant theoretical concern. Low-lying coastal and small island States, such as itself, were particularly vulnerable to the effects of sea level rise. As they could not afford to mitigate the effects of sea level rise on their own, the cooperation of the international community was essential to ensure adequate, predictable and accessible assistance to those States'. Two States echoed the importance of cooperation in this respect.[2]

The commonality of the issue is not the only important feature of sea level rise; scientific predictions also forecast that its local impact will always be different in each case, depending on a number of geographical factors. Therefore, it just seems reasonable that local or regional responses may effectively supplement global cooperation efforts. It is reflected in the second issue paper, namely in its acknowledgement of the need for regional cooperation (ILC, 2022b, p. 15, para. 49(g)).

While sea level rise is a global issue, it is obviously the most pressing for small island nations facing the possibility of disappearance and a potential loss of statehood. The question of maritime borders is therefore adjacent to statehood. There seems to be an agreement that statehood is persistent, and even after a loss of territory, elements of the state's international juridical personality survive. There are historical examples to prove this that may also serve as models for regulatory solutions. International cooperation is vital in this regard as well (ILC, 2022b p. 29, para. 112).

The loss of statehood is not just an abstract legal problem. As territorial States are under an obligation to protect the people living in their territory, a loss of territory affects first and foremost the people living in that area. Therefore, the principles regarding the protection of persons affected by sea level rise must also be developed. Many States have submitted statements in that regard, and the second issue paper highlights the one from the Solomon Islands, which forwards an interpretation of cooperation in the context of human rights (ILC, 2022b,

---

[2] Cuba and Argentina.

p. 106, para. 436). Connecting sea level rise to human rights can significantly contribute to the enhancement of the duty of States to cooperate (Bellinkx et al., 2022, p. 74).

More recently, the report of the Study Group identified some elements that serve to fulfil the duty to cooperate. Technical assistance was elaborated in this regard: 'the provision of technical or logistical assistance, qualified human resources or financial assistance to States particularly affected by the phenomenon that lacked sufficient capacity of their own was considered essential' (ILC, 2024, p. 10, para. 43).

3   Conclusions

Sea level rise is putting the existing system of international cooperation to a previously unseen stress-test. The frequency and intensity of climate-related disasters are on the rise; therefore, the need to stop global warming, and protect the atmosphere and biodiversity, is probably obvious; however, the identification of the underlying values may not be possible when looking at only strictly legal sources of international law (Brunnée, 2008, p. 554). Other approaches that are value-based and less individualistic may assist such efforts, like the African ubuntu approach, or Catholic social teachings, particularly the Laudato Si' encyclical letter of Pope Francis, where he appeals 'for a new dialogue about how we are shaping the future of our planet. We need a conversation which includes everyone, since the environmental challenge we are undergoing, and its human roots, concern and affect us all'.

An identification of the missing values would surely assist the interpretation of the existing rules on cooperation, but the lack of a joint understanding should not necessarily hamper the observation of the duty of States to cooperate. The cooperation framework exists, and we do not need to wait for the creation of new institutions or principles. Instead, a major improvement can be reached by improving the actual content of the existing system of cooperation, and clarifying the actual obligation deriving from the generic 'duty to cooperate'.

There are several good examples of that approach, e.g. those in the 2015 Paris Agreement. As a further promising development in the collective action against sea level rise, an agreement seems to be forming in the ILC 'that it [has been] essential for the Commission to focus on the duty of cooperation, whether as a general principle of law or as a rule of customary international law' (ILC, 2024, pp. 10–11, para. 47). This cooperation is not just among the affected nations, but it is also a duty of other members of the international community (ILC, 2024, p. 14, para. 62). The difficulty of the discussions on the merits lies in the necessary combination of a needs-based and a rights-based approach (ILC, 2024, p. 15, para. 66). The examples of already existing regional cooperations of this sort include the Latin American model of humanitarian visas, although 'it was noted that matters of admission of foreign nationals fell under the purview of domestic authorities' (ILC, 2024, p. 18, para. 80).

Further improvements towards a mandatory cooperation can be achieved by the emergence of a new peremptory norm of international law; however, there are virtually no traces of such tendencies for the time being. A more realistic approach can be the identification of the connection between sea level rise and human rights. This could help to point out that no State can hope to adapt to the rising sea levels or stop global warming on its own. Therefore, the existing duty to cooperate should be understood as covering positive action: this is a necessary, but, in itself, not necessarily sufficient requirement of adaptation.

References

Arangio-Ruiz, G. (1979). *The United Nations Declaration on Friendly Relations and the System of the Sources of International Law*. Brill.

Bellinkx, V., Casalin, D., Türkelli, G. E., Scholtz, W., & Vandenhole, W. (2022). Addressing Climate Change Through International Human Rights Law: From (Extra)Territoriality to Common Concern of Humankind. *Transnational Environmental Law, 11*(1), 69–93. https://doi.org/10.1017/S204710252100011X

Bowling, C., Pierson, E., & Ratté, S. (2016). *The Common Concern of Humankind: A Potential Framework for a New International Legally Binding Instrument on the Conservation and Sustainable Use of Marine Biological Diversity in the High Seas (White Paper)*. United Nations. https://www.un.org/depts/los/biodiversity/prepcom_files/BowlingPiersonandRatte_Common_Concern.pdf

Brunnée, J. (2008). Common Areas, Common Heritage, and Common Concern. In L. Rajamani & J. Peel (Eds.), *The Oxford Handbook of International Environmental Law* (pp. 550–573). Oxford University Press.

Carli, E. (2023, December 29). Community Interests Above All: The Ongoing Procedural Effects of *Erga Omnes Partes* Obligations Before the International

Court of Justice. *EJIL: Talk!* https://www.ejiltalk.org/community-interests-above-all-the-ongoing-procedural-effects-of-erga-omnes-partes-obligations-before-the-international-court-of-justice/

De Feyter, K. (2017). *Globalization and Common Responsibilities of States.* Routledge.

Dronkers, J., Gilbert, J. T. E., Butler, L. W., Carey, J. J., Campbell, J., James, E., McKenzie, C., Misdorp, R., Quin, N., Ries, K. L., Schroder, P. C., Spradley, J. R., Titus, J. G., Vallianos, L., & Von Dadelszen, J. (1990, November). *Strategies for Adaptation to Sea Level Rise: Report of the Intergovernmental Panel on Climate Change (IPCC) Coastal Zone Management Subgroup.* http://papers.risingsea.net/federal_reports/IPCC-1990-adaption-to-sea-level-rise.pdf

Houben, P.-H. (1967). Principles of International Law Concerning Friendly Relations and Co-operation Among States. *American Journal of International Law, 61*(3), 703–736. https://doi.org/10.2307/2197463

Intergovernmental Panel on Climate Change (IPCC). (1990). *Climate Change: The IPCC Scientific Assessment.* https://www.ipcc.ch/site/assets/uploads/2018/03/ipcc_far_wg_I_full_report.pdf

Intergovernmental Panel on Climate Change (IPCC). (2019). *Special Report on the Ocean and Cryosphere in a Changing Climate.* https://www.ipcc.ch/srocc/

International Law Commission (ILC). (2020). *Sea Level Rise in Relation to International Law: First Issues Paper by Bogdan Aurescu and Nilüfer Oral, Co-Chairs of the Study Group on Sea Level Rise: Seventy-Second Session* (A/CN.4/740). United Nations. https://documents.un.org/doc/undoc/gen/n20/053/91/pdf/n2005391.pdf

International Law Commission (ILC). (2021). *Report of the International Law Commission: Seventy-Second Session (26 April–4 June and 5 July–6 August 2021)* (A/76/10). United Nations. https://documents.un.org/doc/undoc/gen/g21/224/41/pdf/g2122441.pdf

International Law Commission (ILC). (2022a). *Draft Conclusions on Identification and Legal Consequences of Peremptory Norms of General International Law (Jus Cogens).* United Nations. https://legal.un.org/ilc/texts/instruments/english/draft_articles/1_14_2022.pdf

International Law Commission (ILC). (2022b). *Sea Level Rise in Relation to International Law: Second Issues Paper by Patrícia Galvão Teles and Juan José Ruda Santolaria, Co-Chairs of the Study Group on Sea Level Rise: Seventy-Third Session* (A/CN.4/752). United Nations. https://documents.un.org/doc/undoc/gen/n22/276/29/pdf/n2227629.pdf

International Law Commission (ILC). (2024). *Study Group on Sea Level Rise in Relation to International Law: Seventy-Fifth Session* (A/CN.4/L.1002). United Nations.

Kaul, I., Grunberg, I., & Stern, M. A. (1999). Defining Global Public Goods. In *Global Public Goods: International Cooperation in the 21st Century* (pp. 3–21). Oxford University Press.

Paris Agreement. (2015, December 12). United Nations Framework Convention on Climate Change.

Sanchez Castillo-Winckels, N. (2017). Why Common Concern of Humankind Should Return to the Work of the International Law Commission on the Atmosphere. *Georgetown Environmental Law Review, 29*(1), 131–152.

Service de l'information des Nations Unies. (1960). *L'ONU pour tous - La structure et les activitiés de l'organisation des nations unies et des institutions qui lui sont rattachées au cours des années 1945 à 1958*. United Nations.

Shelton, D. (2009). Common Concern of Humanity. *Iustum Aequum Salutare, 5*(1), 33–40. https://ias.jak.ppke.hu/20091sz/05.pdf

Tung, W. L. (1968). *International Law in an Organizing World*. Thomas Y. Crowell Co.

United Nations (UN). (1970). *Report of the Special Committee on Principles of International Law Concerning Friendly Relations and Co-Operation Among States* (A/8018).

United Nations General Assembly (UNGA). (1970a, October 24). *Declaration on Principles of International Law Concerning Friendly Relations and Cooperation Among States in Accordance with the Charter of the United Nations* (A/RES/2625(XXV)). https://documents.un.org/doc/resolution/gen/nr0/348/90/pdf/nr034890.pdf

United Nations General Assembly (UNGA). (1970b, October 24). *Resolutions Adopted on the Reports of the Sixth Committee: Twenty-Fifth Session* (A/RES/2625(XXV)). https://documents.un.org/doc/resolution/gen/nr0/348/90/pdf/nr034890.pdf

Open Access This chapter is licensed under the terms of the Creative Commons Attribution 4.0 International License (http://creativecommons.org/licenses/by/4.0/), which permits use, sharing, adaptation, distribution and reproduction in any medium or format, as long as you give appropriate credit to the original author(s) and the source, provide a link to the Creative Commons license and indicate if changes were made.

The images or other third party material in this chapter are included in the chapter's Creative Commons license, unless indicated otherwise in a credit line to the material. If material is not included in the chapter's Creative Commons license and your intended use is not permitted by statutory regulation or exceeds the permitted use, you will need to obtain permission directly from the copyright holder.

CHAPTER 10

The Advisory Opinion of the International Tribunal for the Law of the Sea

Curtis F. J. Doebbler

> Rising seas are amplifying the frequency and severity of storm surges and coastal flooding. These floods swamp coastal communities. Ruin fisheries. Damage crops. Contaminate fresh water.
> Antonio Guterres, United Nations Secretary-General. (Needham, 2024)

1 THE BACKGROUND TO THE ADVISORY OPINION

The International Tribunal for the Law of the Sea ('ITLOS' or 'Tribunal') is created by Part VI of the United Nations Convention on the Law of Sea ('UNCLOS'). It deals with problems related to the sea, including the marine environment. These problems are closely related to more general environmental issues, especially climate change. And climate change is related to sea rise. While the ITLOS deals with marine pollution in the Advisory Opinion discussed in this contribution, many of the findings are

C. F. J. Doebbler (✉)
Department of Law, University of Makeni, Makeni, Sierra Leone
e-mail: cdoebbler@gmail.com; cdoebbler@unimak.edu.sl

© The Author(s) 2026
E. Fornalé (ed.), *Sea Level Rise*,
https://doi.org/10.1007/978-3-031-89171-7_10

of relevance to sea rise. The connection between sea level rise has been well documented by the IPCC (2022).

In accordance with Article 21 of the Statute of the Tribunal, it has jurisdiction over 'all applications submitted to it in accordance with this Convention [on the Law of the Sea] and all matters specifically provided for in any other agreement which confers jurisdiction on the Tribunal' (Statute of the ITLOS).

This Request did not materialize out of the blue. The relationship between climate change and sea rise has been long recognized (Palm & Bolsen, 2020 and IPPC, 2023 at 98 and 106). Our oceans are sinks that absorb greenhouse gases and thereby protect the planet against some of the worse adverse effects of climate change (IPCC, 2022). The Intergovernmental Panel on Climate Change ('IPCC') had noted that the importance of the oceans and ice caps stating that

> [a]ll people on Earth depend directly or indirectly on the ocean and cryosphere. The global ocean covers 71% of the Earth surface and contains about 97% of the Earth's water ... within the climate system, [oceans provide] ... the uptake and redistribution of natural and anthropogenic carbon dioxide (CO2) and heat, as well as ecosystem support, services provided to people ... including food and water supply, renewable energy, and benefits for health and well-being, cultural values, tourism, trade, and transport. (IPCC, 2022, p. 5)

Furthermore, the 'State of the ocean and cryosphere interacts with each aspect of sustainability reflected in the United Nations Sustainable Development Goals (SDGs)' (*id*.). According to the Food and Agricultural Organization ('FAO'), the 'climate interconnections between atmospheric pollution from GHG emissions and marine waters' were 'well-documented' (FAO, 2023, p. 6, n. 30). Thus, failure to address the deleterious consequences of climate change for our oceans affects our human development more generally. And these effects of climate change do not harm all countries equally as Sierra Leone noted in its Written Statement to the Tribunal stating that '[d]eveloping countries are already ten times more likely to be affected by a climate disaster, as compared to developed countries' (Sierra Leone, 2023, p. 4, para. 4, citing Notre Dame Global Adaptation Initiative). The Commission of Small Island States in their Submission to the Tribunal explained how the impact of climate change on the oceans will adversely effect fishing, coral reefs, ocean acidification, and sea coasts (Commission of Small Island States on Climate Change and International Law, 2023). This will cause, among other adverse effects, loss of employment and food insecurity from lost fish stocks and displacement of persons as the oceans encroach on low-lying coasts.

On 25 September 2024, during the High-Level Segment of the General Assembly, the United Nations held a High-level plenary meeting on addressing the existential threats posed by sea level rise. The meeting was authorized by UNGA Resolution 78/544 of 16 January 2024 and the modalities defined in UNGA Resolution 78/319 of 1 August 2024. As part of this meeting, a panel was held on 'Sea level rise and its legal dimensions' (the General Assembly was only one of several international entities concerned with sea level rise). On 25 September 2024, for example, for the first time the UN General Assembly addressed sea level rise as part of a High-Level Meeting during the High-Level Segment of the 79th Session of the UN General Assembly. At the meeting, Greek Prime Minister Kyriakos Mitsotakis recognized that '[g]lobal sea levels are undoubtedly rising because of anthropogenic global warming' (Office of the Greek Prime Minister).

The International Law Commission is also considering the legal consequences of sea rise (UN GAOR, 2018, para. 369), which was put on its agenda in 2018. Other international bodies that are considering sea level rise caused by climate change, include, among others, the Intergovernmental Panel on Climate Change, the Conference of the Parties of the United Nations Framework Convention on Climate Change, the Intergovernmental Oceanographic Commission of the United Nations Educational, Scientific and Cultural Organization, the International Atomic Energy Agency and the United Nations Environment Programme, the International Partnership for Blue Carbon, the Blue Carbon Initiative, the Global Decade for Blue Carbon, the Conference of the Parties of the Convention on the Conservation of Migratory Species of Wild Animals, Joint Group of Experts on the Scientific Aspects of Marine Pollution, the United Nations Development Programme, the United Nations Human Settlements Programme (UN-Habitat), and the United Nations Conference on Trade and Development.

The UN Secretary-General notes, '[r]ising seas are a crisis entirely of humanity's making. A crisis that will soon swell to an almost unimaginable scale, with no lifeboat to take us back to safety' (Needham, 2024).

Despite these obvious concerns, the legal obligations of States in relation to the oceans and climate change had previously received limited attention from international Tribunals. This ITLOS Advisory Opinion not only clarified those obligations, but was the forerunner of the Advisory Opinions provided by the Inter-American Court of Human Rights (IACtHR Advisory Opinion 2025), which elaborated human rights obligations, and the International Court of Justice's Advisory Opinion (ICJ Advisory Opinion, 2025), elaborating general legal obligations, the principles of States responsibility, and reiterating the ITLOS holdings in relation to climate change and sea rise. The ICJ notes that '[a]lthough the [ICJ] is not obliged, in the exercise of its judicial functions, to model its own interpretation of UNCLOS on that of ITLOS, it considers that, in so far as it is called upon to interpret the Convention, it should ascribe great weight to the interpretation adopted by the Tribunal' (ICJ Advisory Opinion, 2025, p. 101, para. 338).

2 The Request for an Advisory Opinion

The submission of the Questions for the Advisory Opinion was done on 12 December 2022, when the intergovernmental organization called the Commission of Small Island States on Climate Change and International Law (COSIS) submitted a request for an Advisory Opinion to the Registry of the Tribunal. The Commission had been created on 31 October 2021 by the Caribbean and Pacific States of Antigua and Barbuda, Tuvalu, Palau, Niue, Vanuatu, and Saint Lucia.Later Saint Vincent and the Grenadines, Saint Christopher (Saint Kitts) and Nevis, and the Bahamas acceded to the Commission constitutive treaty (Commission of Small Island States Decision, 2022). All of these States are parties to the United Nations Convention on the Law of the Sea. The submission of a request for an Advisory Opinion to the Tribunal was one of the primary reasons why the Commission had been created (Freestone, et al., 2022).

On 18 June 2022, the Sub-Committee on Protection and Preservation of the Marine Environment of the COSIS approved the request for an Advisory Opinion from ITLOS consistent with Article 2(2) of the Agreement (COSIS, 2022). The question asked was stated as follows:

What are the specific obligations of State Parties to the United Nations Convention on the Law of the Sea ('UNCLOS'), including under Part XII:
(a) to prevent, reduce, and control pollution of the marine environment in relation to the deleterious effects that result or are likely to result from climate change, including through ocean warming and sea level rise, and ocean acidification, which are caused by anthropogenic greenhouse gas emissions into the atmosphere?
(b) to protect and preserve the marine environment in relation to climate change impacts, including ocean warming and sea level rise, and ocean acidification? (See para. 3 of the ITLOS Advisory Opinion).

Thirty-one States, eight intergovernmental organizations submitted written statements to the Tribunal by the 15 February 2023 deadline. Three States made submissions after the deadline set by the Tribunal, but the Tribunal accepted them for consideration albeit listing them on their website as having been submitted late. Of the States about half were developing countries Pursuant to Article 138, paragraph 3, and Article 133, paragraph 3, of the Rules of the Tribunal, nine intergovernmental organizations were invited and submitted written statements. In addition, ten non-State actors made submissions as allowed by the Rules of the Tribunal.

3 The Submission of States

As an organ of an intergovernmental organization the Tribunal must take into account both written and oral account submissions made by States and intergovernmental organizations. As this is an intergovernmental process, the submission by non-State actors does not have to be taken into account and only States participate in the oral proceedings that took place from 11 to 25 September 2023.

Perhaps the most striking aspect of State representations is that few States argued that the Tribunal should refrain from giving the ITLOS Advisory Opinion. China was a significant exception in arguing that the word 'matters' in Article 21 of the Tribunal's Statute does not include Advisory Opinions (China, 2023, paras. 10–17) and therefore the Tribunal has no jurisdiction to respond to the Request. Interestingly, China did not object to the International Court of Justice's jurisdiction to give its Advisory Opinion indicating that the ICJ could find obligations existed under the UNCLOS (China, ICJ, 2024). China and virtually all

States do acknowledge that climate change is impacting the oceans in a manner that causes significant harm. Nevertheless, there was a clear divide between States (for example, the United Kingdom and Australia) who argued for an opinion that was limited in scope and those (for example, Sierra Leone and Egypt) who argue for recourse to a broad realm of law or for extensive obligations to be drawn from the applicable law.

As has been the case in the annual climate negotiations that have been taking place since the mid-1990s there was a division between developed States and developing States. Developed States, which are Annex I States in terms of the UNFCCC, generally argued that if they met their obligations under the UNFCCC and the Paris Agreement they were taking sufficient action to satisfy any obligations they might have under the UNCLOS (Written Statement of Australia, Written Statement of China) or that the jurisdiction of the Tribunal was significantly limited and perhaps should not be exercised at all (China, 2023; France, 2023; Guatemala, 2023; United Kingdom, 2023). The developing States generally argued that the obligations under the UNCLOS co-existed with those under the instruments of international climate change law, but were not identical. Therefore, according to most developing States, all States had even more extensive obligations under the UNCLOS to ensure action to address the adverse effects of climate change for sea rise and the marine environment.

One of the major actors in the global climate change forums, the United States did not participate as it has not ratified the UNCLOS and does not recognize the jurisdiction of the Tribunal.

In addition to COSIS, eight other intergovernmental organizations submitted Written Statements. The African Union, for example, provided an almost two-hundred-page Written Statement that argued that the Tribunal has jurisdiction to grant Advisory Opinions under Article 21 of its Statute and should do so because the COSIS met the requirements of that article (African Union, 2023).

Finally, ten non-State actors made submissions which were not part of the official records of the case. Although the Tribunal does not have to take these into account it may do so. It does appear that the Advisory Opinion was informed by the representations of the non-State actors, especially as concerns the relevance of international human rights law. The Inter-American Court of Human Rights, in its Advisory Opinion, of course focused on human rights obligations (IACtHR Advisory Opinion, 2025).

4 The Submission of Non-State Actors

The Tribunal may also have recourse to the views expressed by non-State actors. Unlike the views of States, however, the Tribunal may also ignore the views of non-State actors.

Three UN Special Rapporteurs on Climate Change, Environment, and Toxic Wastes submitted an 'Amicus Brief' (UNSR's Amicus) arguing that UNCLOS must be interpreted consistent with international human rights treaties (UN Special Rapporteurs Amicus Brief, 2023, pp. 5–8), that climate change-related pollution constitutes pollution of the marine environment (UN Special Rapporteurs Amicus Brief, 2023, pp. 8–9), and that climate change-related pollution threatens human rights (UN Special Rapporteurs Amicus Brief, pp. 9–20). The Special Rapporteurs went on to explain the consequences of these three observations for the interpretation and application of the UNCLOS. Notable among the observations was the conclusion that governments have '[t]he obligation to respect, protect and fulfil this right to science requires that governments respond to the climate crisis using the best available science' (UN Special Rapporteurs, p. 20, para. 66). Although the Special Rapporteurs report to the UN Human Rights Council, they are independent insofar as their reports express their independent expert views. An Amicus Brief by CIEL and Greenpeace also discussed the parallel application of international human rights law extensively (CIEL & Greenpeace Amicus Brief, pp. 13–40). In fact, it was non-governmental organizations (NGOs) that highlighted human rights most succinctly. In this regard, it is notable that while 16 non-governmental organizations or coalitions of non-governmental organizations (NGOs) submitted Amicus Briefs to the ITLOS, 57 NGOs made written submissions and 38 NGOs made oral submissions to the IACtHR for its Advisory Opinion which was focused on human rights.

5 Description and Discussion of the Advisory Opinion

The Tribunal's Advisory Opinion interpreted the obligations of States under UNCLOS in relation to climate change. It focused on UNCLOS articles 192 and 194 as those are the articles pertinent to the questions that had been asked by COSIS. The Tribunal interpreted Article 192 to include the obligation to protect and preserve the marine environment. The Tribunal interpreted Article 194 to include obligations to prevent reduce and control marine pollution. Before considering either of these articles the Tribunal had to determine if it had jurisdiction to give the Advisory Opinion, whether even if it had jurisdiction there were still compelling reasons for it to refuse to give an Advisory Opinion, and the preliminary issue of whether the anthropogenic greenhouse gas emissions (GHGs) constituted pollution of the marine environment.

The Tribunal's Advisory Opinion is divided into nine parts. Generally, these nine parts address two general issues. First, the context in which the decision is given and the sources of law to be applied (Parts I–VI, paras. 1–153), and second, the questions themselves that the Tribunal has been asked (Parts VII–IX). What follows in this part of this contribution is a description of Advisory Opinion with some intermittent comments.

Part I of the ITLOS Advisory Opinion is merely an introduction that states the questions that the Tribunal was asked and outlines the procedure that has been followed (paras. 1–44). This has already been discussed above. Below the other parts of the Advisory Opinion are described and commented upon. Where pertinent, the Tribunal's Advisory Opinion is compared to the Advisory Opinions of the ICJ and IACtHR.

5.1 *Background*

After describing the procedure, somewhat unusually, before addressing its own jurisdiction, the Tribunal describes the background against which the Request has been made. It does so in two parts, A and B. In part A it addresses the science and then the law concerning climate change.

In paragraphs 46–66, the Tribunal reiterates the importance of science to the question of legal obligations under the UNCLOS and describes the harmful effects of climate change on the oceans. It also notes that the UN General Assembly in its resolution 76/296 of 25 July 2022 (UN Doc. A/RES/76/296) endorsed the Declaration adopted by the 2022 United Nations Ocean Conference and expressed that it was 'deeply alarmed by the adverse effects of climate change on the ocean and marine life' (para. 47). It is notable that in these paragraphs the work of the Intergovernmental Panel on Climate Change (IPCC) is relied upon. The IPCC itself does not actually do research. Instead, the IPCC reviews hundreds of already peer-reviewed research articles and reports, and extracts the most reliable evidence for its reports.

5.2 The Best Available Science

Considering the crucial question of which science to rely upon, the Tribunal notes that most States have admitted that the reports of the IPCC contain the best available science concerning climate change (para. 51). The Tribunal has already noted (para. 48) several relevant reports of the IPCC including its 2019 *Special Report on Oceans and the Cryosphere in a Changing Climate* (IPCC Oceans Report, 2022). It then goes on to describe the IPCC's findings concerning the harmful effects of climate change on oceans in some detail (paras. 52–65). The Tribunal notes the IPCC was established by States acting through the cooperation of a United Nations specialized agency and a United Nations programme and currently has 193 state members (para. 47). The Tribunal also notes the IPCC's conclusions about the detrimental effect of climate change on our oceans (paras. 57–58). The Tribunal notes how climate change creates sea level rises (para. 59), as well as the harmful effects of acidification (paras. 61–62) as well as other climate-related risks (para. 63). The Tribunal points out that the IPCC has indicated that the risks to humans can be mitigated by maintaining global temperature rises to under 1.5 °C (paras. 63–65) and that failure to achieve this goal will be a 'threat to human well-being and planetary health' (para. 66, quoting IPCC, 2023, p. 89).

In paragraph 66—unusually placed at the very end of Section III.A, entitled 'Scientific aspects', and which mainly discusses the best available science—the Tribunal concludes '[t]hat climate change represents an existential threat and raises human rights concerns' (para. 66). This conclusion is reiterated in the IACtHR Advisory Opinion, which found obligations for Member States of the Organization of American States to ensure a healthy environment for present and future generations as well as the rights to life, personal integrity, health, private and family life, property and housing, freedom of residence and movement, water and food, work and social security, culture and education (IACtHR Advisory Opinion, 2025). The ICJ Advisory Opinion found that the legal obligations in human rights treaties and customary international law more broadly were relevant to States' responsibility for the adverse effects of climate change (ICJ Advisory Opinion, 2025, pp. 51–52, paras. 143–145).

5.3 Sources of Law

The Tribunal describes the instruments of international climate change law (paras. 67–82). It notes first the United Nations Framework Convention on Climate Change (paras. 67–69) and highlighted this treaty's expression of the principles of common but differentiated responsibilities, the precautionary principle, the principle of sustainable development, and the principle of cooperation (para. 69).

The Tribunal goes on to find relevant the Kyoto Protocol (paras. 70–71), the Paris Agreement (paras. 72–76), and numerous decisions of the conferences of the parties to these treaties (para. 77). The Tribunal also notes the relevance of the instruments adopted under the auspices of the International Maritime Organization (paras. 78–80), the International Civil Aviation Organization (para. 81), and the Montreal Protocol on protection of the ozone layer (para. 82). In comparison to the ICJ, which found a broad corpus of international law to be applicable to determining States' legal obligations in respect of climate change (ICJ Advisory Opinion, 2025, pp. 45–58, paras. 113–173), the Tribunal focused only on the UNCLOS, which it was required to do, and international climate change instruments (paras. 123–137, discussed in more detail below in Sect. 5.5). The IACtHR, which was restrained by its mandate to address the legal obligations of States members of the Organization of American States (OAS), limited its review to the international human rights law of the OAS while finding the 'Legal Frameworks on Climate and Environment', as well as several other legal regimes to be relevant to interpretation of the Inter-American human rights obligations (IACtHR Advisory Opinion, pp. 44–65, paras. 120–171).

5.4 Jurisdiction

Only after this lengthy recital of climate change science and law does the Tribunal turn to the consideration of its own jurisdiction. This is somewhat unusual in part because several States had made Written and Oral Statements that challenged the Tribunal's jurisdiction. The Tribunal, however, dispenses with these challenges by reiterating its finding in its SRFC Advisory Opinion. In that case, the Tribunal opined that Article 21 of its Statute provides the competence to respond to requests for an Advisory Opinion on "all matters specifically provided for in any other agreement which confers jurisdiction on the Tribunal" (Art. 21

of the Statute; ITLOS, 2015, p. 21, para. 54). The Tribunal found the COSIS Agreement to be an 'other agreement' that 'confers jurisdiction on the Tribunal' within the meaning of Article 21 of the Statute (para 88). Although some States had tried to distinguish the SRFC Advisory Opinion by claiming that the current request did not relate to the purpose of the COSIS Agreement, the Tribunal rejected this line of reasoning (paras. 105–109) The Tribunal also noted that it was not necessary for the request for an advisory opinion to be related to the other agreement (paras. 106–108). And on this basis the Tribunal concluded that it had jurisdiction (para. 109).

Considering its discretion as concerns replying to the Request the Tribunal considered different concerns but rejected them all in favour of the exercise of its discretion to give an Advisory Opinion (paras. 110–122). These concerns included, for example, the fact that because not all States were party to the COSIS Agreement, it should not give an Advisory Opinion. The Tribunal rejects this reasoning finding that 'the fact that the advisory opinion has been requested by some States Parties to the Convention, and not by all, cannot be a reason for the Tribunal to refrain from giving the opinion' (para. 113). The Tribunal also concludes that there is no pending legal dispute that might make providing an Advisory Opinion inappropriate (para. 116), that the fact that the request came from COSIS did not distract from its importance to the interpretation of UNCLOS (paras. 117–118), and the general nature of the Request was not a concern as 'the questions raised by the Commission are clear and specific enough to enable it to give an advisory opinion' (para. 120). The ICJ and the IACtHR also rejected challenges to their jurisdiction and requests for them to exercise their discretion to refuse to give an advisory opinion.

5.5 The Applicable Law

The Tribunal then defines what international law it should apply by referring to Article 138, paragraph 3, of its Rules of Procedure that itself refers to articles 130 to 137 (para. 123). These articles note that in giving an Advisory Opinion the Seabed Disputes Chamber shall apply the Convention as well as 'other rules of international law not incompatible with this Convention' (para. 126, referring to Article 293 of the Convention).

It is noteworthy that while discussing the science in Section II.A, the Tribunal refers to 'human rights concerns' (para. 66), but it does not draw on its earlier reference in this section to include human rights treaties among the applicable law. At the same time, the Tribunal does not exclude the application of human rights treaties as relevant to its interpretation and several human rights bodies have already made it clear that they do apply to restrain and guide State action on climate change. Human rights bodies applying international human rights bodies to the State responsibilities to adequately address the adverse effects of climate change on human rights include the UN Human Rights Council (for example, Human Rights Council Res. 10/4, HRC 41st meeting (25 March 2009), the UN Human Rights Committee (*Daniel Billy and others v Australia (Torres Strait Islanders Petition)*, UN Doc. CCPR/C/135/D/3624/2019 (2019)), the European Court of Human Rights (*Verein KlimaSeniorinnen Schweiz and Others v. Switzerland*, ECtHR Appl. No. 53600/20 (9 April 2024)), and the African Commission on Human and Peoples' Rights ('ACHPR'), (ACHPR Res. A/HRC/28/27 (2008)). The entire IACtHR Advisory Opinion, 2025, and the significant sections of the ICJ Advisory Opinion, 2025, (pp. 108–116, para. 369–404) both indicated the important relevance of human rights as legal obligations that must be respected by States both for ensuring the protection of human rights and for ensuring adequate climate action. The ICJ emphasized human rights in the two international covenants (ICCPR, 1966 and ICESCR, 1996) and under customary international law (see ICJ Advisory Opinion, 2025, pp. 108–116, paras. 369–404, especially p. 108, para. 369 and p. 115, para. 402). It focused on three themes. First, that adverse effects of climate change impact the enjoyment of human rights (pp. 109–112, paras. 372–386), including 'the impact on the health and livelihoods of individuals through events such as sea level rise' (p. 109, para. 376). The Court found that among the effected rights may be the right to life (pp. 109–110, paras. 377–378), the right to health (p. 110, para. 379), the right to an adequate standard of living (p. 111, para. 380), the right to privacy, family and home (p. 111, para. 381), the rights of women, children and indigenous peoples (p. 111, para. 382), and the right to water (pp. 111–112, para. 384). Second, the ICJ surveys several sources of State practice and *opinio juris* (pp. 112–113, paras. 387–393) to find that there is 'under international law, the human right to a clean, healthy and sustainable environment' that 'is essential for the enjoyment of other human rights' (p. 113, para. 393). While the right to a clean, healthy and sustainable environment has been widely accepted, as the ICJ survey indicates, it was not without controversy for a small minority of States. For example, eight States abstained from the adoption of UN General Assembly Resolution 76/300 of 28 July 2022 recognizing this right, including China and Russia, and the United States only voted for the resolution voicing serious reservations (Spijkers, 2025).

5.6 The Interpretation of the Law

In Part V of its Advisory Opinion the Tribunal considers 'the interpretation of the Convention and the relationship between the Convention and other relevant rules of international law' (para. 128). Relying on the rules of treaty interpretation (paras. 128–129), especially those found in Articles 31–33 of the Vienna Convention on the Law of Treaties, 1969, the Tribunal concludes that it should have recourse to external rules of international law, especially international climate change law, to interpret the UNCLOS (paras. 136–137). The Tribunal notes that the provisions of the Convention refer to external rules (para. 131) and that the International Court of Justice has referred to the broader context of a treaty, especially 'any relevant rules of international law applicable in the relations between the parties' as pertinent to interpreting a treaty (para. 135). The Tribunal then concludes that the "extensive treaty regime addressing climate change" is applicable (para. 137). Like the Tribunal, the ICJ (ICJ Advisory Opinion, 2025, p. 63, para. 190) and the IACtHR (IACtHR Advisory Opinion, p. 13, para. 35) both drew on the Vienna Convention on the Law of Treaties, 1969, for the rules regarding the interpretation of treaty obligations.

5.7 The Two Questions and Their Relationship

In Part VI of its Advisory Opinion the Tribunal considers the scope of the Request and the relationship between the two questions. Although no State had seriously challenged the Tribunal's jurisdiction to provide an Advisory Opinion, several States cautioned about too broad an opinion. The Tribunal concludes that the Request concerns 'the specific obligations of States Parties under the Convention' and in order to answer the two questions it has been asked 'the Tribunal will have to interpret the Convention and, in doing so, also take into account external rules, as appropriate' (para. 142). The Tribunal also determines that it has been requested 'to provide guidance as to the specific obligations of the States Parties under Part XII as well as other relevant provisions of the Convention' (para. 144). In this part, the Tribunal also opines that it has only been asked to interpret primary obligations (para. 148), which are the specific international legal obligations that require States to act in a certain way or refrain for acting in a certain way. The Tribunal somewhat confusingly refers to 'responsibility' and 'liability' as opposed to 'primary obligations' when the former two terms are not usually considered the opposite of primary obligations. The Tribunal also notes that the Request focuses on 'protection and preservation of the marine environment' and not 'sea rise' (para. 150). This is important because some States had argued that the mention of sea level rise in the Request, 'invited the Tribunal to deal with the issue of the relationship between sea level rise and existing maritime claims or entitlements' (para. 149). Doing so would have involved the Tribunal in the complex issues of determining 'base points, baselines, claims, rights or entitlements to maritime zones … or maritime boundaries, and the corresponding obligations in the context of "physical changes connected to climate change-related sea-level rise"' (para. 150). The Tribunal's restrictive interpretation of the Request leaves this thorny issue for the ICJ. The ICJ Advisory Opinion, 2025, does address the issue of sea rise (pp. 105–107, paras. 355–365). It describes the issue generally as whether the complete submergence of a State's territory can deprive that State of its maritime entitlements (p. 105, para. 355). It notes that the IPCC has 'described sea level rise as 'unavoidable' and has concluded with a high level of confidence that, as a result, the risks for coastal ecosystems, people and infrastructure will continue to increase' (p. 105, para. 356, citing the IPCC, 2023 at p. 15, B.2.2 and p. 18, B.3.1). The ICJ notes that sea rise is likely to result in forced displacement of persons, 'as well as affecting the territorial integrity of States and their permanent sovereignty over their natural resources' (p. 105, para. 357). And Court notes that 'these principles are closely connected with the right to self-determination' and that 'sea level rise is not without consequences for the exercise of this right' *Ibid*. The Court then notes that the issue of sea rise is linked to the UNCLOS provisions requiring States to establish, publicize, and maintain charts or lists of geographical co-ordinates that show the limits of States maritime zones (p. 105–106, para. 358). The Court further notes that the UNCLOS obligations do not require that once the breadth of maritime zones measured from the baselines has been duly established and the State has given due publicity to the charts or lists of geographical co-ordinates in accordance with UNCLOS there is no provision in the

Convention requiring States parties to update them' (p. 106, paras. 362) after it described how this is consistent with States *opinio juris* (p. 106, para. 360) and the views expressed in final report of the Study Group on sea level rise in relation to international law, adopted by the UN's International Law Commission (p. 106, para. 361, citing UN ILC, 2025). In an apparent nod to the right of self-determination in accordance with the position taken by small island States, some of which sea level rise threatens their very existence, the Court finds that there is 'a strong presumption in favour of continued statehood' and that 'once a State is established, the disappearance of one of its constituent elements would not necessarily entail the loss of its statehood' (ICJ Advisory Opinion, 2025, p. 107, para. 363).

Considering the relationship between the two questions on which the Advisory Opinion was requested the Tribunal recognized that the second question was broader because '[t]he obligation to protect and preserve the marine environment encompasses the obligation to prevent, reduce and control marine pollution' (para. 152).

5.8 The First Question: What Are States Obligations Under the Convention to Prevent, Reduce, and Control Pollution of the Marine Environment

In part VII the Tribunal responds to the first question. It begins by clarifying terms. The Tribunal looks to the relevant legal instruments relating to climate change or in authoritative scientific works such as in the IPCC reports to define 'climate change', 'greenhouse gas emissions', and 'ocean acidification' (para. 157). Its first task, according to the Tribunal, is to '[w]hether anthropogenic GHG emissions fall within the definition of marine pollution under the Convention' (paras. 159–179). It answers this question in the affirmative based on its review of the scientific evidence, jurisprudence, relevant treaties, and the views of the participants in the proceedings. Although it does not clarify if by participants in the proceedings, it means only States and intergovernmental organizations or also non-State actors, the view that GHG emissions constitute marine pollution was supported by both types of actors.

Focusing on Article 1, paragraph 1, subparagraph 4, of the Convention, the Tribunal sets out the criteria for determining what 'pollutants or forms of pollution' can constitute pollution of the marine environment (para. 161). These criteria are '(1) there must be a substance or energy;

(2) this substance or energy must be introduced by humans, directly or indirectly, into the marine environment; and (3) such introduction must result or be likely to result in deleterious effects' (para. 161). The Tribunal then examines GHG emissions (paras. 162–178). It so concludes, because it finds 'substance' and 'energy' should have a broad meaning (para. 163), that energy includes heat (para. 163), and that ILC agreed that atmospheric pollution included the introduction of energy and that the ILC also agreed energy includes heat (para. 163, quoting ILC, 2021, subpara. (b)).

The Tribunal then continues finding 'gas' must be given its ordinary meaning which clearly qualifies it as a 'substance' (para. 164). The Tribunal turns to the Convention and the IPCC's classifications to find that the introduction of GHGs into the atmosphere that are of concern are those done by human beings (para. 165). Marine environment is interpreted (para. 166) as a combination of both 'spatial and material components' based on the 'the context in which the term is used in the Convention, in light of its object and purpose, by the relevant subsequent practice of the States Parties to the Convention regarding its interpretation, and by the corresponding international jurisprudence' citing examples of each in prior and subsequent paragraphs. As an example of jurisprudence, the Tribunal cites ICJ's decision in the *Nuclear Weapons Advisory Opinion*, several of its own decisions, and the *Arbitration regarding the Chagos Marine Protected Area*. As examples of subsequent practice, the Tribunal cites the Regulations on Polymetallic Sulphides and the Convention on Biological Diversity.

In oral argument, COSIS, Argentine, Micronesia, Sierra Leone (ITLOS, 2023a, pp. 23–29/92/175/293/391) had argued unequivocally that GHG emissions constitute pollution in the terms of the Convention. Professor Margaretha Wewerinke-Singh, for example, appearing for COSIS, argued that it is 'straightforward and uncontroversial' that GHG emissions constitute pollution under the Convention (ITLOS, 2023a, September 11, pp. 23–29). The Tribunal agreed with that argument stating 'that anthropogenic GHG emissions into the atmosphere constitute pollution of the marine environment' within the meaning of the Convention (para. 179).

The finding that GHG emissions constitute pollution establishes the connection between climate change and the marine environment. It also establishes that action to prevent marine pollution helps to combat sea level rise and the consequential harm to the coastal areas of States and marine life. Although perhaps an uncontroversial finding by the Tribunal, once this has been found the application of obligations in the Convention becomes much more straightforward. Subsequently, in Section VII.C, the Tribunal is able to apply Part XII of the Convention to the actions of States. Through this transition from VII.B to VII.C, the Tribunal has used a rhetorical argument to match the words of the Request to those of the Convention. This enables the Tribunal to avoid addressing the lex specialis arguments of countries such as India.

The Tribunal also concludes that States have obligations to protect the marine environment from pollution. These obligations flow from Article 194, paragraph 1, of the Convention (paras. 193–197). The aim of these obligations is that all necessary measures are taken to 'prevent, reduce and control pollution' (art. 194 of the Convention; para. 198 of the Advisory Opinion). The Tribunal's emphasis is on effective action. Nevertheless, the Tribunal somewhat confusingly finds that the obligation 'to take all necessary measures … does not entail the immediate cessation of marine pollution from anthropogenic GHG emissions' (para. 199). And even perhaps more disappointingly is the Tribunal reference to the weak language of the Paris Agreement (para. 200), which contains aspirational language and virtually no concrete obligations (see Doebbler, 2015; Doebbler & Wewerinke-Singh, 2016). Nevertheless, the Tribunal finds that the UNFCCC, the Paris Agreement (and instruments of international human rights law (see para. 66) must be applied to interpret the obligations in UNCLOS. This determination by the Tribunal was important because some States, for example Australia (Australia Written Statement), had argued that abiding by their obligations under the Paris Agreement satisfied any obligation States had under any other source of international law to protect the marine environment from climate change. The Tribunal reiterates that 'it does not consider that Paris Agreement modifies or limits the obligation under the Convention' (para. 224). Furthermore, the Tribunal emphasized that 'treaties do not apply in isolation but are "interpreted and applied within the framework of the entire legal system prevailing at the time of interpretation"' quoting the International Court of Justice's statement in its Advisory Opinion on the 'Legal Consequences for States of the Continued Presence of South Africa in Namibia (South West Africa) notwithstanding Security Council Res. 276', 1971 I.C.J. Rep. 16, para. 53. This conclusion was echoed by the ICJ Advisory Opinion, 2025, and the IACtHR Advisory Opinion, 2025.

In describing the general obligations of States 'to take all necessary measures to prevent, reduce and control marine pollution' under Article 194 (para. 193) the Tribunal divides its discussion of the legal obligation into (a) the scope and content, (b) nature, and a conclusion.

The Tribunal indicates that this provision applies future or potential pollution as well as controlling current pollution. The Tribunal thus concludes that obligation is one 'to take all necessary measures with a view to reducing and controlling existing marine pollution from such emissions and eventually preventing such pollution from occurring at all' (para. 199), but then cautions that 'this obligation does not entail the immediate cessation of marine pollution from anthropogenic GHG emissions' (para. 199), curiously citing the weak provisions of Paris (para. 200).

Addressing the modalities the Tribunal first notes that actions must be understood as 'all measures necessary' either jointly or by individual States (para. 201), but it also recognizes that preventing transboundary pollution may require joint action (para. 202). These measures must be understood broadly as '"indispensable", "requisite" or "essential"' (para. 203) and as part of an expansive obligation (para. 203) 'to prevent, reduce and control marine pollution but also … [to take] … other measures which make it possible to achieve that objective' (para. 203). The Tribunal then also notes that such measures may be limited by other UNCLOS obligations such as obligations related to exploitation or control of a State's Exclusive Economic Zone (EEZ) or the free passages of ships that is provided for in UNCLOS (para. 204 citing art. 194(1) of the UNCLOS). But the Tribunal makes it clear that such measures include mitigation action to reduce 'anthropogenic GHG emissions into the atmosphere' (para. 205).

Such measures must not merely be determined by the whims of States (para. 206), but must be determined objectively based on international climate change law (para. 207) and the 'best available science is found in the works of the IPCC which reflect the scientific consensus of the majority of States' (para. 208; *see also* paras. 206–224 elaborating the standards of science and law). Contrary to the arguments made by a number of developed States, the Tribunal opined that it 'does not consider that the obligation under Article 194, paragraph 1, of the Convention would be satisfied simply by complying with the obligations and commitments under the Paris Agreement' (para. 223), and that the Tribunal 'does not consider that the Paris Agreement modifies or limits the obligation under the Convention' (para. 224).

In paragraphs 225–230 the Tribunal emphasizes the applicability of the principle of common but differentiated responsibilities (CBDR), which requires States who have contributed to climate change through their greater GHG emissions must do more to combat the adverse consequences of climate change. The Tribunal notes that 'in the context of marine pollution from anthropogenic GHG emissions, States with greater means and capabilities must do more to reduce such emissions than States with less means and capabilities' (para. 227). The Tribunal also notes that 'the reference to available means and capabilities should not be used as an excuse to unduly postpone, or even be exempt from, the obligations under Article 194(1) of the UNCLOS' (para. 226). The principle of CBDR was also considered relevant to the obligations of States in both the ICJ Advisory Opinion, 2025 (p. 53–54, paras. 148–151), and IACtHR Advisory Opinion, 2025 (p. 72, para. 192; p. 75, para. 201; p. 84, para. 237; p. 89, para. 250; pp. 90–91, paras. 254–255; p. 92, para. 258; p. 108, para. 309; p. 111, paras. 324 and 327; and, p. 145, para. 431).

Under Article 194(1), the Tribunal also notes there is a duty to make an honest effort to align their policies with preventing, reducing, and controlling marine pollution (para. 230). There is additionally a duty under Article 195 to ensure that marine geoengineering does not have 'the consequence of transforming one type of pollution into another' (para. 231). This reflects the principle of good faith as reflected in the Vienna Convention on the Law of Treaties that is stated in Article 26 as '[e]very treaty in force is binding upon the parties to it and must be performed by them in good faith' and applies equally to customary international law rules (Nuclear Tests Case, 1974, p. 268, para. 46).

In eleven paragraphs (paras. 232–242), the Tribunal opines, in what may be viewed as a weakening of the obligations of States, that the obligation to prevent, reduce and control pollution of the marine environment is not one of result. In other words, States have an obligation of conduct (para. 237), 'it is an obligation "to deploy adequate means, to exercise best possible efforts, to do the utmost" to obtain the intended result' (Seabed Disputes Chamber, 2011, pp. 10/41, para. 110). The Tribunal notes that this due diligence applies in relation to a State's duty to regulate non-State actors (para. 236).

In paragraph 241, the Tribunal opines that due to the serious nature of the harm it 'considers that the standard of due diligence States must exercise in relation to marine pollution from anthropogenic GHG emissions needs to be **stringent**' [emphasis added] (para. 241) and links this heightened level of due diligence to the precautionary principle (para. 241).

5.9 The First Question: What Are States Obligations Under the Convention to Prevent, Reduce, and Control Pollution of the Marine Environment

As concerns the obligations of States for transboundary pollution paragraph 2 of Article 194, the Tribunal reiterates the long-established obligation of all States not to cause harm to other States (paras. 244–258, at para. 246 citing, among other sources, ICJ, 1996, p. 242, para. 29; and Trail Smelter, 1952). The Tribunal finds that '[w]ith respect to transboundary pollution affecting the environment of other States, the standard of due diligence can be even **more stringent**' [emphasis added] (para. 256), while also finding that this standard applies both to already occurred harm as well as harm that is likely to occur (para. 248) (Desierto, 2024). Again, the Tribunal finds the principle of CBDR to be applicable to determining the exact nature of 'stringent due diligence' (para. 249). In paragraph 252, the Tribunal addresses the difficult question of the 'diffused and cumulative causes and global effects of climate change' and the difficulty in determining 'how anthropogenic GHG emissions from activities under the jurisdiction or control of one State cause damage to other States'. The Advisory Opinion avoids answering this question. It is, however, an answerable question. For example, the International Law Commission has pointed out in Article 47(1) of its Draft Articles on State Responsibility that '[w]here several States are responsible for the same internationally wrongful act, the responsibility of each State may be invoked in relation to that act' (Draft Articles on State Responsibility, p. 124). In other words, although a State may not claim more reparations for harm than it has actually suffered, it does not have to consider how much of that harm has been caused by one particular State against which it is making a claim, as long as it can prove that that State made some contribution to the harm the claiming State has suffered.

As it did in relation to Article 194(1), once again, the Tribunal found that because anthropogenic gases constitute pollution, 'there appears to be no convincing reason to exclude the application of Article 194, paragraph 2, to such pollution' and its transboundary effects (paras 252–253).

In Section VI.E of its Advisory Opinion, the Tribunal addresses obligations applicable to specific sources of pollution (paras. 259–291) by considering both obligations to adopt national and international laws (paras. 265–280) as well as obligations to enforce these standards (paras. 281–291). It finds these obligations apply primarily to land-based pollution in Article 207, pollution ships in Article 211, and pollution from or through the atmosphere in Article 212 (paras. 260/263). The enforcement obligations relevant to each of these forms of pollution are found in articles 213 in respect of pollution from land-based sources, Article 217 in respect of pollution from ships, and Article 222 in respect of atmospheric pollution (para. 264). For each of these types of pollution, the Tribunal finds obligations to abide by treaty obligations and make an effort to adopt domestic laws (paras. 265–280). For atmospheric pollution the treaty obligations include those found in 'climate change treaties' as well as aviation treaties 'establishing a carbon offsetting and reduction scheme' (para. 277). In respect of GHG emissions from vessels or ships, Article 211, paragraph 2 imposes upon States the obligation to adopt 'laws and regulations' that have 'at least have the same effect as that of generally accepted international rules and standards' (para. 279). The Tribunal distinguishes this obligation from the obligation to merely 'take into account' rules of international law such as is provided for in paragraphs 1 of articles 207 (concerning land-based pollution) and 212, paragraph 1 (pollution 'from or through the atmosphere applicable to the airspace' under a State's sovereignty by vessels and aircraft) of the Convention (para. 279 of the Advisory Opinion). Finally, the Tribunal then considers obligations of enforcement in relation to specific types of pollution (paras. 281–291). It notes that the obligations of States are similar in respect of land-based and atmospheric pollution, but distinguishes pollution from vessels (para. 281). In respect of land-based and atmospheric pollution (arts. 213–222), the Tribunal finds in respect of the types of pollution described in Articles 213 and 222, States have obligations both to adopt legislation and ensure that the legislation is effectively applied (paras. 282–283). Furthermore, States must 'ensure that recourse is available in accordance with their legal systems for prompt and adequate compensation or other relief in respect of damage caused by pollution of the marine environment by natural or juridical persons under their jurisdiction' (para. 284, quoting art. 235(2)). The Tribunal also finds that States have obligations to 'take other measures that are necessary to implement applicable international rules and standards' (para. 285). The Tribunal stresses that this refers to treaties to which a State is a party and applicable rules of customary international law that reflect evolving standards of care (*id.*). This obligation has been long established as part of international law (UNGA Declaration on International Law). It reflects a fundamental tenant of State responsibility under international law, which is that '[e]very internationally wrongful act of a State entails the international responsibility of that State' (ILC, 2001, art. 1) and the Rainbow Warrior Case Arbitral Tribunal noted that 'any violation by a State of any obligation, of whatever origin, gives rise to State responsibility' (Rainbow Warrior Case, p. 251, para. 75). The Tribunal notes that when 'State Party to the Convention, which is bound by those rules and standards, fails to take such measures, its international responsibility would be engaged for breach of the obligations under Article 213 or 222 of the Convention' (para. 286). Tribunal finds similar obligations of States to control pollution in relation to obligations under Article 217 pertaining to pollution from 'vessels flying their flag or of their registry' (paras. 287–291), again finding that implementation includes 'laws and regulations, and other necessary measures ... [including those which] ... may be wide-ranging and include administrative and judicial measures' (para. 290).

In the final part of its answer to question (a) in VI.F of the Advisory Opinion, the Tribunal considers the obligations of States to cooperate, to share technical assistance, and to monitoring and the conduct of environmental assessments (para. 292). The Tribunal prefaces its findings by noting that its findings in regard to the first question also apply to the second question (para. 293).

As concerns global cooperation, the Tribunal notes the consensus of States that protecting the marine environment from consequences caused by climate change requires international cooperation (para. 295), citing its finding in an earlier case that 'the duty to cooperate is a fundamental principle in the prevention of pollution of the marine environment under Part XII of the Convention and general international law' (para. 296, quoting the *MOX Plant Case* at p. 110, para. 82). The Tribunal reiterates the general importance of the duty to cooperate, how it has been found to be essential by the IPCC (para. 297), and is found in treaties on climate change (para. 298) as well as in the specific provisions of the Convention (para. 299). The Tribunal addresses the duty to cooperate under Article 197, which is to create standards (paras. 300–311), and under articles 200 and 201, which is to ensure the standards are based on comparative data (paras. 312–320). It concludes that 'articles 197, 200 and 201, read together with articles 194 and 192 of the Convention, impose specific obligations on States Parties to cooperate, directly or through competent international organizations, continuously, meaningfully and in good faith in order to prevent, reduce and control marine pollution from anthropogenic GHG emissions' (para. 321). The Tribunal then describes three obligations incumbent on States. First, the obligation to formulate standards and take action to reduce GHG emissions (para. 321). Second, the obligation to promote research and data sharing on the impact of marine pollution (para. 321). And third, the establishment of scientific criteria for standard setting (para. 321). As concerns technical assistance the Tribunal interprets Article 202 together with Article 203 to find that States have an obligation to cooperate on technical assistance with particular emphasis on providing technical assistance to developing States (paras. 322–340). While the Tribunal recognizes that the principles of common but differentiated responsibilities are not expressly stated in the Convention (para. 326) it finds that the language of Articles 202 and 203 implicitly incorporate the principle stating 'that articles 202 and 203 of the Convention set out specific obligations to assist developing States, in particular vulnerable developing States, in their efforts to address marine pollution from anthropogenic GHG emissions' (para. 339). Both the ICJ and the IACtHR also emphasized obligations of cooperation throughout their advisory opinions (for example, ICJ Advisory Opinion, 2025, pp. 50–51, paras. 140–142; pp. 70–71, paras. 214–215; pp. 81–84, paras. 160–270; and IACtHR Advisory Opinion, 2025, pp. 88–93, paras. 247–265).

Finally, the Tribunal addresses the specific obligations of monitoring pollution in Article 204, reporting on pollution in Article 205, and undertaking environmental assessments in Article 206 (paras. 340–368). As concerns monitoring the Tribunal points to the precedent of the *Chagos Marine Protected Area Arbitration*, at p. 500, para. 322 (which itself refers to the ICJ judgement in the *Pulp Mills Case*, at p. 83, para. 205) to observe that the procedural requirement of monitoring is as important as any substantive legal obligations (para. 345). Again, this finding is consistent with long-established international law indicating that all legal obligations entered into by States are legally binding on them. Although the Tribunal finds that the duty of monitoring under Article 204 allows States some discretions over the means used (para.

348), the monitoring must be continuous (para. 346), based upon 'recognized scientific methods' (para. 347), and include the monitoring of actions by non-State actors operating under the State's broad jurisdiction (para. 349). As a complement to the duty to monitor the Tribunal finds Article 205 imposes a duty of transparent reporting (para. 350–351). The Tribunal however stops short of stating that the reports must be made public, again apparently deferring to the discretion of States. Finally, the Tribunal finds there is an obligation on States to conclude environmental impact assessments (EIA) 'to assess the potentially harmful effects of a planned activity prior to its execution and to disseminate the obtained results thereafter' (para. 352). Notably, the Tribunal finds 'obligation to conduct an environmental impact assessment [exists] under the Convention and customary international law' (paras. 353–356; para. 356 citing Seabed Disputes Chamber, 2011, pp. 50–51, paras. 145/147; ICJ, 2010, p. 83, para. 204). The obligation of conducting EIA is triggered, according to the Tribunal, when a contemplated activity is under the 'jurisdiction or control' of a State and the States has 'reasonable grounds for believing' that these activities 'may cause substantial pollution of or significant and harmful changes to the marine environment' (para. 359, quoting art. 206). The concepts of jurisdiction and control (paras. 359–360) and of reasonable grounds for believing harm might occur for an activity to be undertaken (para. 361) are to be interpreted broadly in favour of the obligation to undertake an EIA as is required by 'the precautionary approach [which] may restrict the margin of discretion on the part of the State concerned' (para. 361). Although mentioning the 'precautionary approach' in four different places (paras. 69/213/242/353), the Tribunal does not use the terms 'precautionary measures' (the term used in Article 3(3) of the UNFCCC) or 'precautionary principle' or discuss the responsibility to take precautionary measures as part of customary international law. The ICJ had already found that the obligation of States to undertake environmental impact assessments (EIA) when 'a proposed industrial activity may have significant adverse impact in a transboundary context, in particular, on a shared resource' is 'a requirement under general international law' (ICJ Pulp Mills, 2010 at para. 204). The ICJ Advisory Opinion, 2025 (p. 49, para. 136; p. 69, para. 210; and p. 92, para. 298), reiterated this holding. Finally, the Tribunal acknowledges the discretion of States in determining the content of an EIA (para. 363), the impact of the different capacities of States (paras. 364), the possibility of including cumulative impacts (para. 365), and new (but not yet entered into force) BBNJ Agreement's detailed provisions on EIAs (para. 366). The Tribunal concludes Section VII.F by stating that

> [i]n light of the foregoing, the Tribunal is of the view that articles 204, 205 and 206 of the Convention impose specific obligations on States Parties to monitor the risks or effects of pollution, to publish reports and to conduct environmental impact assessments as a means to address marine pollution from anthropogenic GHG emissions. Both the ICJ and the IACtHR also indicated that environmental impact assessments (EIA) may be part of States legal obligations under general international law and under international human rights law, respectively. (ICJ Advisory Opinion, 2025, p. 69, para. 210, and IACtHR Advisory Opinion, 2025, p. 83, para. 230, p. 120–121, paras. 358–362).

5.10 The Second Question: What Are States Obligations Under the Convention to Protect and Preserve the Marine Environment?

In Section VIII, the Tribunal addresses the second question concerning the protection and preservation of the marine environment through a similar approach to the first question. It begins by noting that this question is broader than the first question (para. 370) and that the definitions it stated when answering the first question apply equally to this question (para. 371). It must be remembered that the Tribunal already found that anthropogenic GHG emissions fall within the definition of marine pollution under the Convention and have harmful effects on the marine environment.

The Tribunal then clarifies relevant terms and expressions, noting that the definitions already arrived for the first question apply equally to the second question (paras. 373–376). The Tribunal notes however that the term 'climate change impacts' … 'is used in relation to circumstances in which drivers of climate change cause deleterious effects to the marine environment' … [and therefore] … '[t]he Tribunal is of the view that Question (b) concerns the negative impacts of climate change and ocean acidification on the marine environment' (para. 375).

The Tribunal defines the relevant provisions to be applied in Section VIII.B as those in Part XII of the Convention (paras. 380–383), especially Articles 192–196, as well as Articles 61, 63, and 64 in Part V, and Articles 117, 118, and 119 in Part VII (para. 382). The Tribunal determines that replying to this question requires it to set out the specific obligations of States Parties under the Convention to protect and preserve the marine environment against climate change impacts and ocean acidification (para. 378).

Addressing the obligations in the just-mentioned articles is done in four sections starting with the obligations under Article 192 of the Convention (paras. 384–400). As concerns Article 192, the Tribunal rejects the argument that the climate change treaties are *lex specialis* and notes that it was not an argument favoured by the 'vast majority of participants' (para. 384). Instead, the Tribunal finds that 'the obligation contained in article 192 of the Convention has a broad scope, encompassing any type of harm or threat to the marine environment' (para. 385). Moreover, the obligations to protect and preserve 'can be invoked to combat any form of degradation of the marine environment, including climate change impacts, such as ocean warming and sea level rise, and ocean acidification' (para. 388). The Tribunal is of the opinion that both article 192 and the climate change instruments require 'States to implement measures to protect and preserve the marine environment in relation to climate change impacts and ocean acidification that include resilience and adaptation actions…' (para. 391).

Second, addressing the obligation under article 194, paragraph 5 (paras. 401–406), the Tribunal finds that this provision must be understood broadly to include 'the measures necessary to protect and preserve rare or fragile ecosystems as well as the habitat of depleted, threatened or endangered species and other forms of marine life are those which make it possible to achieve that objective' (para. 402). The Tribunal also notes that the definition of rare or fragile ecosystems may change over time, pointing to the example of ice-covered areas mentioned in Article 234 of the Convention (para. 403). The Tribunal also notes that the phrase 'depleted, threatened or endangered species' can be understood in terms of the widely-ratified CITES, which 'classifies species threatened with extinction and those likely to become endangered in the absence of trade regulations' (para. 404). Finally, the Tribunal notes 'that the obligation imposed by Article 194, paragraph 5, of the Convention may call for specific measures, such as the enactment and enforcement of laws and regulations or the undertaking of monitoring and assessment' (referring to is paras. 340–367) and that '[t]hese measures are context-specific and call for objectively reasonable approaches to be taken on the basis of the best available science … [although t]heir implementation depends on the relevant domestic legal system and allows for the exercise of [limited] discretion' (para. 405) the limits on a State's discretion, according to the Tribunal, can be found in the requirement that a 'State must take into account, objectively, the relevant options in a manner that is reasonable, relevant and conducive to the benefit of mankind as a whole' (para. 405; Seabed Disputes Chamber, [2011], p. 71, para. 230). And a State 'must act in good faith, especially when its action is likely to affect prejudicially the interests of mankind as a whole' (Seabed Disputes Chamber, 2011, p. 71, para. 230).

Third, the obligations under other provisions of the Convention are addressed (paras. 407–436). The Tribunal finds a number of obligations to exist under Article 192, read in conjunction with the rest of the Convention. There is an obligation to conserve living resources and marine life (para. 409). There is an obligation for States to take into account the economic and cultural impacts of climate change and ocean acidification on fishing when taking measures to protect and preserve the marine environment (para. 410, citing para. 66). And there is an obligation to conserve living resources in a State's exclusive economic zone (EEZ) (referring to art. 61) and in the high seas (referring to arts. 117/119; para. 411). Each of these obligations is further elaborated in paragraphs 412–436. As concerns the conservation in the EEZ, the Tribunal notes that the 'purpose of conservation and management measures under Article 61 of the Convention is to ensure that the maintenance of the living resources in the exclusive economic zone is not endangered by overexploitation' (para. 414) and concerning protection and preservation of fish stocks on the high seas under Articles 117 and 119 (para. 415). The Tribunal goes on to explain what this duty means by reiterating what is determined in its SRFC Advisory Opinion at page 60, paragraph 214, that regional and global efforts are both encouraged and that conservation efforts 'should concern the whole stock unit over its entire area of distribution or migration routes' (paras. 415/417). In paragraphs 419–428, the Tribunal emphasizes duties of international cooperation found under Articles 63, 64, and 118 of the Convention, again reminding States of the 'duty of diligence' and 'good faith' (paras. 422–423) in undertaking their obligations of cooperation, 'directly or through appropriate international organizations, in implementing conservation and management measures with regard to straddling and highly migratory species and other living resources of the high seas' (para. 428). And finally, the Tribunal considers the obligation of States to act to mitigate and adapt to climate change to be able to meet their obligations under Article 196, which obliges States to control the use of new technologies and harmful introduction of new or alien species to the marine environment (paras. 429–436). The Tribunal again finds this to require a precautionary approach (para. 434). The recognition that Article 196 extends to action to address climate change is important for many Small Island States whose coastal communities depend on small-scale fishing for their livelihood. Similarly, the Tribunal's generally confirming the obligations of States to protect and preserve fishing stocks from the adverse effects of climate change will be welcome by States with significant small-scale fishing communities, which are predominately developing States.

Finally, in Section VIII.C.4 the Tribunal considers how area-based management tools (ABMTs) for achieving the goals of protection and preservation of the marine environment (paras. 437–440). The Tribunal notes that the IPCC has recognized that ABMTs might be valuable for protecting and preserving the marine environment (para. 438, citing IPCC, 2022, p. 483), Convention on Biodiversity (para. 439, citing art. 2), and the OSPAR Convention (para. 439, citing the 11th Preambular para.) as evidencing the value of regional international cooperation. The Tribunal concludes that although regional cooperation and agreements are within the remit of States, they 'must be consistent with the Convention and other rules of international law', noting that the BBNJ expresses the need for a 'global framework' (para. 440).

5.11   Operative Clause

The Tribunal concludes its Advisory Opinion operative clauses that summarize its findings and indicating that they have been agreed unanimously. The operative clause are contained in paragraph 441, which is the final paragraph of the Advisory Opinion and runs for almost seven pages. The operative clauses are essentially a summary of the procedural and substantive findings of Tribunal that have been described above.

6   Declarations of Individual Judges

The Tribunal's Advisory Opinion was unanimously adopted by all 21 judges, although five judges attached declarations.

Judge Infante Caffi supported the Tribunal's Advisory Opinion but called for a greater elaboration of human rights concerns. He emphasized that this 'is an instance in which regimes on human rights require law of the sea principles to be applied, and likewise, in which the law of the sea requires States to consider the human implications of regulatory measures, policies and enforcement actions' (Caffi, p. 2, para. 4).

Judge Pawlak also called for greater focus on human rights lamenting that the Tribunal had not taken into account the recent UN Human Rights Committee's *Torres Strait Islanders case* and the Swiss Constitutional Court's holding in *Verein KlimaSeniorinnen Schweiz and Others v. Switzerland*. In both of these cases international human rights law was found relevant in the context of climate change.

Judge José Luis Jesus disagreed with the characterization of the obligation in Article 194, paragraph 2 of UNCLOS as one of only due diligence. In his view, it was more. It was, he opined in his Declaration, an obligation of result. This was because '[e]ven when a State has taken all measures of due diligence … but damage has nevertheless occurred, the responsibility of the former State may still be engaged on the basis of strict responsibility for the environmental damage caused' (Jesus, p. 5, para. 15). He also understood Article 194, paragraph 1 to apply to the joint and several action of States, while he understood Article 194, paragraph 2 to apply only to individual States or bilateral relations in general (Jesus, pp. 2–3, paras. 6–8).

Judge Markiyan Kulyk expressed concern about the nuances of the Advisory Opinion in relation to its denying a second round of written statements (Kulyk, 2024, p. 1, sec. I), its failure to further interpret COSIS (Kulyk, 2024, pp. 1–2, sec. II), and the failure to sufficiently consider the balance explicit in Article 193 between States right to use resources and their duty to protect them (Kulyk, 2024, pp. 2–3, sec. III). In the final sections IV.1 and IV.2 Judge Kulyk draws attention to the principle of 'common but differentiated responsibilities' referring to it as the distinction 'between the obligation to take all necessary measures and how it is actually implemented', noting all States have obligations, but some are conditional (Kulyk, 2024, pp. 3–4, sec. IV.1/IV.2).

Judge Kittichaisaree declaration considered jurisdiction and discretion to refuse to provide an Advisory Opinion, obligations of conduct and result, human rights, and the distinction between primary and secondary obligations. An unusual take on the technical assistance provisions of UNCLOS expressed with emotion and a quote to Shakespeare's Macbeth and Dante's Inferno the hope that the Advisory Opinion might be a guiding light in the effort to address the adverse consequences of climate change. In relation to jurisdiction, he addresses the concern by some States (Guatemala, 2023, p. 6, para. 23) claim that COSIS is not an appropriate international organization to ask an Advisory Opinion because it was established for that purpose. He points out that it has the intention to contribute to other advisory opinions, for example, that before the International Court of Justice, and therefore, it cannot be said to have been established merely for the purpose of this Request.

Although the separate Declarations raised important points, some critical of the Advisory Opinion, none of them distracted significantly from the core findings of the Advisory Opinion.

7   The Advisory Opinion in Context

The Tribunal Advisory Opinion significantly influenced the advisory opinions of both the International Court of Justice and the Inter-American Court of Justice in several ways noted above and it is likely to also influence the forthcoming Advisory Opinion of the African Court of Human and Peoples Rights. Having set the scene for these other advisory opinions the Tribunal has been an important trailblazer.

The Tribunal framed its opinion as one seeking consensus, often referring to the views of the majority of States. It also emphasized the importance of science for dealing with issues of climate change. It emphasized cooperation. And it established significant stringent standards for the duty of care States must abide by in relation to their acts or omissions that have consequences for the planet's atmosphere and its oceans. A critical question is whether these Advisory Opinions will be taken to heart by States in the global climate negotiations to overcome political differences and the intransigent efforts of some developed States to maintain advantages they have accumulated from the overexploitation of the earth's atmosphere and aquatic resources. If States do not take the interpretations of international law to heart, the advisory opinions suggest that injured States and non-States actors must be allowed to have recourse to national and international courts to enforce their rights.

More specifically, the Tribunal took advantage of the majority of States supporting its jurisdiction to reiterate its earlier determination in its SRFC Advisory Opinion. This likely will fairly definitively put to rest challenges to its authority to give Advisory Opinions. This is a welcome reaffirmation of the Tribunal's *compétence de la compétence* for jurists and others seeking to promote the rule of international law.

The finding that the IPCC's studies constituted the best available science is also notable for confirming what many, but not all, climate scientists and activists had argued for decades, which is that climate change is causing harm to the planet and those who live on it. If this finding is followed by other judicial bodies, especially the ICJ, it will significantly enhance consensus about climate science.

The Tribunal in its substantive holdings interprets Articles 194 and 192 of the Convention to create specific obligations for States. Under Article 194, paragraph 1, of the Convention, the Tribunal found that 'States Parties to the Convention have the specific obligations to take all necessary measures to prevent, reduce and control marine pollution from anthropogenic GHG emissions and to endeavour to harmonize their policies in this connection' (Operative Clause, para. 3(b)). And under 192 of the Convention the Tribunal found that State Parties have the broad obligation 'to protect and preserve the marine environment …[from]…any type of harm or threat to the marine environment' (Operative Clause, para. 4(b)). Both these findings provide a basis for strengthening climate action, but will either finding go far enough in creating the political will that is needed for State action and which to date has been sorely lacking.

Moreover, although it would be significant if all the States who weighed-in on the Advisory Opinion took climate action, if the United States, which was absent, does not, the inequalities and inequities in the world are likely to impose even greater burdens on developing States. In addition, both China and India objected to the Tribunal's Advisory Opinion jurisdiction (China, 2023; India, 2023) neither the United States nor Russia participated in the proceedings when both likely would have also objected to the Tribunal's jurisdiction. And some States, for example, Indonesia, although submitting that the Tribunal had jurisdiction, noted that the Tribunal's Advisory Opinion 'shall have no binding force, and that it is only used to assist the requesting body in its activities' (Indonesia, 2023, p. 7, para. 32).

Similarly, likely to face political opposition is the Tribunal's finding that the principle of common but differentiated responsibilities (CBDR) applies to action required by the UNCLOS (see, e.g., paras. 226/229/ 327), although the principle is not expressly endorsed in the treaty. While this finding is a reasonable reading of the clear words of the UNCLOS articles, developed States continue to object to the application of this principle in the UNFCCC where it is expressly stated. The Advisory Opinion may encourage States to accept this reasonable interpretation of the clear words of the UNCLOS, but it is still to be seen if it contributes towards encouraging States to respect this principle in the regime of international climate change law.

The Tribunals also find the duty to cooperation to be a duty that 'permeates the entirety of Part XII of UNCLOS' (para. 297). The Tribunal reiterates this duty repeatedly (see, e.g., paras. 202–203/302/307–308/312–320/423). Indeed, cooperation may be the most important obligations of States in relation to adequately addressing climate change, but it has also been one of the main political obstacles. This is problematic because international cooperation between States requires political will.

One of the most important parts of the Tribunal's Advisory Opinion is its finding of a 'stringent' standard of 'due diligence' in environmental matters, as indicated above by its frequent mention by the Tribunal. While the determination that States owed a duty of due diligence was neither progress nor controversial, the articulation of a more 'stringent' standard of due diligence is somewhat unusual. It has left jurists speculating as to what this enhanced standard of due diligence actually is and how a State can achieve it beyond due diligence (see, e.g., Desierto, 2024). The Tribunal did go to some lengths to explain it. For example, the Tribunal noted that it requires States to 'put in place a national system, including legislation, administrative procedures and an enforcement mechanism necessary to regulate the activities in question, and to exercise adequate vigilance to make such a system function efficiently, with a view to achieving the intended objective' (para. 235). And later that '[t]he standard of due diligence under Article 192 is ... stringent given the high risks of serious and irreversible harm to the marine environment by climate change impacts and ocean acidification' (para. 399). This standard must be met by action based on 'scientific and technological information, relevant international rules and standards, the risk of harm and the urgency involved' (para. 239). The Tribunal also indicates that stringent due diligence may apply in respect of States action to control the activities of non-State actors (for example, para. 236). The obligation to control non-State actors operating on a State's territory is not new in international law. It was established more than eighty years ago and has since been often repeated (Trail Smelter, 1952). However, given some recent efforts to demand more freedom for private profit-making entities, this reiteration of the law comes as a poignant reminder, perhaps even a warning, that this rule of international law applies.

The Tribunal referred to the Seabed Disputes Chamber (2011, p. 47, para. 135) in finding that the precautionary principle was part of the duty of due diligence States owed to each other in respect of the marine environment. And while referencing due diligence to the UNCLOS, it appears to accept that other rules of international law treaties may be relevant to interpreting UNCLOS obligations, including international climate change and international human rights law (para. 66). Indeed, both the ICJ and the IACtHR provided additional content to the obligation of stringent due diligence. The ICJ indicated that it is not enough to merely abide by the international legal obligations in environmental or climate change treaties; States must also respect and fulfill their legal obligations under other treaties they have ratified, including human rights treaties, in order to meet their burden of due diligence (ICJ Advisory Opinion, 2025, para. 393).

It is noteworthy that throughout its Opinion the Tribunal appears to be seeking an understanding of the law that is consistent with the views of the majority of States. In this respect, the Advisory Opinion has taken perhaps too optimistic a view of the consensus of States. This aspect of the Advisory Opinion may be used by States, especially the United States that did not participate in its proceedings, to challenge the Tribunal's authority and the authority of its Opinion. However, seeking to distil the consensus views of States is perhaps the wisest and safest route the Tribunal could take.

With an increasing number of sirens going off about the adverse effects of climate change for life on our planet, the Tribunal's Advisory Opinion should be welcomed as a courageous step into what was a relatively disputed and developing area of international law: the responsibility of States for climate change. It should be welcomed that a body with authority to interpret law related to our most plentiful natural resource has had the integrity and courage to step up to the challenge. The International Court of Justice in its Advisory Opinion ascribed 'great weight' to the Tribunal's Opinion and agreed with its interpretation of the UNCLOS insofar as it considered it. The Inter-American Court of Human Rights considered the effect of sea level rise in numerous parts of its Advisory Opinion and generally followed the reasoning established by the Tribunal. Perhaps most importantly, both of these latter judicial bodies followed the Tribunal's finding of a more stringent duty of care. Considering how Africans are among the people most adversely affected by climate change, it remains to be seen how the African Court of Human and Peoples' Rights will respond to the questions put to it about the legal obligations of States to address climate change while respecting their obligations under the African Charter on Human and Peoples' Rights, a question currently before the African Court (ACHPR, 2025).

The ITLOS Advisory Opinion affirmed the role of the law of the sea in addressing climate change. The subsequent advisory opinions of the International Court of Justice and Inter-American Court of Human Rights confirmed most of the Tribunal's reasoning and advanced it by firmly anchoring climate obligations in human rights law, clarifying the content of stringent due diligence, and affirming the applicability of the precautionary principle and State responsibility in the context of transboundary environmental harm. Going forward, international judicial and quasi-judicial bodies—including the Tribunal itself—should view the Tribunal's Advisory Opinion not as a final word, but as an important initial interpretation of the legal obligations of States for the adverse effects of climate change. The integration of human rights, science, and equity into the core of environmental law is no longer a matter of academic debate but has become a judicial imperative. As the adverse effects of climate change intensify and cause increasing loss and damage to the most vulnerable States, the international community, human rights defenders, and legal representatives must look to all three advisory opinions not merely as interpretive tools, but as catalysts for achieving adequate legal and policy action to protect individuals, peoples, and communities from these adverse effects.

References

Although citations below are to the official UN document numbers and dates, references to the Advisory Opinion, the separate opinions of individual judges the written and oral submissions can be located online at https://www.itlos.org/en/main/cases/list-of-cases/request-for-an-advisory-opinion-submitted-by-the-commission-of-small-island-states-on-climate-change-and-international-law-request-for-advisory-opinion-submitted-to-the-tribunal/#c8453 (last visited on 10 October 2024).

African Charter on Human and Peoples' Rights, 1520 UNTS 218, Adopted 27 June 1981, Entered Into Force 21 October 1986 (1986).

African Commission on Human and Peoples' Rights. (2008). ACHPR Res. A/HRC/28/27.

African Convention on the Conservation of Nature and Natural Resources (African Convention), 1001 UNTS 3 (1976).

African Union. (2023, June 16). *Written Statement of the African Union* (ITLOS Doc. C31-WS-2-7-African-Union).

Agreement for the Establishment of the Commission of Small Island States on Climate Change and International Law (COSIS Treaty). UNTS Vol. 3444, No. 56940, Edinburgh, 31 October 2021. Note: This Intergovernmental Organization Was Officially Registered With the Secretary-General of the United Nations in Accordance With Article 102 of the Charter of the United Nations on 3 February 2022. United Nations, Note by the UN Secretary-General Dated 3 February 2022. https://www.itlos.org/fileadmin/itlos/documents/cases/31/Certificate_of_Registration__E_Fr_.pdf

Agreement Under the United Nations Convention on the Law of the Sea on the Conservation and Sustainable Use of Marine Biological Diversity of Areas Beyond National Jurisdiction ("BBNJ Agreement"), UNTS Cataloguing Number C.N.203.2023.TREATIES-XXI.10, 20 July 2023 (105 States have signed, 13 have ratified, the treaty requires 60 ratification to enter into force).

Arbitration Regarding the Chagos Marine Protected Area Between Mauritius and the United Kingdom of Great Britain and Northern Ireland (Arbitration Regarding the Chagos Marine Protected Area), Award of 18 March 2015, Reports of International Arbitral Awards, Vol. XXXI, pp. 359/580, para. 538.

Australia. (2023, June 16). *Written Statement of Australia* (ITLOS Doc. C31-WS-1-11-Australia).

Center for International Environmental Law (CIEL), Pacific Island Students Fighting Climate Change, ClientEarth, & World's Youth for Climate Justice. (2024, July). *Legal Memorandum: Advisory Opinion on Climate Change Delivered by the International Tribunal for the Law of the Sea: Relevance for the International Court of Justice Climate Advisory Proceedings*.

China. (2023, June 16). *Written Statement of China* (ITLOS Doc. C31-WS-1-8-China-transmission-ltr).

Commission of Small Island States on Climate Change and International Law. (2023, June 16). *Written Statement of the Commission of Small Island States on Climate Change and International Law: Written Submission of the Pacific Community* (ITLOS Doc. C31-WS-2-5-SPC).

Committee of Legal Experts of the Commission. (2022, June 18). *Recommendation CLE (1/2022/Rec COSIS Sub-Committee Recommendation)*.

Commission of Small Island States on Climate Change & International Law, Decision Establishing the Sub-Committee on Protection and Preservation of the Marine Environment, Doc. COSIS/SC-PPME/2022/1 (18 June 2022).

Convention for the Prevention of Marine Pollution From Land-Based Source (Paris Convention), 1546 UNTS 119, Entered Into Force May 6, 1978.

Convention on Biological Diversity, 1760 UNTS 79, Adopted on June 5, 1992, Entered Into Force on December 29, 1993.

Diane A. Desierto, The "Stringent" Standard of Due Diligence in ITLOS' Climate-Change Advisory Opinion, EJIL Talk! (Mar. 5, 2024).

Doebbler, C. F. J. (2015). Ensuring Consistency with Existing International Law of Another Climate Change Agreement. *Journal of the South Pacific Law, 1*, A1–A22.

Doebbler, C. F. J., & Wewerinke-Singh, M. (2016). The Paris Agreement: Some Critical Reflections on Process and Substance. *University of New South Wales Law Journal, 39*(4), 1486.

Food and Agriculture Organization of the United Nations (FAO). (2023, June 16). *Written Statement of FAO*. https://www.itlos.org/fileadmin/itlos/documents/cases/31/written_statements/3/C31-WS-3-2-FAO.pdf

France. (2023, June 16). *Written Statement of France* (ITLOS Doc. C31-WS-1-19-France-translation-ITLOS).

Freestone, D., Barnes, R., & Akhavan, P. (2022). Agreement for the Establishment of the Commission of Small Island States on Climate Change and International Law (COSIS). *The International Journal of Marine and Coastal Law, 37*(1), 1–13. https://doi.org/10.1163/15718085-bja10087

Fry, I., Orellana, M., & Boyd, D. (2023, May 23). *Amicus Brief Submitted to the International Tribunal for the Law of the Sea by the UN Special Rapporteurs on Human Rights & Climate Change, Toxics & Human Rights, and Human Rights & the Environment* (ITLOS Doc. C31-WS-4-1_Amicus_Brief_UN_Special_Rapporteurs).

Guatemala. (2023, June 16). *Written Statement of Guatemala* (ITLOS Doc. C31-WS-1–26-Guatemala-01).

India. (2023, September 8). *Written Statement of India*. https://www.itlos.org/fileadmin/itlos/documents/cases/31/written_statements/3/C31-WS-3-4-India.pdf

Indonesia. (2023, June 15). *Written Statement of Indonesia*. https://www.itlos.org/fileadmin/itlos/documents/cases/31/written_statements/1/C31-WS-1-13-Indonesia.pdf

Infante Caffi, J. (2024, May 21). *Declaration of Judge Infante Caffi* (ITLOS Doc. C31-Adv-Op-21.05.2024-decl-Infante-Caffi-orig).

Inter-American Court of Human Rights (IACtHR). (2025). *Advisory Opinion on the Climate Emergency and Human Rights (IACtHR Advisory Opinion)*,

Advisory Opinion OC-32-25 (29 May 2025), https://www.corteidh.or.cr/docs/opiniones/seriea_32_en.pdf.

International Court of Justice (ICJ). (1996). *Advisory Opinion: Legality of the Threat or Use of Nuclear Weapons*. I.C.J. Reports.

International Covenant on Civil and Political Rights (ICCPR, 1966), Dec. 16, 1966, 999 U.N.T.S. 171.

International Covenant on Economic, Social and Cultural Rights (ICESCR, 1966), Dec. 16, 1966, 993 U.N.T.S. 3.

International Court of Justice (ICJ). (2010, April 20). *Pulp Mills on the River Uruguay (Argentina v. Uruguay)*. https://climatecasechart.com/non-us-case/pulp-mills-on-the-river-uruguay-argentina-v-uruguay/

International Court of Justice (ICJ). (2025, 25 July). *Advisory Opinion on the Obligations of States in Respect of Climate Change (ICJ Advisory Opinion), General List No. 187*, https://www.icj-cij.org/sites/default/files/case-related/187/187-20250723-adv-01-00-en.pdf.

International Law Commission (2018). Report on the Work of Its Seventieth Session, U.N. GAOR, 73d Sess., Supp. No. 10, U.N. Doc. A/73/10.

International Panel on Climate Change (IPCC). (2022). *The Ocean and Cryosphere in a Changing Climate: Special Report of the Intergovernmental Panel on Climate Change (IPCC Oceans Report)*. Cambridge University Press. https://www.ipcc.ch/srocc/

Intergovernmental Panel on Climate Change, Climate Change 2023: Synthesis Report (R. Pidcock et al. eds., 2023) (IPCC, 2023).

International Seabed Authority. (2010, May 7). *Regulations on Prospecting and Exploration for Polymetallic Sulphides in the Area* (ISBA/16/A/12/Rev.1).

International Tribunal for the Law of the Sea (ITLOS). (2015, April 2). *Request for Advisory Opinion Submitted by the Sub-Regional Fisheries Commission (SRFC)*. ITLOS Reports 2015.

International Tribunal for the Law of the Sea (ITLOS). (2022a, December 12). *Request for an Advisory Opinion Submitted on by the Commission of Small Island States on Climate Change and International Law*. ITLOS Doc.

International Tribunal for the Law of the Sea (ITLOS). (2022b, December 16). *Request for Advisory Opinion submitted by the Commission of Small Island States on Climate Change and International Law*. ITLOS Reports 2022–2023.

International Tribunal for the Law of the Sea (ITLOS). (2023a). *Minutes of the Public Sittings Held From September 11 to 25, 2023*. https://www.itlos.org/fileadmin/itlos/documents/cases/31/Oral_proceedings/C31_Minutes.pdf

International Tribunal for the Law of the Sea (ITLOS). (2023b, February 15). *Request for Advisory Opinion submitted by the Commission of Small Island States on Climate Change and International Law*. ITLOS Reports 2022–2023.

International Tribunal for the Law of the Sea (ITLOS). (2023c, June 30). *Request for Advisory Opinion submitted by the Commission of Small Island States on Climate Change and International Law*. ITLOS Reports 2022–2023.

International Tribunal for the Law of the Sea (ITLOS). (2023d, September 11). *Public Sitting in Advisory Opinion Held* (ITLOS Doc. ITLOS/PV.23/C31/2/Rev.1).

International Tribunal for the Law of the Sea (ITLOS). (2024, May 21). *Request for Advisory Opinion submitted by the Commission of Small Island States on Climate Change and International Law*. ITLOS Reports 2024.

Jesus, J. (2024, May 21). *Declaration of Judge Jesus* (ITLOS Doc. C31-Adv-Op-21.05.2024-decl-Jesus-rev).

Kittichaisaree, J. (2024, May 21). *Declaration of Judge Kittichaisaree* (ITLOS Doc. C31-Adv-Op-21.05.2024-decl-Kittichaisaree-orig).

Kulyk, J. (2024, May 21). *Declaration of Judge Kulyk* (ITLOS Doc. C31-Adv-Op-21.05.2024-decl-Kulyk-orig).

Needham, K. (2024, August 26). *UN Chief Issues 'SOS' for Pacific Islands Worst Hit by Warming Oceans*. Reuters News Agency. https://www.reuters.com/business/environment/un-chief-issues-sos-pacific-islands-worst-hit-by-warming-ocean-2024-08-26

Notre Dame Global Adaptation Initiative. (n.d.). *About*. Retrieved at October 7, 2024, from https://gain.nd.edu/about/

Nuclear Tests (Austl. v. Fr.) (Nuclear Test Cases, 1974), Judgment, 1974 I.C.J. Rep. 253 (Dec. 20).

Palm, R., & Bolsen, T. (2020). The Science of Climate Change and Sea-Level Rise. In *Climate Change and Sea Level Rise in South Florida: The View of Coastal Residents* (pp. 5–13). Springer Verlag. https://doi.org/10.1007/978-3-030-32602-9

Paris Agreement, UN Treaty Series No. 54113 (2015).

Pawlak, J. (2024, May 21). *Declaration of Judge Pawlak* (ITLOS Doc. C31-Adv-Op-21.05.2024-decl-Pawlak-orig).

Pulp Mills (Arg. v Uru.), Judgment, 2010 I.C.J. Rep. 14.

Rainbow Warrior Case, XX UNRIAA p. 215 (1990).

Request for Advisory Opinion by the Pan African Lawyers Union (ACHPR, 2025), No. 001/2025, Afr. Ct. H.P.R. (pending).

Request for an Advisory Opinion Submitted by the Commission of Small Island States on Climate Change and International Law (CIEL & Greenpeace), Case No. 31, Memorial Filed on Behalf of Center for International Environmental Law and Stichting Greenpeace Council (June 15, 2023) (The brief was drafted on behalf of the Center for International Environmental Law, Washington, D.C., and Greenpeace International, Amsterdam by Kristin Casper, Louise

Fournier, Upasana Khatri, Tamara Morgenthau, Maria Alejandra Serra, and Nikki Reisch).
Seabed Disputes Chamber. (2011, February 1). *Advisory Opinion: Responsibilities and Obligations of States With Respect to Activities in the Area.* ITLOS Reports.
Sierra Leone. (2023, June 16). *Written Statement of Sierra Leone* (ITLOS Doc. C31-WS-1-29-Sierra_Leone).
Spijkers, Otto: The Status of the Right to a Clean, Healthy and Sustainable Environment Under Customary International Law, VerfBlog, 2025/4/04, https://verfassungsblog.de/ip-hr2he-customary-international-law/, https://doi.org/10.59704/b4113d1f4013cea9
Statute of the International Tribunal for the Law of the Sea (Statute), Annex VI to the United Nations Convention on the Law of the Sea (UNCLOS), 1833 UNTS 3 (1994).
Torres Strait Islanders Case, UN Doc CCPR/C/135/D/3624/2019 (18 September 2024).
Trail Smelter Case (United States, Canada), 3 UNRIAA, p. 1905 (1952).
United Kingdom. (2023, June 16). *Written Statement of the United Kingdom* (ITLOS Doc. C31-WS-1-27-UK).
United Nations Convention on the Law of the Sea (UNCLOS), Opened for Signature December 10, 1982, 1833 UNTS 397 (Entered Into Force November 16, 1994).
United Nations Framework Convention on Climate Change (UNFCCC), 1771 UNTS 107 (1993).
United Nations General Assembly (UNGA). (1990). *UNGA Resolution 45/94: The Need to Ensure a Healthy Environment for the Well-Being of Individuals* (UN Doc. A/45/40).
United Nations General Assembly, Official Records, 73rd Sess., Supp. No. 49, U.N. Doc. A/73/49 (2018).
United Nations International Law Commission (UN ILC). (2001). *Draft Articles on Responsibility of States for Internationally Wrongful Acts: With Commentaries. Yearbook of the International Law Commission*, 2, Part Two. https://legal.un.org/ilc/texts/instruments/english/commentaries/9_6_2001.pdf
United Nations International Law Commission (UN ILC). (2021). Draft Guidelines on the Protection of the Atmosphere: With Commentaries. *Yearbook of the International Law Commission*, 2, Part Two.
United Nations International Law Commission, Sea-level rise in relation to international law: Final consolidated report of the Co-Chairs of the Study Group on sea-level rise in relation to international law, Patrícia Galvão Teles, Nilüfer Oral and Juan José Ruda Santolaria (ILC, 2025), UN Doc. A/CN.4/783 (3 February 2025).

UN Special Rapporteurs Ian Fry, Marcos Orellana, and David Boyd, Amicus Brief Submitted to the International Tribunal for the Law of the Sea in Response to Request for an Advisory Opinion from the Commission of Small Island States on Climate Change and International Law, (UN Special Rapporteurs Amicus Brief, 2023), Case No. 31 (May 30, 2023).

Vienna Convention on the Law of Treaties, May 23, 1969, 1155 U.N.T.S. 331.

Verein KlimaSeniorinnen Schweiz and Others v. Switzerland, Application No. 53600/20, European Court of Human Rights, Grand Chamber Judgment (9 April 2024). https://hudoc.echr.coe.int/eng/?i=001-233206

Written Submission of China to the ICJ for its consideration of its Advisory Opinion on the Obligations of States in Respect of Climate Change (China ICJ, 2024), 22 March 2024 accessed at https://www.icj-cij.org/sites/def ault/files/case-related/187/187-20240322-wri-19-00-en.pdf (last accessed 28 July 2025).

Open Access This chapter is licensed under the terms of the Creative Commons Attribution 4.0 International License (http://creativecommons.org/licenses/by/4.0/), which permits use, sharing, adaptation, distribution and reproduction in any medium or format, as long as you give appropriate credit to the original author(s) and the source, provide a link to the Creative Commons license and indicate if changes were made.

The images or other third party material in this chapter are included in the chapter's Creative Commons license, unless indicated otherwise in a credit line to the material. If material is not included in the chapter's Creative Commons license and your intended use is not permitted by statutory regulation or exceeds the permitted use, you will need to obtain permission directly from the copyright holder.

CHAPTER 11

Assessing Innovative Sources for the Loss and Damage Mechanism: The Role and Prospective Regulation of Climate-Friendly Foreign Investment

Federica Cristani

1 INTRODUCTION

On the very first day of the UN Framework Convention on Climate Change's (UNFCCC) 28th Conference of the Parties (COP28), the participating States decided,[1] without any objection, to operationalize the Loss and Damage Fund to 'provide finance for addressing a variety of challenges associated with the adverse effects of climate change, such as climate-related emergencies, *sea level rise*, [...] and the need for climate-resilient reconstruction and recovery' (Hemingway Jaynes,

---

[1] This Chapter is based on the work carried out within the framework of the Horizon Europe HRJust project (States' Practice of Human Rights Justification: a study in civil society engagement and human rights through the lens of gender and intersectionality), GA No. 101094346 and the SERI (Swiss State Secretariat for Education, Research and Innovation—SERI) under grant agreement n. 23.00131/ 1,010,943,546.

F. Cristani (✉)
Institute of International Relations Prague, Prague, Czech Republic
e-mail: cristani@iir.cz

© The Author(s) 2025
E. Fornalé (ed.), *Sea Level Rise*,
https://doi.org/10.1007/978-3-031-89171-7_11

2023; UNFCCC COP, 2024b; United Nations, 2023). The decision 'has set much needed momentum for addressing loss and damage in the most vulnerable nations', as stated by Ms. Rabab Fatima, the Under-Secretary-General and High Representative for the Least Developed Countries, Landlocked Developing Countries and Small Island Developing States, who, during COP28, recalled that '[t]he world's most vulnerable nations—in particular, the Small Island Developing States (SIDS)—are among the nations who have contributed the least to the climate crisis, yet they bear the brunt of the impacts' (UN, 2023). For small island States, indeed, the Loss and Damage Fund represents a crucial mechanism—suffice to recall that '[o]ver the past fifty years, small island developing States lost $153 billion due to weather, climate, and water related hazards' (UN, 2023).

The decision taken at COP28 came after a long path to the establishment of the loss and damage mechanism within the UNFCCC. As the first milestone, COP19 established the *Warsaw International Mechanism for Loss and Damage*, while COP25 established the *Santiago Network* to channel technical assistance from all relevant stakeholders. At COP26, the *Glasgow Dialogue* was initiated to serve as a platform among the parties and relevant stakeholders to address loss and damage associated with the adverse impacts of climate change. At COP27, an agreement was reached to 'establish new funding arrangements for assisting developing countries that are particularly vulnerable to the adverse effects of climate change [including sea level rise], in responding to loss and damage, including with a focus on addressing loss and damage by providing and assisting in mobilizing new and additional resources' (UNFCCC COP, 2023, p. 12). This agreement was welcomed by the Chair of the Alliance of Small Island States (AOSIS) with satisfaction, as the Chair claimed '[a] mission 30 years in the making [was] accomplished'—but at the same time this was accompanied by the warning that States 'must work even harder [...] to operationalize the loss and damage fund' (AOSIS, 2022). Also at COP27 a Transitional Committee on the operationalization of the new funding arrangements was set up with a mandate to provide recommendations for consideration and adoption by the forthcoming COP meetings (UNFCCC COP, 2023).

One of the tasks of the Transitional Committee has been the identification of potential sources of funding, including *innovative sources* (UNFCCC COP, 2023). In this regard, climate-friendly FDI can be part of the strategies that help developing countries like SIDS address climate change. However, the current international legal framework for FDI is not well equipped to protect climate-positive foreign investments. The 2022 UNCTAD report on 'Investment Policy Trends in Climate Change Sectors' has highlighted that 'policy measures to attract investment in climate change adaptation sectors still need to be developed and implemented in developing countries to respond to the growing financing needs in those sectors' (UNCTAD, 2022a, p. 1). As discussed also in the following paragraphs, the discussion on climate-friendly FDI, and relevant data, has mainly been focused on mitigation and adaptation measures. When it comes to climate-friendly FDI as a source of finance for the Loss and Damage Fund, the picture becomes quite blurred, with virtually no available data on whether and to what extent FDI has been involved in loss and damage mechanisms.

The discussion on the loss and damage mechanism has also addressed other interlinked issues, such as gender and human rights. This chapter will offer the first overview of this topic with the aim to answer the research question of whether the current international (investment) legal framework is well equipped for promoting and regulating climate-friendly FDI for the loss and damage mechanism, which would also reinforce human rights protection in the field. The first part will provide the background analysis of the loss and damage mechanism under the UNFCCC, with an insight into the importance of the mechanism for small island developing States affected by the consequences of, among other things, sea level rise, while the second part will focus on an analysis of the international regulation of climate-friendly FDI, which comprises an assessment of the current legal framework, its gaps and possible proposals for a new generation of investment-related agreements, with an additional focus on how to include a gender and a human rights perspective in the picture.

2 The Establishment of the Loss and Damage Mechanism Within the Framework of the UNFCCC

Though an agreed upon and uniform definition of 'loss and damage' is still lacking within the UNFCCC, it 'can generally be understood as the negative impacts of climate change that occur despite, or in the absence of, mitigation and adaptation' (Burkett, 2014; UNEP, 2024a; United Nations (UN) Climate Action Thought Leaders, 2022). As highlighted in the Final Report of the ILA International Law and Sea Level Rise Committee, '[l]oss generally refers to climate-related impacts for which restoration is not possible' while damage 'refers to negative impacts for which restoration is possible' (International Law Association, 2024, p. 12). Different types of loss and damage exist, but they are usually categorized as economic and non-economic. Economic damages are those which are quantifiable, such as damage to infrastructure, agriculture, and property, whereas non-economic losses include, for example, the degradation of ecosystems, loss of cultural heritage, and displacement of communities—which are more challenging to measure (United Nations Framework Convention on Climate Change (UNFCCC), Executive Committee of the Warsaw International Mechanism for Loss and Damage, 2024).[2]

Loss and damage do not affect countries in the same way: for the most part, developing countries are disproportionately impacted by climate disasters, which often overwhelm the adaptive capacities of these nations, leading to unavoidable losses. When talking about developed/developing countries, we refer primarily to the classification elaborated by the World Bank, which distinguishes between low, lower-middle, upper-middle and high-income groups of countries according to the annually updated threshold level of gross national income (GNI) per capita (World Bank, 2024a, 2024b). Accordingly, the World Bank has generally referred to 'low and middle-income countries' as 'developing countries' (Khokhar & Serajuddin, 2015). Other international organizations have elaborated slightly different categorizations, though it led to the same countries always being included in the 'developing' group—e.g. the International Monetary Fund classifies 37 countries as 'advanced economies', while all others are considered 'emerging and developing economies' (IMF, 2015); the United Nations Development Programme has instead elaborated the Human Development Index as a way to measure 'very high', 'high', 'medium' and 'low' levels of human development (UNDP, 2024). Within the United Nations, while a formal definition of developing countries has not been adopted there so far, the term is used to identify 159 countries (i.e. all countries except for those in Europe and North America along with Japan, Australia, and New Zealand) (United Nations (UN), 1999). Moreover, the UN maintains a list of 'least developed countries', which are defined as 'countries that have low levels of income and face severe structural impediments to sustainable development' (United

---

[2] With special regard to the challenges related to the quantification of non-economic losses, see also UNFCCC (2013).

Nations, 2024). Again a very unique group within the UN is the group of the Small Island Developing States (SIDS), which is identified as 'a distinct group of 39 States and 18 Associate Members of United Nations regional commissions that face unique social, economic and environmental vulnerabilities' (United Nations, 2024). The SIDS group includes countries from the Caribbean, the Pacific, the Atlantic, the Indian Ocean, and the South China Sea. They have been recognized as 'a special case both for their environment and development' since the 1992 United Nations Conference on Environment and Development in Rio de Janeiro (United Nations, 2024). SIDS are home to 65 million people, and due to their remote geography, they are particularly prone to the adverse effects of climate change, especially hurricanes and sea level rise (United Nations, 2024), thus 'represent[ing] some of the most vulnerable and marginalized populations on the planet' (UNDP, 2024). This makes the SIDS a very particular group of vulnerable countries (Benjamin & Thomas, 2023; UNEP, 2024b). According to the data elaborated by the Sendai Framework Monitor, 'on average, 18% of the total population is affected after each disaster in SIDS, compared to 6% in non-SIDS countries' (United Nations Office for Disaster Risk Reduction (UNDRR, 2024). This in turn generates a great amount of costs associated with the adverse effects of climate change (Warner et al., 2012). A 2023 academic research report published in *Nature Communications* found that between 2000 and 2019, the world suffered a loss of at least $2.8 trillion in losses and damages from climate change—costing around $16 million per hour (Newman & Noy, 2023). And SIDS pay the highest price: indeed, '[o]ver the past fifty years, SIDS have lost $153 billion due to weather, climate, and water-related hazards [and o]n average, SIDS experience 2.1% of GDP loss due to disasters, whereas other countries face an average of 0.3% of GDP' (United Nations Office for Disaster Risk Reduction (UNDRR), 2024). As rightly highlighted, 'losses and damages are disproportionately experienced by developing countries and by vulnerable groups, such as people of low socio-economic class, migrant groups, the elderly, women and children' (United Nations (UN) Climate Action Thought Leaders, 2022).

Accordingly, SIDS have always been at the frontline in pushing for the establishment of a loss and damage mechanism during international climate negotiations. Back in 1989, the Male Declaration on Global Warming and Sea Level Rise, which was adopted in the framework of the Small States Conference on Sea Level Rise that took place between 14 and

18 November 1989, already 'call[ed] upon all State[s] [...] to consider ways and means of protecting the small States of the world which are most vulnerable to sea level rise' (Small States Conference on Sea Level Rise, 1989). The first-ever express reference to 'loss and damage' in climate negotiations appeared in 1991, when Vanuatu, acting on behalf of the Alliance of Small Island States (AOSIS), proposed an 'insurance mechanism' for inclusion in the forthcoming UNFCCC, with an express request to 'industrialised' countries to pay for the 'loss and damage' that would harm SIDS as a result of the sea level rise. According to the submission by Vanuatu, '[t]he financial burden of *loss and damage* suffered by the most vulnerable small island and low-lying developing countries (Group 1 countries) as a result of sea level rise shall be distributed in an equitable manner amongst the industrialized developed countries (Group 2 countries) by means of an Insurance Pool' [emphasis added] (Intergovernmental Negotiating Committee for a Framework Convention on Climate Change Working Group II, 1991). Actually, this approach took inspiration from already existing international conventions that were designed to provide compensation for damage resulting from nuclear incidents and oil spills (Appadoo, 2021).[3]

Later, at COP13 in Bali, the 'loss and damage' reference was entered into a negotiated text. The Bali Action Plan indeed aimed to provide 'enhanced action on adaptation', establishing adaptation as a separate 'pillar' of negotiations that is independent from mitigation, and this included not only strategies to help countries reduce climate risk, but also, more explicitly, 'means to address *loss and damage* associated with climate change impacts in developing countries that are particularly vulnerable to the adverse effects of climate change' (UNFCCC COP, 2008; Warner & Zakieldeen, 2012).

A landmark milestone for the loss and damage mechanism came in 2013, when the Warsaw International Mechanism for Loss and Damage associated with Climate Change Impacts (WIM) was set up at COP19 with the aim to 'address loss and damage associated with the impacts of climate change, including extreme events and slow-onset events, in developing countries that are particularly vulnerable to the adverse effects

---

[3] For example, the 1963 Brussels Supplementary Convention on Third Party Liability in the Field of Nuclear Energy, which was adopted in order to provide additional funds to compensate damage resulting from a nuclear incident where other funds would be insufficient.

of climate change', especially SIDS (Tietjen & Gopalakrishnan, 2023; UNFCCC COP, 2014, Preamble). Alongside the WIM, an executive committee of the mechanism was set up as well, with its tasks including, among others, reporting annually to the UNFCCC Conference of the Parties and making recommendations (UNFCCC COP, 2014, paras. 2–3). The WIM was assigned the functions of '[e]nhancing knowledge and understanding of comprehensive risk management approaches to address loss and damage [...; taking a]ction to address gaps in the understanding of and expertise in approaches to address loss and damage [...; c]ollect[ing...], shar[ing...], manag[ing...] and us[ing...] relevant data and information, including gender-disaggregated data; [providing] overviews of best practices, challenges, experiences and lessons learned in undertaking approaches to address loss and damage; [s]trengthening dialogue, coordination, coherence, and synergies among relevant stakeholders [...; and e]nhancing action and support, including finance, technology and capacitybuilding, to address loss and damage' (UNFCCC COP, 2014, para. 5). However, the decision just '[r]equests developed country Parties to provide developing country Parties with finance, technology and capacity-building' (UNFCCC COP, 2014, para. 14).

In 2015, at COP21, the Paris Agreement was adopted (UNFCCC, 2015); 'loss and damage' were included in Article 8, according to which the 'Parties recognize the importance of averting, minimizing and addressing loss and damage associated with the adverse effects of climate change, including extreme weather events and slow onset events [...]'. However, at the same time the parties '[a]gree[d] that Article 8 of the Agreement does not involve or provide a basis for any liability or compensation.' (Franczak, 2022; UNFCCC COP, 2016).

Later, in 2019, at COP25, the Santiago Network for Loss and Damage was established with the aim to contribute to the implementation of the WIM through data gathering and technical assistance (Franczak, 2022; UNFCCC COP, 2022b).[4] At the subsequent COP26, the developing countries led by the SIDS pushed for a new loss and damage fund. In particular, a proposal from the Alliance of Small Island States (AOSIS) requested "establish[ing] the 'Glasgow Loss and Damage Facility' as

---

[4] For a general overview, see also UNFCCC (2024a).

a standalone facility under the financial mechanism of the Convention".[5] However, the proposal was not successful—instead, the 'Glasgow Dialogue' was established with the Glasgow Climate Pact, which further emphasized the need for increased financial resources to address loss and damage (Gabbatiss, 2022; UNFCCC COP, 2022a).[6]

Another important milestone came in 2022, namely at COP27, where an agreement was reached to establish new funding arrangements, including a loss and damage fund, with the newly established Transitional Committee tasked with, among other things, preparing recommendations for the operationalization of these new funding arrangements, including the fund (UNFCCC COP, 2023; UNFCCC COP, 2022c). This agreement attracted global attention due to the devastating climate events at the time, including the 2022 floods in Pakistan, which directly impacted millions of people and led to an emphasis on the need for immediate action (Bhandari et al., 2023).

At COP28, the Loss and Damage Fund was operationalized as an entity entrusted with the operation of the Financial Mechanism of the UNFCCC Convention, which would also serve the Paris Agreement. In particular, Decisions 5/CMA.5 (UNFCCC COP, 2024b) and 1/CP.28 (UNFCCC COP, 2024b) on the operationalization of the new fund established that the fund would be serviced by a new secretariat (UNFCCC COP, 2024b, para. 3), and governed and supervised by a board (UNFCCC COP, 2024b, para. 4). They also endorsed the recommendations of the Transitional Committee on funding arrangements, including establishing an annual high-level dialogue (Tietjen & Gopalakrishnan, 2023; UNFCCC COP, 2024b, para. 1). To date, the Board of the Fund has met three times, first in Abu Dhabi (30 April–2 May 2024)—where more administrative-like issues related to the practical organization of the fund were discussed (UNFCCC, 2024b), then in Songdo–Incheon (9–12 July 2024) (UNFCCC, 2024c)—where the Philippines was selected to host the Board of the fund (UNFCCC, 2024e), and most recently in Baku (18–20 September 2024) (UNFCCC,

---

[5] As reported in Franczak (2022), p. 4.

[6] The establishment of the Glasgow Dialogue was seen as 'another excuse for delaying the process of taking real action to address loss and damage', as declared by Ineza Umuhoza Grace, co-director of the Loss and Damage Youth Coalition, and reported by Carbon Brief.

2024d)—where Ibrahima Cheikh Diong was selected as the first Executive Director of the fund for a four-year term beginning on 1 November 2024 (UNFCCC, 2024f).

Discussions on the fund will carry on also in COP29, which will take place between 11 and 22 November 2024, and where the WIM will be further revised as well (Mwale et al., 2024).[7]

2.1 The Puzzle of Climate Finance and the Challenge of Securing Funding Sources

The Loss and Damage Fund forms part of a puzzle of climate financial mechanisms (Warner & Weisberg, 2023). Generally, when talking about 'climate finance', we refer to 'local, national or transnational financing—drawn from public, private and alternative sources of financing—that seeks to support mitigation and adaptation actions that will address climate change' (UNFCCC, 2024g). The UNFCCC Convention,[8] as well as the Kyoto Protocol[9] and the Paris Agreement,[10] has made an express call for financial assistance from State parties with more financial resources to the more vulnerable countries, especially SIDS. In accordance with the principle of 'common but differentiated responsibility and respective capabilities' set out in the UNFCCC Convention,[11] developed country Parties are to provide financial resources to assist developing country Parties in

---

[7] The present chapter covers the discussions that have taken place at the time of writing (October 2024), before COP29.

[8] According to Article 11 of the UNFCC Convention, '[a] mechanism for the provision of financial resources on a grant or concessional basis, including for the transfer of technology, is hereby defined. [...] The developed country Parties may also provide and developing country Parties avail themselves of, financial resources related to the implementation of the Convention through bilateral, regional and other multilateral channels'.

[9] According to article 11 of the Kyoto Protocol, 'developed country Parties and other developed Parties in Annex II to the Convention may also provide, and developing country Parties avail themselves of, financial resources for the implementation of Article 10, through bilateral, regional and other multilateral channels'.

[10] According to article 9 of the Paris Agreement, 'developed country Parties shall provide financial resources to assist developing country Parties with respect to both mitigation and adaptation in continuation of their existing obligations under the Convention'.

[11] See, among other similar texts, the Preamble of the UNFCCC Convention, which '[a]cknowledg[es] that the global nature of climate change calls for the widest possible

implementing the objectives of the UNFCCC. To facilitate the provision of climate finance, the UNFCCC Convention established a financial mechanism under Article 11 to provide financial resources to developing country Parties, including SIDS. The financial mechanism also serves the Kyoto Protocol and the Paris Agreement, while the COP decides on its policies, programme priorities, and eligibility criteria for funding. The UNFCCC Convention states that the 'operation [of the financial mechanism] shall be entrusted to one or more existing international entities'.[12] In this respect, the Global Environment Facility has served as an operating entity of the financial mechanism since the Convention's entry into force in 1994—it comprises a partnership of 186 member governments as well as civil society organizations, Indigenous Peoples, women, and youth, with a focus on integration and inclusivity.[13]

In 2001, States parties to the UNFCCC also established the Special Climate Change Fund and the Least Developed Countries Fund (LDCF),[14] both managed by the Global Environment Facility, as well as the Adaptation Fund under the Kyoto Protocol.[15]

At COP16 in 2010, the Green Climate Fund was then established[16] as 'the world's largest climate fund, mandated to support developing

cooperation by all countries and their participation in an effective and appropriate international response, in accordance with their common but differentiated responsibilities and respective capabilities and their social and economic conditions'.

[12] See article 11, para. 1 of the UNFCCC Convention.

[13] For more details see the official website of the Global Environment Facility. (n.d.). *Who We Are*. Retrieved September 23, 2024, from https://www.thegef.org/who-we-are.

[14] The Special Climate Change Fund was established to finance projects relating to adaptation, technology transfer and capacity building, energy, transport, industry, agriculture, forestry and waste management, and economic diversification. On the other hand, the Least Developed Countries Fund was created with the aim to assist least developed countries in the preparation and implementation of the relevant national adaptation programmes of action (UNFCCC COP, 2001b).

[15] The aim of the Adaptation Fund was to finance concrete adaptation projects and programmes in developing countries under the Kyoto Protocol (UNFCCC COP, 2001a).

[16] According to para. 102 of the Decision, the 'Green Climate Fund, [...is] designated as an operating entity of the financial mechanism of the Convention under Article 11, with arrangements to be concluded between the Conference of the Parties and the Green Climate Fund to ensure that it is accountable to and functions under the guidance of the Conference of the Parties, to support projects, programmes, policies and other activities in developing country Parties using thematic funding windows' (UNFCCC COP, 2011).

countries [especially SIDS in] rais[ing] and realiz[ing] their Nationally Determined Contributions (NDC) ambitions towards low-emissions, climate-resilient pathways' (Green Climate Fund, 2024).

The above-mentioned funds have been designed to address mitigation and adaptation rather than loss and damage. Indeed, one challenge related to the Loss and Damage Fund has been to make sure that it does not divert resources from adaptation and mitigation efforts. Climate adaptation and mitigation measures remain a critical priority, and there is a need to distinguish and secure adequate funding for both adaptation/mitigation measures and measures to address loss and damage in order to avoid any overlapping between them (Bhandari et al., 2023). Instead, efforts to address loss and damage should be part of a broader strategy that would include stronger mitigation and adaptation measures (UNEP, 2024).

Determining and securing financial sources for the Loss and Damage Fund has been another major challenge in the negotiations: on the one hand, developing countries, and especially SIDS, which contribute the least to global emissions but are the most affected by the adverse impacts of climate change, have been arguing that high-emitting nations should bear the financial responsibility for the related loss and damage; on the other hand, developed countries have been reluctant to introduce binding commitments in this regard due to concerns over their potential liability for historical emissions (Tietjen & Gopalakrishnan, 2023). Consequently, funding pledges have been slow (Franczak, 2022; Perkins, 2023).[17]

Taking into account the above challenges, the Transitional Committee has explored *innovative funding sources* for the Loss and Damage Fund, including carbon taxes, levies on international aviation and shipping, fossil fuel extraction taxes, and FDI in climate-friendly sectors (UNEP, 2024a).

---

[17] The question of which states should contribute to the funds is crucial and debated also in connection with funding for adaptation and mitigation measures. In this respect, the World Resources Institute has just released its climate finance calculator, where one can create different scenarios that take into account—to varying degrees—both historical greenhouse gas emissions and the economic maturity of countries while focusing on climate justice by showing which countries bear the greatest historical responsibility for climate change. It shows, for example, that Somalia, since the 1950s, emitted roughly as much CO2 by burning fossil fuels as the US economy emits in 3 days on average. See more information about the climate finance calculator at Qinan Zou et al. (2024).

What has seemed to be a relatively understudied question is how to efficiently include climate-friendly FDI in the discussion, in particular, whether the current international investment legal framework is well equipped to promote and regulate such FDI.

3   Climate-Friendly FDI: Opportunities and Challenges for the Loss and Damage Fund

Since the 2010s, there has been a growing debate among stakeholders on climate-friendly FDI; an internationally agreed definition of 'climate-friendly' FDI is not yet in place—however, it is usually 'broadly defined as investment in mitigation and adaptation: the former involves investment in cleaner and more energy-efficient technology supporting the reduction of greenhouse gas emissions; the latter involves investment in critical infrastructure, technology and activities to increase resilience and help adapt to the consequences of climate change' (UNCTAD, Trade and Development Board, 2022).

Since 2010, the above-mentioned Green Climate Fund has mobilized billions of dollars for funding projects in, among others, the renewable energy sector, sustainable agriculture, and ecosystem restoration.[18] Also, the World Bank has been a major player in climate finance through its lending and investment activities in developing countries, including SIDS. In particular, in 2017, the World Bank has been promoting the first prototype of 'blue bonds' in the Seychelles to improve the marine resources of the region (with a $15 million package). It is also worth mentioning that in 2007, the World Bank established two ad hoc regional risk insurance schemes for SIDS, namely the Caribbean Catastrophe Risk Insurance Facility[19] and the Pacific Catastrophe Risk Assessment and the Financing Initiative (Mahul et al., 2016), which have contributed to, among other things, issuing recovery packages for the Bahamas following Hurricane Dorian in 2019 and for Tonga after Tropical Cyclone Gita in 2018 (World Bank Group, 2019).

---

[18] In order to de-risk the delivery of capital flows from the private sector, the Green Climate Fund has also established the Private Sector Facility, a division designed to fund and mobilize private sector actors, including institutional investors, project sponsors and financial institutions. For more information see Green Climate Fund (n.d.b).

[19] Caribbean Catastrophe Risk Insurance Facility. Company Overview. Retrieved October 16, 2024, from https://www.ccrif.org/about-us.

However, resources from the private sector are still far from being adequate to fulfil the needs of climate finance. Already in 2014, UNCTAD highlighted (with the first analysis of this kind being undertaken by an international organization) the investment gap in climate change mitigation and adaptation in its annual World Investment Report (UNCTAD, 2014). When analysing the resource implications for the implementation of the Sustainable Development Goals (SDGs) for the period 2015–2030, the report observed that '[e]stimates for total investment needs in developing countries alone range from $3.3 trillion to $4.5 trillion per year, for basic infrastructure (roads, rail and ports; power stations; water and sanitation), food security (agriculture and rural development), *climate change mitigation and adaptation*, health and education. Reaching the SDGs will require a step-change in both public and private investment […]. However, today, the participation of the private sector in investment in these sectors is relatively low. […] At today's level of investment—public and private—in SDG-related sectors in developing countries, an annual funding shortfall of some $2.5 trillion remains' [emphasis added] (UNCTAD, 2014, p. xxiv).

Though climate-friendly FDI has grown exponentially in the last decade, still the climate finance gap is wide, as it has been estimated to be in the order of $5.8–5.9 trillion for developing countries, including SIDS, in the pre-2030 period (WEF, 2024). As highlighted in the last COP, COP28, 'the adaptation finance needs of developing countries are estimated at USD 215–387 billion annually up until 2030, and [it is also estimated] that about USD 4.3 trillion per year needs to be invested in clean energy up until 2030, increasing thereafter to USD 5 trillion per year up until 2050, to be able to reach net zero emissions by 2050' (UNFCCC COP, 2024a).

While the relevant discussions have been largely focused on climate-friendly FDI for mitigation and adaptation measures, it should be noted that FDI can be a valid source of funding also for the Loss and Damage Fund. This has been very well expressed by the Technical Support Unit of the Transitional Committee in its 2023 report, where it observed that 'philanthropic investors and foundations could be a significant source of funds that has been largely untapped so far' (Technical Support Unit of the Transitional Committee. Working Group 5(c), 2023, July, para. 8), adding that '[w]hile there are barriers, such as […] *regulation uncertainty*, the potential is huge and should be explored through individual or hybrid (public–private) investment and solutions' [emphasis added]

(Technical Support Unit of the Transitional Committee. Working Group 5(c), (2023, July, para. 38).

Boosting climate-friendly FDI in the context of the loss and damage mechanism can bridge the gap between the funding needs of developing countries, especially SIDS, and the financial resources available through traditional channels. At the same time, climate-friendly FDI also poses some challenges, including the risk that foreign private investors may prioritize their own financial interests over the needs of local communities and the environment of the host countries or that existing inequalities and injustices can be exacerbated, as vulnerable communities, like the ones in SIDS, may not have a say in the decision-making process or benefit from the profits generated by the related projects in the host State (Loungani & Razin, 2001).

Accordingly, it is important to rely on a robust regulatory framework for climate-friendly FDI and to overcome the 'regulation uncertainty' referred to by the Technical Support Unit in the above-mentioned 2023 report. We should thus investigate whether the current international (investment) legal framework is well equipped to promote and regulate climate-friendly FDI.

3.1 Regulatory Gaps and Some Initial Proposals

FDI is traditionally regulated by international investment agreements (IIAs), including bilateral investment treaties (BITs) and treaties with investment provisions (TIPs). To date, there are 2615 IIAs in force.[20] While old-generation IIAs (concluded in the period from the 1980s until the early 2010s) usually made no reference to climate concerns, new-generation IIAs (concluded in the 2010s and 2020s) have included climate change-related provisions (UNCTAD, 2022c). Indeed, there has been a growing tendency to incorporate provisions promoting and facilitating climate-friendly FDI in the relevant environment chapters of international trade agreements,[21] as well as in investment facilitation

---

[20] See the relevant database of the UN Trade and Development. International Investment Agreements Navigator. Retrieved September 10, 2024, from https://investmentpolicy.unctad.org/international-investment-agreements.

[21] As, for example, in the case of the EU–New Zealand FTA, art. 19.11 (Trade and investment supporting sustainable development), paras. 4–5, according to which '[...] each Party shall promote and facilitate trade and investment in: (a) environmental goods

agreements[22] and in a number of non-binding international instruments focused on the transition to the green economy.[23]

---

and services; (b) goods that contribute to enhanced social conditions; and (c) goods subject to transparent, factual and non-misleading sustainability assurance schemes such as fair and ethical trade schemes and ecolabels. 5. Activities to promote and facilitate trade and investment as referred to in paragraph 4 may include: (a) awareness-raising actions and information and public education campaigns; (b) adoption of policy frameworks conducive to the deployment of best available technologies; (c) encouraging the uptake of transparent, factual and non-misleading sustainability schemes, especially for SMEs; (d) addressing related non-tariff barriers; and (e) reference to relevant international standards, such as the ILO conventions and guidelines or MEAs' (Free Trade Agreement Between the European Union (EU)-New Zealand Free Trade Agreement, 2023, July 9). See also the UK–Australia FTA, art. 22.6 (Environmental Goods and Services), according to which '[t]he Parties recognise the importance of trade and investment in environmental goods and services as a means of improving environmental and economic performance, contributing to clean growth, and addressing global environmental challenges. 2. Accordingly, each Party shall facilitate and promote, as appropriate, trade and investment in environmental goods and services, including environmental and low emissions technologies, clean and renewable energy and enabling infrastructure, and energy-efficient goods and services. 3. The Environment Working Group shall consider issues identified by a Party or the Parties related to trade in environmental goods and services, including issues identified as potential non-tariff barriers to that trade. The Parties shall endeavour to address any potential barrier to trade in environmental goods and services that may be identified by a Party, including by working through the Environment Working Group and in conjunction with other relevant committees established under this Agreement, as appropriate. 4. The Parties shall cooperate bilaterally and in international fora, as appropriate, on ways to enhance trade and investment in environmental goods and services' (Free Trade Agreement Between the United Kingdom of Great Britain (UK) and Northern Ireland and Australia, 2021, December 16).

[22] See the first Sustainable Investment Facilitation Agreement between the EU and Angola that entered into force on 1 September 2024, which aims to promote FDI to achieve the sustainable development goals. See, in particular, article 32 (Investment and climate change), para. 2, according to which '[e]ach Party shall: [...] promote the mutual supportiveness of investment and climate policies and measures, thereby contributing to the transition to a low greenhouse gas emission, resource-efficient economy and to climate-resilient development', and article 33 (Investment contributing to sustainable development), para 1, according to which '[i]n accordance with their commitment to enhance the contribution of investment to the goal of sustainable development, the Parties shall facilitate and encourage investment in sustainable production and consumption, in environmental goods and services, and investment of relevance for climate change mitigation and adaptation' (Sustainable Investment Facilitation Agreement Between the European Union (EU) and the Republic of Angola, 2023, November 17).

[23] See for example, the Chair's Statement of the 13th Asia–Pacific Economic Cooperation Energy Ministerial Meeting (2023).

Especially investment facilitation agreements' provisions are quite an interesting case for climate-friendly FDI, since they tend to include also commitments to an ongoing cooperation between the parties—rather than simply relying on the regulatory framework within each party's domestic legal order or containing non-enforceable commitments to facilitate FDI (Paine & Sheargold, 2024a). In particular, some IIAs provide for the creation of 'platforms for lasting cooperation' between States (UNCTAD, 2023).[24] Cooperation can operate on different levels and in different fields of policy action, such as technological transfer or knowledge sharing. Usually, states identify specific areas for cooperation in the negotiation plan—in this framework, they clarify which sectors and types of investment they intend to promote as instances of 'climate-friendly' FDI (Paine & Sheargold, 2024b). For example, the Australia–Singapore Green Economy Agreement includes seven priority areas for cooperation and identifies specific actions to be taken, such as 'identify[ing] a comprehensive list of environmental goods and services [...] in supporting the transition to sustainable economic growth and facilitating trade and investment flows [...or] facilitat[ing] and promot[ing] investment that will support decarbonisation efforts and open up new green economy opportunities' (Green Economy Agreement Between Singapore & Australia, 2022, para. 9).

However, IIAs are still not well equipped to protect climate-friendly FDI and encourage States to adopt measures that might attract more FDI of this kind. A major obstacle has been the constant tensions between the States' right to regulate and provisions protecting FDI in the relevant treaties—the so-called regulatory chill effect, according to which States are reluctant to adopt climate policies due to fears of compensation claims from foreign investors protected under IIAs that might appear whenever the new climate policies adversely impact their investments (Lieser, 2024). Indeed, some countries have had to rethink their national climate change measures in the light of possible investor-State arbitrations (Sharma, 2022); in the past decade, there have been around 190 Investor-State arbitration cases initiated by foreign investors in the fossil fuel and renewable energy sectors (UNCTAD, 2022d). Foreign investors have put forward investment claims in order to obtain compensation for the value reductions of their investments due to new climate policies that

---

[24] See, for example, the Kenya–United Kingdom Economic Partnership Agreement (UK Government, 2020).

the host States have introduced (RWE v. Kingdom of the Netherlands, 2021; *Uniper v. Netherlands*, ICSID Case No. ARB/21/22, 2021) and that have aimed, for example, to restrict coal-fired power plants (*Vattenfall AB and others v. Federal Republic of Germany*, ICSID Case No. ARB/12/12, 2021), ban offshore oil drilling (*Rockhopper v Italy*, ICSID Case No. ARB/17/14, 2022) or introduce incentives for the renewable energy sector (Sharma, 2022).[25] The 2022 Report of the UN Special Rapporteur on the Promotion and Protection of Human Rights in the Context of Climate Change has 'estimated that legal claims by oil and gas investors against those States that impose laws to limit fossil fuel activities could reach a total cost of $340 billion' (United Nations (UN) Special Rapporteur on the Promotion and Protection of Human Rights in the Context of Climate Change, 2022).

In light of this, States are increasingly securing their regulatory powers in climate action when negotiating IIAs; accordingly, the new generation of treaties include express safeguards for the State's right to regulate in some areas.[26]

---

[25] See, for the example, the series of cases that have been brought against Spain that relate to a series of governmental measures aimed at imposing a 7% tax on power generators' revenues and a reduction in subsidies for renewable energy producers. See, among others, *Charanne B.V. and Construction Investments S.a.r.l. v. Spain*, SCC, Award (2016, January 21); *BayWa r.e. Renewable Energy GmbH and BayWa r.e. Asset Holding GmbH v. Spain*, ICSID Case No. ARB/15/16 (2019, December 2), *Isolux Infrastructure Netherlands B.V. v. Spain*, SCC Arb No. 2013/153, Award (2016, July 17); *Eiser Infrastructure Limited and Energía Solar Luxembourg S.à.r.l. v. Kingdom of Spain*, ICSID Case No. ARB/13/36, Award (2017, May 4). See also Prabhu and Staffer (2023).

[26] For example, the 2019 Model Dutch BIT acknowledges that '[t]he provisions of th[e] Agreement shall not affect the right of the Contracting Parties to regulate within their territories necessary to achieve legitimate policy objectives such as the protection of public health, safety, environment, public morals, labor rights, animal welfare, social or consumer protection or for prudential financial reasons. The mere fact that a Contracting Party regulates, including through a modification to its laws, in a manner which negatively affects an investment or interferes with an investor's expectations, including its expectation of profits, is not a breach of an obligation under this Agreement'. This practice has been labelled as 'sovereignty reassertion' (Sharma, 2022). For example, Article 10.10 of the Free Trade Agreement Between the Oman and the United States (US) (2009, January 1) provides that '[n]othing in this Chapter shall be construed to prevent a Party from adopting, maintaining, or enforcing any measure otherwise consistent with this Chapter that it considers appropriate to ensure that investment activity in its territory is undertaken in a manner sensitive to environmental concerns'.

These kinds of provisions open the way for host States to adopt measures that can attract and incentivize climate-friendly FDI. Additionally, there seems to be a new approach of States that involves progressively introducing investment-related obligations to foreign investors. For example, the Morocco-Nigeria BIT (not yet in force) requires foreign investors to comply with the home or host State's rigorous environmental impact assessments prior to establishing the proposed investment[27] (Lieser, 2024).

Several international initiatives aim to contribute to the reform of the IIA regime in the light of climate change objectives, such as the OECD work programme on the future of investment treaties (Organisation for Economic Co-operation and Development (OECD), 2024), the UN Working Group on Business and Human Rights' reports on human rights and IIAs (United Nations (UN) Working Group on the Issue of Human Rights and Transnational Corporations and Other Business Enterprises, 2021; United Nations (UN) Working Group on the Issue of Human Rights and Transnational Corporations and Other Business Enterprises, 2024) and the works of the UNCITRAL Working Group III: Investor-State Dispute Settlement Reform (United Nations Commission on International Trade Law (UNCITRAL), 2024).

To date, provisions aimed at promoting climate-friendly FDI have been largely focused on FDI in mitigation and adaptation measures. On the other hand, IIAs seem silent regarding climate-friendly FDI related to loss and damage. Apart from the recent report from the Technical Support Unit of the Transitional Committee, which has made a clear reference to the need to promote FDI for loss and damage, investment reports

---

[27] See Reciprocal Investment Promotion and Protection Agreement between the Government of the Kingdom of Morocco and the Government of the Federal Republic of Nigeria (2016, December 3), Article 14 (Impact Assessment): '(1) Investors or the investment shall comply with environmental assessment screening and assessment processes applicable to their proposed investments prior to their establishment, as required by the laws of the host State for such an investment or the laws of the home State for such an investment, whichever is more rigorous in relation to the investment in question. (2) Investors or the investment shall conduct a social impact assessment of the potential investment. The Parties shall adopt standards for this purpose at the meeting of the Joint Committee. (3) Investors, their investment and host State authorities shall apply the precautionary principle to their environmental impact assessment and to decisions taken in relation to a proposed investment, including any necessary mitigation or alternative approaches of the precautionary principle by investors[,] and investments shall be described in the environmental impact assessment they undertake'.

and the scholarly debate seem to focus more on a specific kind of FDI, namely FDI for the post-disaster recovery phase. As already mentioned, loss and damage mechanisms should address the negative impacts of climate change that occur despite, or in the absence of, mitigation and adaptation measures. This would include impacts following sudden-onset events—such as heat waves, flooding, and storms—and also slow-onset events—such as sea level rise or ocean acidification, and encompass both economic and non-economic losses and damages (United Nations (UN) Climate Action Thought Leaders, 2022). And FDI should be channelled to address all aspects of loss and damage. However, when it comes to loss and damage in cases of slow-onset events and non-economic loss, data and debate on the relevant FDI seem absent.

Also when it comes to FDI in the loss and damage area in cases of sudden-onset events (referred to as FDI in disaster recovery), only sectorial studies are available, like the 2022 research on disaster management financing in pre- and post-disaster phases in the period from 2013 to 2019 in Pakistan (Khan et al., 2022) or the 2020 World Bank report on 'Private Sector Participation in Disaster Recovery and Mitigation' (Marcelo Gordillo et al., 2020). At the same time, the attention to the need to promote FDI in SIDS is increasing, as testified, for example, by the discussions that took place at the workshop on 'Facilitating investment in SDG projects: spotlight on small island developing states' organized in 2019 by UN Trade and Development (UNCTAD) in partnership with the Caribbean Association of Investment Promotion Agencies and the Caribbean Export Development Agency, where a strong emphasis was put on the necessity to strengthen investment promotion and facilitation in the region (UNCTAD, 2019). This was echoed in the 'Call to Action on Mobilization of Resources for SIDS', which was jointly issued by the United Nations Secretary-General and the Prime Minister of Antigua and Barbuda on 28 May 2024 as an outcome of the 4th International Conference on Small Island Developing States. It called 'on leaders of international financial institutions, development banks, UN entities, the private sector, as well as donor partners, to [...s]cale-up climate finance to SIDS and urgently capitalize the new fund for responding to loss and damage, ensuring that SIDS priorities and needs are considered in the context of access and resource allocation' (UNSG Prime Minister of Antigua and Barbuda, 2024). Also the recent UNCTAD report on the 'Strategy to support Small Island Developing States' highlights the need to '[r]eview investment policies and international investment agreements to attract higher levels of FDI [to SIDS] for sustainable development [and f]acilitate FDI attracting to SIDS through promotion activities, cooperation and partnerships' (UNCTAD, 2024). Data is generally missing on the impact of climate-friendly FDI in the host States and, especially when it comes to FDI for loss and damage, on how and to what extent the FDI contributed to decreasing losses in the given region affected by sudden-onset disasters. In this respect, more research is surely needed, also to help policymakers and all relevant stakeholders to understand how to efficiently regulate this kind of FDI.

Generally, the reported FDI for disaster recovery includes that which is used for restoration of objects damaged by floods, rehabilitation of water supply and sanitation schemes, and reconstruction of schools or health facilities damaged by disasters (Khan et al., 2022). To date, no international legal framework (not even a non-binding one) has been dedicated to this kind of FDI, and IIAs, to date, have not provided any specific regulation in this respect—including the ones which SIDS are parties of.[28] Usually, FDI for disaster recovery is mainly based on solidarity-based donor contributions, and 'only a marginal proportion of [the] claimed financial compensation is covered. [...] It is evident that losses and damages will continue to affect the most vulnerable across and within countries, fostering [...] *climate investment traps* as climate impacts limit the availability of financial resources even more' [emphasis added] (Shahud et al., 2023).[29]

In order to secure a regulatory framework that would promote and protect climate-friendly FDI (the investments in mitigation and adaptation measures as well as those directed towards disaster recovery and funding the loss and damage mechanism), a holistic climate-responsive IIA reform is needed (Burke-White, 2022; Dietrich Brauch, 2023). Under this reform, States, with the consultation and cooperation of relevant stakeholders, could re-design the scope and purpose of IIAs, so that they could 'contribute[...] to sustainable development, [...be] coherent

---

[28] To the best knowledge of the author at the time of writing (October 2024) and according to the public data of IIAs collected in the United Nations Trade and Development (UNCTAD) Investment Policy Hub. Retrieved October 17, 2024, from https://investmentpolicy.unctad.org.

[29] A 'climate investment trap occurs when climate-related investments remain chronically insufficient' (Ameli et al., 2021).

with domestic policies, and [...be] consistent with international obligations, including those relating to climate action' (UNCTAD, 2022b). This would guarantee reliance on a transparent regulatory framework (Paine & Sheargold, 2024b; Stephenson & Zhan, 2022) and overcoming the regulatory uncertainties referred to in the above-mentioned 2023 report of the Technical Support Unit of the Transitional Committee—especially when it comes to the regulatory framework for the promotion and protection of climate-friendly FDI pertaining to loss and damage, in particular that targeting SIDS. In this respect, cooperation among relevant stakeholders, as mentioned, would be key to boosting such a reform. More generally, '[c]ooperation is central in tackling the threat posed by climate change' (Rudall, 2021), as this is also true when it comes to implementing the loss and damage mechanism (Freestone & Çiçek, 2021). In this respect, it is worth recalling that according to the Paris Agreement, the '[p]arties should enhance understanding, action and support, including through the Warsaw International Mechanism, as appropriate, on a *cooperative* and facilitative basis with respect to loss and damage associated with the adverse effects of climate change' [emphasis added] (Paris Agreement, 2015, Article 8(3)).

4 Integrating a Gender Perspective into the Loss and Damage Mechanism

As already mentioned, the promotion and protection of FDI pertaining to loss and damage require a new approach to the existing regulatory framework. It should be a holistic approach which would take into account all the relevant stakeholders that would benefit from this source of finance, including women. In this respect, the discussion on women and climate change is already in place (Fornalé, 2023a; Fornalé & De Vido, 2023).[30] While engaging in this discussion is outside the scope of this chapter, what is worth briefly recalling is the recent report presented at COP28 by UN Women on 'feminist climate justice', according to which 'by 2050 climate change may push up to 158 million more women and girls into poverty, and cause 232 million to face food insecurity' (United Nations (UN) Women, 2023).

---

[30] See the chapter by De Vido in this book.

The report makes it clear that 'the new loss and damage fund [...] should [...] address *gendered* economic and non-economic losses' [emphasis added] (Turquet et al., 2023, p. 31). UN Women provides the example of *gendered* loss and damage in the food systems, highlighting how 'specifically due to the climate crisis, [it has] disproportionately affected women and gender-diverse people [...]. Such losses and damages are often unaccounted for because they either impact the informal economy or unpaid care workers or they are intangible in nature, such as emotional or cultural loss[,] and as such not captured by traditional econometric measures' (Turquet et al., 2023, p. 31). This is the case, for example, of women owning small farms and affected by crop loss and land dispossession or women in certain regions who are more likely to be affected by climate-related disasters due to their roles in agriculture, water collection, and caregiving (Tewari et al., 2023). Additionally, non-quantifiable losses should also be taken into account, especially in the case of sea level rise—indeed, one of the side effects of sea level rise is the risk that the affected communities may increase their saltwater consumption due to the infiltration of salt water into freshwater drinking supplies. Studies have shown that 'saltwater consumption has negative, long-lasting effects on nearly every stage of a woman's reproductive cycle' (Teirstein & Al Hasnat, 2024) and this issue 'has become one of the main threats to the safety of freshwater supply in coastal zones' (Cao et al., 2021), especially for women.

In this respect, introducing a gender perspective into the loss and damage mechanism is crucial for ensuring that the specific needs of women are adequately addressed.

A first step would be to collect comprehensive data on the impacts of sudden- and slow-onset events, such as sea level rise, on women; to date, disaggregated data on this subject are missing,[31] and this makes it more challenging for all the stakeholders to have a clear picture of the relevant phenomena. As highlighted in the above-mentioned UN Women report, '[t]he large gaps in gender statistics on this critical issue are holding

---

[31] But there have been already some reports with disaggregated data on the topic, like the one on Asia by Oxam (Tewari et al., 2023), one on Africa (Chakma et al., 2022) and the above-mentioned recent one presented at COP28 by UN Women. However, further research is still needed, as highlighted by the UN Women report itself: '[c]apturing the intersections of gender and environment in quantitative data is challenging and has long been neglected' (Turquet et al., 2023, p. 14).

back progress on gender-responsive policymaking' (Turquet et al., 2023, p. 14).

In parallel, there is the need to keep raising attention to the topics at hand and translate all the efforts that have been made by relevant stakeholders in the field into programmatic policy actions. This includes, for example, introducing a clear reference to gender-related questions in, among other things, the loss and damage mechanism (Tewari et al., 2023; Wijenayake, 2013).

Some first steps (though quite isolated) in the direction of including a gender dimension in climate finance have already been taken. Worth recalling is the Gender Action Plan adopted in the framework of the Green Climate Fund, which requires that all projects include gender assessments and promote gender equality (Green Climate Fund Board, 2014; Green Climate Fund Board, 2019).

However, significant barriers remain: accessing these funds requires a technical expertise and an institutional capacity that are lacking in several situations, especially those in developing countries, or more particularly SIDS, and this is coupled with bureaucratic hurdles and a lack of gender-related policies (Chakma et al., 2022; Price, 2021). More needs to be done to mainstream gender equality across all climate finance mechanisms. This holds true also for the loss and damage mechanism, and in the way sources of funding (including FDI) are promoted and channelled[32] so that they can be gender-responsive.

5 Adding a Human Rights Perspective to the Loss and Damage Mechanism

The loss and damage mechanism entails a human rights dimension, among others. This has recently been very clearly expressed in the first analytical study on the impacts of loss and damage on human rights, which was carried out by the UN Secretary-General and presented to the UN Human Rights Council on 28 August 2024 (HRC, 2023; HRC, 2024). It calls for 'human rights and equity-based approaches to loss and damage', which would imply 'anchoring relevant policies and measures [...] in human rights' (HRC, 2024, para. 30).

---

[32] See, in this respect, the first guide developer by the World Bank on how to mainstream gender issues in investment climate policies, authored by Blackden et al. (2010) and the World Bank Group (2024).

The UN Secretary-General made it clear that a *'paradigm shift* from contemporary economic and governance systems that threaten humanity's future towards a human rights economy that is fair, equitable, inclusive and sustainable, creates decent work opportunities, reduces inequalities and poverty and upholds human rights is needed', adding that '[p]olicies and programmes [...] need to be tailored to national circumstances within a global economy where businesses and *investors* respect the rule of law, including international human rights and labour law' [emphasis added] (HRC, 2024, para. 36). The mentioned 'paradigm shift' to a human rights-based approach in climate finance in the field of loss and damage is required not only for states, but also for investors—which seems to further imply that the FDI in the loss and damage mechanism should also be anchored in the rule of law.

Considering that to date, as mentioned, this kind of FDI seems to lie in an almost unregulated limbo, the call of the UN Secretary-General can be interpreted as encouraging a reinforcement of the relevant regulatory framework—as suggested in the previous paragraph. Though not explicitly mentioning it, the report seems to suggest that FDI pertaining to loss and damage should also be based on a human rights approach. Indeed, the Secretary-General warns that '[b]usinesses involved in loss and damage responses must respect human rights' (HRC, 2024, para. 34).

The report of the Secretary-General also makes a call for the need to adopt a gender and, more broadly, an intersectionality perspective when dealing with loss and damage, highlighting that 'responses and policies should take into account impacts on disproportionately affected groups, including [...] *women* [...]. The robust collection of disaggregated data and intersectional, gender-responsive [...] analysis can help to ensure that the diverse experiences of loss and damage beyond marketable interests are effectively addressed' (HRC, 2024, para. 31).

Throughout the report, there is also a strong call to guarantee 'access to judicial remedy [...and] developing [of] transitional justice approaches' (HRC, 2024, para. 30).

This reference comes at a timely moment when climate litigation seems to have been gaining *momentum*, with manifold climate litigation cases grounded on human rights arguments being brought against States at the national level, as well as before the European Court of Human Rights (Arling & Taghavi, 2023; Banda & Scott, 2017; Beauregard et al., 2021;

Fornalé, 2023b; Iyenga, 2023; Savaresi & Setzer, 2021). Also, international courts have been called to reflect on climate change-human rights issues, with three requests for Advisory Opinions in this regard that have been filed before the Inter-American Court of Human Rights (IACHR, 2023), the International Court of Justice (ICJ, 2023) and the International Tribunal of the Law of the Sea, respectively (ITLOS, 2022). The ITLOS has issued its Advisory Opinion on 21 May 2024 (ITLOS, 2024), the IACHR on 29 May 2025 (IACHR, 2025), and the ICJ on 23 July 2025 (ICJ, 2025).

Climate litigation has been subject to several analyses, coupled with the establishment of various databases aimed at collecting the relevant case law (Urgenda, 2024; University of Zurich (UZH), 2024; Sabin Center for Climate Change Law, 2024). While a detailed analysis of such case law is outside the scope of this chapter,[33] it is worth recalling that courts in several jurisdictions are now recognizing the linkages between human rights and climate change—though the outcomes vary, so it still seems difficult to draw a uniform pattern in this respect (UNEP, 2023).

For the scope of this chapter, we can briefly highlight how the human rights and climate change arguments have also entered into the argumentation of international investment arbitration cases. There has been a tendency for States to rely on human rights and environmental arguments when trying to defend or justify their actions involving alleged breaches of their treaty obligations towards foreign investors before investment arbitral tribunals. When relying on environmental arguments, states have generally highlighted that when faced with the need to balance competing interests, i.e. protection of foreign investments according to their obligations under the relevant IIAs and compliance with international environmental-related agreements, they had to prioritize their environmental-related obligations. This is what was claimed, for example, by Canada in the *S D. Myers Inc. v Canada* NAFTA/UNCITRAL case, where it argued that its obligations towards foreign investors were 'inconsistent with Canada's other international [environmental] obligations, including the Basel Convention and Transboundary Agreement[,] and

---

[33] For more detail, see the chapter by Fornalé and Cristani in this book: Fornalé E. Cristani F. Appendix. Typologies of human rights justifications in climate litigation. A first attempt at conceptualization through a case-law-mapping exercise with a zoom in on the KlimaSeniorinnen case.

that these prevail over [its] obligations' under the applicable IIA (*S.D. Myers Inc. v. Canada*, 2000, November 13, para. 150).

States have also used human rights justifications when trying to justify the breaches of their obligations towards foreign investors under the relevant IIAs (for example, in cases of expropriation), particularly as part of their defences grounded in the police power doctrine (Gavriil, 2024).[34] This was very clear in the landmark case *Philip Morris Brands Sàrl, Philip Morris Products SA, and Abal Hermanos SA v. Uruguay*, where Uruguay claimed that it acted against its obligations under the relevant IIA in order 'to protect public health in fulfilment of its national and international obligations' (*Philip Morris Brands Sàrl, Philip Morris Products SA, and Abal Hermanos SA v. Uruguay*, 2016, July 8). In this case, the arbitral tribunal sided with the State's argumentation, stating that the measures 'were a valid exercise by Uruguay of its police powers for the protection of public health' (*Philip Morris Brands Sàrl, Philip Morris Products SA, and Abal Hermanos SA v. Uruguay*, 2016, July 8, paras. 306–307). However, sometimes arbitral tribunals do not uphold states' arguments based on human rights justification due to a lack of a relevant regulatory framework in international investment law. This was well illustrated in the *Pezold v. Zimbabwe* case; Zimbabwe relied on human rights justifications to support its defence, also relying on the case law of the European Court of Human Rights on the margin of the appreciation doctrine. However, the arbitral tribunal dismissed the arguments, affirming that '[b]alancing competing (and non-absolute) human rights and the need to grant States a margin of appreciation when making those balancing decisions is well established in human rights law, but the Tribunal is not aware that the concept has found much support in international investment law' (*Bernhard von Pezold and Others v. Zimbabwe*, 2015, July 28, para. 465).

Despite the diverse outcomes of investment case law as to (the success of) States' reliance on human rights and environmental arguments in their defences, it is worth noting how such kinds of defences are becoming more common (Gismondi, 2023)—this has also led to a call for a

---

[34] Though there is no uniform understanding of what the policy power doctrine exactly entails, it is commonly understood as a doctrine according to which states maintain sovereign regulatory powers over core issues such as public policy, including the maintenance of public order, and the protection of public health and the environment (Titi, 2018).

'climate-responsive reform of investor-State dispute settlement (ISDS) and international investment agreements (IIAs)' (Scherer & Reichenbach, 2023, p. 105). It is also worth recalling that climate litigation is also being used to seek remedies for damages caused by climate change, though climate loss and damage cases have not been so frequent. According to a research conducted in 2023, among the 160 rights-based climate change cases (31 at the international level and 129 at the domestic level), only 24 (10 at the international level and 14 at the domestic level), i.e. around 15% of all the rights-based cases, related to claims of loss and damage by the claimants (Shrivastava & Franz Derler, 2024; Wewerinke-Sing, 2023). Keeping the focus on investment in the loss and damage mechanism, it is interesting to note that to date no case has related to an alleged misuse of FDI pertaining to loss and damage in the host country. Since the regulation of such FDI is almost non-existent, and also given the fact that it is not even very clear how such FDI is channelled, it seems quite difficult to predict on which basis and how potentially affected claimants might initiate a claim against foreign investors for misuse of FDI for loss and damage and before which jurisdiction. This adds up to the call to put in place a transparent regulation of FDI for loss and damage. In light of the above, such a regulation should be human rights-based. This approach is further reflected in the most recent report issued by the UN Working Group on Business and Human Rights on the use by investors of environmental, social, and sustainability approaches, according to which investors are '[r]equir[ed...] to embed human rights into policies and strategies, to undertake the identification and assessment of human rights impacts through ongoing human rights due diligence, to remediate adverse impacts they cause or contribute to, and to disclose these actions' (United Nations (UN) Working Group on the Issue of Human Rights and Transnational Corporations and Other Business Enterprises, 2024).

6 Some Concluding Remarks: Towards an Inclusive and Effective Loss and Damage Mechanism to Foster Cooperation (?)

Just a couple of months ahead of COP29, global leaders adopted the Pact for the Future, which includes Action 9 (we will strengthen our actions to address climate change), according to which, among other things, the parties *decided* to '[f]urther operationalize and capitalize the new funding arrangements, including the Fund, for responding to loss and damage' (UN, 2024). And indeed, the loss and damage mechanism is subject to further discussion also in the framework of COP29 (Barbarà & Hadap, 2024; United Nations (UN) Climate Change Conference Baku, 2024).[35]

As illustrated in the previous paragraph, climate-friendly FDI can provide a valid source of funding for the loss and damage mechanism. However, the relevant stakeholders should coordinate their efforts to put in place a transparent regulation for promoting and protecting this kind of FDI. The relevant regulation should also be human rights-based and include a gender perspective. As highlighted by the above-mentioned 2024 report on the impact of loss and damage on human rights by the UN Secretary-General, it is necessary '[t]o ensure that loss and damage funds integrate gender-responsive, rights-based approaches to make [the] funds directly accessible to those on the front lines of the climate crisis, including environmental human rights defenders' (HRC, 2024, para. 55; Mwale et al., 2024). This would require, as highlighted, a cooperation among all the relevant stakeholders; as pointed out in the Call to Action on Mobilization of Resources for Small Island Developing States issued by the United Nations Secretary-General and the Prime Minister of Antigua and Barbuda at the 2024 Fourth International Conference on Small Island Developing States, '*[c]oncerted* and urgent action is needed' [emphasis added] (United Nations Secretary-General (UNSG) Prime Minister of Antigua and Barbu, 2024) to support SIDS, including in the implementation of the loss and damage mechanism.

References

Alliance of Small Island States (AOSIS). (2022, November 20). *Historic Loss and Damage Fund Established at COP27 in Sharm El-Sheikh*. https://www.aosis.org/historic-loss-and-damage-fund-established-at-cop27-in-sharm-el-sheikh

Ameli, N., Dessens, O., Winning, M., Cronin, J., Chenet, H., Drummond, P., Calzadilla, A., Anandarajah, G., & Grubb, M. (2021). Higher Cost of Finance Exacerbates a Climate Investment Trap in Developing Economies. *Nature Communications, 12*, 4046. https://doi.org/10.1038/s41467-021-24305-3

---

[35] The present chapter gives an account of the discussions at the time of writing (October 2024).

Appadoo, K. (2021). A Short History of the Loss and Damage Principle. *Revue Juridique De L'océan Indien, 31,* 315–323.

Arling, H., & Taghavi, H. (2023, April 6). *KlimaSeniorinnen v. Switzerland. A New Era for Climate Change Protection or Proceeding With the Status Quo?* EJIL:Talk! https://www.ejiltalk.org/klimaseniorinnen-v-switzerland-a-new-era-for-climate-change-protection-or-proceeding-with-the-status-quo/?utm_source=mailpoet&utm_medium=email&utm_campaign=ejil-talk-newsletter-post-title_2

Banda, M. L., & Scott, F. (2017). Litigating Climate Change in National Courts: Recent Trends and Developments in Global Climate Law. *Environmental Law Reporter, 47*(2), 10121–10134.

Barbarà, L., & Hacap, A. (2024, August 12). With Fewer Than 100 Days to COP29: What's on the Agenda? *World Economic Forum.* https://www.weforum.org/agenda/2024/08/what-is-cop29-climate-change-summit-priorities

BayWa r.e. Renewable Energy GmbH and BayWa r.e. Asset Holding GmbH v. Spain, ICSID Case No. ARB/15/16 (2019, December 2).

Beauregard, C., Carlson, D., Robinson, S., Cobb, C., & Patton, M. (2021, May 28). Climate Justice and Rights-Based Litigation in a Post-Paris World. *Climate Policy, 21*(5), 652–665. https://doi.org/10.1080/14693062.2020.1867047

Benjamin, L., & Thomas, A. (2023). The Unvirtuous Cycle of Loss and Damage: Addressing Systemic Impacts of Climate Change in Small Islands from a Vulnerability Perspective. *Review of European, Comparative & International Environmental Law, 32*(3), 390–402. https://doi.org/10.1111/reel.12516

Bernhard von Pezold and Others v. Zimbabwe. (2015, July 28). ICSID Case No. ARB/10/15, Award. https://www.italaw.com/cases/1472

Bhandari, P., Warszawski, N., Javadi-Abhari, N., Simonetti, M., Lechat, T., & Biermann, A. (2023, March 17). 3 Questions on Loss and Damage Funding to Tackle before COP28. *World Resources Institute.* https://www.wri.org/insights/loss-and-damage-funding-questions-transitional-committee

Blackden, C. M., Manuel, C., & Simavi, S. (2010). Gender Dimensions of Investment Climate Reform: A Guide for Policy Makers and Practitioners. *World Bank.* https://documents.worldbank.org/en/publication/documents-reports/documentdetail/260721468321276647/gender-dimensions-of-investment-climate-reform-a-guide-for-policy-makers-and-practitioners

Burkett, M. (2014). Loss and Damage. *Climate Law, 4*(1–2), 119–130. https://doi.org/10.1163/18786561-00402010

Burke-White, W. (2022, November 22). A Green Investment Treaty Can Help Close the Climate Funding Gap. *Bloomberg Law Opinion.* https://news.bloomberglaw.com/us-law-week/a-green-investment-treaty-can-help-close-the-climate-funding-gap

Cao, T., Han, D., & Song, X. (2021, December). Past, Present, and Future of Global Seawater Intrusion Research: A Bibliometric Analysis. *Journal of Hydrology*, *603*, 1–14. https://doi.org/10.1016/j.jhydrol.2021.126844

Chair's Statement of the 13th Asia-Pacific Economic Cooperation Energy Ministerial Meeting. (2023, August 16). *Non-Binding Just Energy Transition Principles for APEC Cooperation: Chair's Statement of the 13th APEC Energy Ministerial Meeting*. https://www.apec.org/meeting-papers/sectoral-ministerial-meetings/energy/13th-apec-energy-ministerial-meeting

Chakma, T., Rigg, S., & Ramsay, A. (2022, November). Women Confronting Loss and Damage in Africa: Feminist Climate Justice Research from Kenya, Nigeria, Rwanda and Zambia. *ActionAid Report*. https://www.actionaid.org.uk/sites/default/files/publications/AAUK%20Loss%20%26%20Damage%20report.pdf

Charanne B.V. and Construction Investments S.a.r.l. v. Spain, SCC, Award (2016, January 21).

De Vido, S., & Fornalé, E. (2023). Achievements and Hurdles Towards Women's Access to Climate Justice. In E. Fornalé & F. Cristani (Eds.), *Women's Empowerment and Its Limits* (pp. 33–51). Springer.

Dietrich Brauch, M. (2023, November 30). Can Existing International Agreements on 'Investment Facilitation' Advance Sustainable Development, Climate Action, and Human Rights? *Columbia Center on Sustainable Investment Blog*. https://ccsi.columbia.edu/news/investment-facilitation-wto-sustainable-development-climate-energy-transition

Eiser Infrastructure Limited and Energía Solar Luxembourg S.à.r.l. v. Kingdom of Spain, ICSID Case No. ARB/13/36, Award (2017, May 4).

Fornalé, E. (2023a). Slow Violence, Gender Equality and Climate Agency in Times of 'Polycrisis'. *Revista General de Derecho Europeo*, *61*. https://www.iustel.com/v2/revistas/detalle_revista.asp?id_noticia=426572

Fornalé, E. (2023b). Vulnerability, Intertemporality, and Climate Litigation. *Nordic Journal of Human Rights*, *41*(4), 357–377. https://doi.org/10.1080/18918131.2023.2225973

Franczak, M. (2022, October 1). *Options for a Loss and Damage Financial Mechanism: Loss and Damage Collaboration*. https://www.lossanddamagecollaboration.org/publication/options-for-a-loss-and-damage-financial-mechanism

Free Trade Agreement Between the European Union (EU)-New Zealand Free Trade Agreement. (2023, July 9). https://policy.trade.ec.europa.eu/eu-trade-relationships-country-and-region/countries-and-regions/new-zealand/eu-new-zealand-agreement/text-agreement_en

Free Trade Agreement Between the United Kingdom of Great Britain (UK) and Northern Ireland and Australia. (2021, December 16). https://www.gov.uk/government/collections/free-trade-agreement-between-the-united-kingdom-of-great-britain-and-northern-ireland-and-australia

Free Trade Agreement Between the Oman and the United States (US) (2009, January 1). https://ustr.gov/trade-agreements/free-trade-agreements/oman-fta/final-text

Freestone, D., & Çiçek, D. (2021). *Legal Dimensions of Sea Level Rise: Pacific Perspectives*. World Bank Group. https://openknowledge.worldbank.org/server/api/core/bitstreams/6ae3f9d7-4d5f-55ce-bfee-1b124561e486/content

Gabbatiss, J. (2022, September 27). *Timeline: The Struggle Over 'Loss and Damage' in UN Climate Talks*. https://interactive.carbonbrief.org/timeline-the-struggle-over-loss-and-damage-in-un-climate-talks/

Gavriil, E. (2024). Protection of Property under Human Rights and International Investment Law: A Case-Law Analysis. *MDPI Laws, 13*(1), 1–20. https://doi.org/10.3390/laws13010006

Gismondi, G. E. (2023). *International Environmental Law and International Human Rights Law in Investment Treaty Arbitration: The Contribution of Argumentation in Reshaping International Investment Law*. Wolters Kluwer.

Green Climate Fund. (2024). *About GCF*. Retrieved September 23, 2024, from https://www.greenclimate.fund/about

Green Climate Fund Board. (2014, October 6). *Decision GCF/B.08/19: Gender Policy and Action Plan*. https://www.greenclimate.fund/document/gcf-b08-19

Green Climate Fund Board. (2019, November 14). *Decision B.24/12. B.24/12: Updated Gender Policy and Action Plan 2020–2023*. https://www.greenclimate.fund/document/gender-action-plan

Green Climate Fund. (n.d.b). *Private Sector Financing*. Retrieved September 10, 2024, from https://www.greenclimate.fund/sectors/private

Green Economy Agreement Between Singapore and Australia. (2022, October 18). https://www.dfat.gov.au/geo/singapore/singapore-australia-green-economy-agreement

Hemingway Jaynes, C. (2023). COP28 Agrees to Establish Loss and Damage Fund for Vulnerable Countries. *World Economic Forum*. https://www.weforum.org/agenda/2023/12/cop28-loss-and-damage-fund-climate-change

Human Rights Council (HRC). (2023, July 19). Resolution 53/6: Human Rights and Climate Change (A/HRC/RES/53/6). https://documents.un.org/doc/undoc/gen/g23/148/70/pdf/g2314870.pdf

Human Rights Council (HRC). (2024, August 28). *Report of the Secretary-General: Analytical Study on the Impact of Loss and Damage From the Adverse Effects of Climate Change on the Full Enjoyment of Human Rights, Exploring Equity-Based Approaches and Solutions to Addressing the Same* (A/HRC/57/30). https://www.ohchr.org/sites/default/files/2024-08/a-hrc-57-30-aev.pdf

IACHR. (2025, May 29). *Advisory Opinion AO-32/25 Requested by the Republic of Chile and the Republic of Colombia. Climate Emergency and Human Rights.* https://jurisprudencia.corteidh.or.cr/en/vid/1084981967

ICJ. (2025, July 23). *Advisory Opinion. Obligations of States in Respect of Climate Change.* https://www.icj-cij.org/case/187

Inter-American Court of Human Rights (IACHR). (2023, January 9). *Request for an Advisory Opinion on the Climate Emergency and Human Rights Submitted by the Republic of Colombia and the Republic of Chile.* https://www.corteidh.or.cr/docs/opiniones/soc_1_2023_en.pdf

Intergovernmental Negotiating Committee for a Framework Convention on Climate Change Working Group II. (1991, December 17). *Vanuatu: Draft Annex Relating to Article 23 (Insurance) for Inclusion in the Revised Single Text on Elements Relating to Mechanisms* (A/AC.237/WG.ii/Misc.13). Submitted by the Co-Chairmen of Working Group II. https://unfccc.int/sites/default/files/resource/docs/a/wg2crp08.pdf

International Court of Justice (ICJ). (2023, April 12). *Request for Advisory Opinion Transmitted to the Court Pursuant to General Assembly Resolution 77/276 of 29 March 2023: Obligations of States in Respect of Climate Change.* https://www.icj-cij.org/case/187

International Law Association (ILA). (2024). *International Law and Sea Level Rise Committee: Final Committee Report Athens May 2024.* https://www.ila-hq.org/en_GB/committees/international-law-and-sea-level-rise

International Monetary Fund (IMF). (2015, October). *World Economic and Financial Surveys. World Economic Outlook: Database WEO Groups and Aggregates Information.* https://www.imf.org/external/pubs/ft/weo/2015/02/weodata/groups.htm#cc

International Tribunal for the Law of the Sea (ITLOS). (2022, December 12). *Request for an Advisory Opinion Submitted by the Commission of Small Island States on Climate Change and International Law (Case No. 31).* https://itlos.org/en/main/cases/list-of-cases/request-for-an-advisory-opinion-submitted-by-the-commission-of-small-island-states-on-climate-change-and-international-law-request-for-advisory-opinion-submitted-to-the-tribunal

Isolux Infrastructure Netherlands B.V. v. Spain, SCC Arb No. 2013/153, Award (2016, July 17).

ITLOS. (2024, May 21). *Request for an Advisory Opinion Submitted by the Commission of Small Island States on Climate Change and International Law (Case No. 31).* https://itlos.org/en/main/cases/list-of-cases/request-for-an-advisory-opinion-submitted-by-the-commission-of-small-island-states-on-climate-change-and-international-law-request-for-advisory-opinion-submitted-to-the-tribunal

Iyenga, S. (2023). Human Rights and Climate Wrongs: Mapping the Landscape of Rights-Based Climate Litigation. *Review of European, Comparative* and *International Environmental Law, 32*(2), 299–309. https://doi.org/10.1111/reel.12498

Khan, I., Ali, A., Waqas, T., Ullah, S., Ullah, S., Ahmad Shah, A., & Imran, S. (2022). Investing in Disaster Relief and Recovery: A Reactive Approach of Disaster Management in Pakistan. *International Journal of Disaster Risk Reduction, 75*. https://doi.org/10.1016/j.ijdrr.2022.102975

Khokhar, T., & Serajuddin, U. (2015, November 16). Should We Continue to Use the Term "Developing World"? *World Bank Blogs*. https://blogs.worldbank.org/en/opendata/should-we-continue-use-term-developing-world

Lieser, A. (2024, September 11). *Bilateral Investment Treaties as a Tool for Global Climate Governance? Bridging the Tension Between International Investment Law and International Climate Change Law by Redesigning Bilateral Investment Treaties*. Völkerrechtsblog. https://vcelkerrechtsblog.org/bilateral-investment-treaties-as-a-tool-for-global-climate-governance

Loungani, P., & Razin, A. (2001). How Beneficial Is Foreign Direct Investment for Developing Countries? *Finance & Development, 38*(2). https://www.imf.org/external/pubs/ft/fandd/2001/06/index.htm

Mahul, O., Cook, S. J., & Prasad, R. (2016, April 1). Enhancing the Financial Resilience of Pacific Island Countries Against Natural Disaster and Climate Risk: Pacific Catastrophe Risk Assessment and Financing Initiative (PCRAFI) Program: Phase II (English). *World Bank Group*. https://documents.worldbank.org/en/publication/documents-reports/documentdetail/716481495545738079/pacific-catastrophe-risk-assessment-and-financing-initiative-pcrafi-program-phase-ii

Marcelo Gordillo, D., Raina, A., & Rawat, S. (2020). Private Sector Participation in Disaster Recovery and Mitigation. *World Bank Group*. https://documents.worldbank.org/en/publication/documents-reports/documentdetail/589341609771914404/private-sector-participation-in-disaster-recovery-and-mitigation

Mwale, B., Sharma, S., White, H., Achampong, L., & Ormond-Skeaping, T. (2024, July 2). *Let's Get to Work, the Loss and Damage Fund in 2024: A Technical Discussion Paper. Loss and Damage Collaboration*. https://www.lossanddamagecollaboration.org/link-page/lets-get-to-work-the-loss-and-damage-fund-in-2024-a-technical-discussion-paper

Newman, R., & Noy, I. (2023). The Global Costs of Extreme Weather That Are Attributable to Climate Change. *Nature Communications, 14*(6103). https://doi.org/10.1038/s41467-023-41888-1

Organisation for Economic Co-operation and Development (OECD). (2024). *The Future of Investment Treaties*. Retrieved September 10, 2024, from https://www.oecd.org/en/topics/sub-issues/the-future-of-investment-treaties.html

Paine, J., & Sheargold, E. (2024a). Research Report. Facilitating Climate Friendly FDI: The Importance of Ongoing Cooperation. *Columbia FDI Perspectives*, 379. https://www.econstor.eu/bitstream/10419/289491/1/1884804330.pdf

Paine, J., & Sheargold, E. (2024b, February 23). *Future of Investment Treaties Track 1: Investment Treaties and Climate Change: Academic Contribution to the OECD 9th Investment Treaty Conference* (DAF/INV/TR1/RD(2024)1). https://one.oecd.org/document/DAF/INV/TR1/RD(2024)1/en/pdf

Paris Agreement. (2015, December 12). https://unfccc.int/sites/default/files/english_paris_agreement.pdf

Perkins, C. (2023). Loss & Damage Needs a New Governance Model. *Journal of Public and International Affairs. Annotations Blog*. https://jpia.princeton.edu/news/loss-damage-needs-new-governance-model

Philip Morris Brands Sàrl, Philip Morris Products SA and Abal Hermanos SA v. Uruguay. (2016, July 8). ICSID Case No. ARB/10/7, Award. https://www.italaw.com/cases/460

Prabhu, A., & Staffer, J. (2023, February 15). Spain's Renewable Energy Disputes: Renewable Energy Needs Reliable Arbitration. *The Arbitration Brief*. https://thearbitrationbrief.com/2023/02/15/spains-renewable-energy-disputes-renewable-energy-needs-reliable-arbitration

Price, R. (2021, May 18). *Access to Climate Finance by Women and Marginalised Groups in the Global South: Helpdesk Report*. Institute of Development Studies. https://opendocs.ids.ac.uk/articles/report/Access_to_Climate_Finance_by_Women_and_Marginalised_Groups_in_the_Global_South/26432386?file=48082189

Qihan Zou, C., Alayza, N., Higgins, H., & Larsen, G. (2024, September 17). *Which Countries Should Pay for International Climate Finance?* World Resource Institute. Insights. https://www.wri.org/insights/international-climate-finance-which-countries-should-pay

Reciprocal Investment Promotion and Protection Agreement Between the Government of the Kingdom of Morocco and the Government of the Federal Republic of Nigeria. (2016, December 3). https://investmentpolicy.unctad.org/international-investment-agreements/treaty-files/5409/download

Rockhopper v. Italy, ICSID Case No. ARB/17/14, Award (2022, August 23).

Rudall, J. (2021). The Obligation to Cooperate in the Fight Against Climate Change. *International Community Law Review, 23*(2–3), 184–196. https://doi.org/10.1163/18719732-12341469

RWE v. Kingdom of the Netherlands, ICSID Case No. ARB/21/4 (2021, February 2).

S.D. Myers Inc. v. Canada. (2000, November 13). NAFTA/UNCITRAL Case, Partial Award. https://www.italaw.com/cases/969

Sabin Center for Climate Change Law. (2024). *Climate Change Litigation Databases*. Retrieved September 26, 2024, from https://climatecasechart.com

Savaresi, A., & Setzer, J. (2021). Rights-Based Litigation in the Climate Emergency: Mapping the Landscape and New Knowledge Frontiers. *Journal of Human Rights and the Environment, 13*(1), 7–34. https://doi.org/10.4337/jhre.2022.01.01

Scherer, M., & Reichenbach, C. (2023). Climate-Related Counterclaims in International Investment Arbitration. In A. Magnusson & A. Ipp (Eds.), *Investment Arbitration and Climate Change* (pp. 105–138). Kluwer Law.

Shahud, M., König, M., Zamarioli, L. H., & Grüning, C. (2023). *Loss and Damage and Climate Litigation: How Can the Maldives and Other Small Island Developing States (SIDS) Position for Greater Climate Action?* Special Edition: UNDP Maldives Economic Bulletin 2023. https://www.undp.org/sites/g/files/zskgxe326/files/2023-11/Loss%20and%20Damage%20and%20Climate%20Litigation%20UNDP%20Maldives%202023.pdf

Sharma, M. (2022). Integrating, Reconciling, and Prioritising Climate Aspirations in Investor-State Arbitration for a Sustainable Future: The Role of Different Players. *Journal of World Investment & Trade, 23*(5-6), 746–777. https://doi.org/10.1163/22119000-12340269

Shrivastava, A., & Franz Derler, R. O. (2024, June 24). *A Global South Perspective on Loss and Damage Litigation*. Verfassungsblog. https://verfassungsblog.de/a-global-south-perspective-on-loss-and-damage-litigation

Small States Conference on Sea Level Rise (1989, November 18). *Male Declaration on Global Warming and Sea Level Rise* (MDV/SLR/15). https://www.islandvulnerability.org/slr1989.html

Stephenson, M., & Zhan, J. (2022). What Is Climate FDI? How Can We Help Grow It? Policy Brief. *T20 Indonesia 2022*. https://www.t20indonesia.org/wp-content/uploads/2022/09/TF1_What-is-Climate-FDI_-How-can-we-help-grow-t_-3.pdf

Sustainable Investment Facilitation Agreement Between the European Union (EU) and the Republic of Angola. (2023, November 17). https://ec.europa.eu/commission/presscorner/detail/en/ip_24_4462

Technical Support Unit of the Transitional Committee. Working Group 5(c). (2023, July). *Identifying and Expanding Sources of Funding*. https://unfccc.int/sites/default/files/resource/Final_Draft_5c_TSU.pdf

Teirstein, Z., & Al Hasnat, M. (2024, May 31). Why Rising Sea Levels Could Trigger a Maternal Health Crisis. *Fast Company*. https://www.fastcompany.com/91133514/why-rising-sea-levels-could-trigger-a-maternal-health-crisis

Tewari, N., Busn, A., Butt, M. N., Stevens, E., & Zafar, S. (2023, December). *Gendered Dimensions of Loss and Damage in Asia*. Oxfam Briefing Paper. https://doi.org/10.21201/2023.000005

Tietjen, B., & Gopalakrishnan, T. (2023, May 9). Loss and Damage Funding in the UN Climate Negotiations: From Dialogue to Reality. *Climate Policy Lab*. https://www.climatepolicylab.org/climatesmart/2023/5/9/loss-and-damage-funding-in-the-un-climate-negotiations-from-dialogue-to-reality

Titi, C. (2018). Police Powers Doctrine and International Investment Law. In A. Gattini, A. Tanzi & F. Fontanelli (Eds.), *General Principles of Law and International Investment Arbitration* (pp. 323–343). Brill. https://doi.org/10.1163/9789004368385_015

Turquet, L., Tabbush, C., Staab, S., Williams, L., & Howell, B. (2023). Feminist Climate Justice: A Framework for Action. *UN Women*. https://www.unwomen.org/sites/default/files/2023-12/Feminist-climate-justice-A-framework-for-action-en.pdf

UNCTAD. (2019, November 26–28). *Facilitating Investment in SDG Projects: Spotlight on Small Island Developing States*. https://unctad.org/meeting/facilitating-investment-sdg-projects-spotlight-small-island-developing-states

UNCTAD. (2022a). *Investment Policy Trends in Climate Change Sectors 2010–2022: UNCTAD Investment Policy Monitor 9*. https://unctad.org/system/files/official-document/diaepcbinf2022d8_en.pdf

UNCTAD. (2022c, September). *The International Investment Treaty Regime and Climate Action: IIA Issues Note, Issue 3*. https://unctad.org/system/files/official-document/diaepcbinf2022d6_en.pdf

UNCTAD. (2022d, September). *Treaty-Based Investor: State Dispute Settlement Cases and Climate Action: IIA Issues Note, Issue 4*. https://unctad.org/system/files/official-document/diaepcbinf2022d7_en.pdf

UNCTAD. (2023, September). *Investment Facilitation in International Investment Agreements: Trends and Policy Options: IIA Issues Note, Issue 3*. https://unctad.org/system/files/official-document/diaepcbinf2023d5_en.pdf

UNCTAD. (2024, April). *UNCTAD Strategy to Support Small Island Developing States*. https://unctad.org/system/files/official-document/aldcinf2024d1_en.pdf

UNCTAD, Trade and Development Board. (2022, August 31). *Investment and Climate Change: Note by the UNCTAD Secretariat*. https://unctad.org/system/files/official-document/ciimem4d25_en_0.pdf

UNDP. (2024). *Small Island Developing States*. Retrieved October 17, 2024, from https://www.undp.org/latin-america/small-island-developing-states

UNEP. (2023, July 27). *Global Climate Litigation Report: 2023 Status Review*. https://www.unep.org/resources/report/global-climate-litigation-report-2023-status-review

UNEP. (2024a). *About Loss and Damage*. https://www.unep.org/topics/climate-action/loss-and-damage/about-loss-and-damage

UNEP. (2024b). *SIDS*. Retrieved October 17, 2024, from https://www.unep.org/topics/ocean-seas-and-coasts/small-island-developing-states

UNFCCC. (2015). *Paris Agreement, Adopted on 12 December 2015 and Entered Into Force on 4 November 2016*. https://unfccc.int/process-and-meetings/the-paris-agreement
UNFCCC. (2024a). *About the Santiago Network*. https://unfccc.int/santiago-network/about
UNFCCC. (2024b). *First Meeting of the Board of the Fund for Responding to Loss and Damage*. https://unfccc.int/event/first-meeting-of-the-board-of-the-fund-for-responding-to-loss-and-damage
UNFCCC. (2024c). *Second Meeting of the Board of the Fund for Responding to Loss and Damage*. https://unfccc.int/event/second-meeting-of-the-board-of-the-fund-for-responding-to-loss-and-damage
UNFCCC. (2024d). *Third Meeting of the Board of the Fund for Responding to Loss and Damage*. https://unfccc.int/event/third-meeting-of-the-board-of-the-fund-for-responding-to-loss-and-damage
UNFCCC. (2024e, July 12). *Philippines Selected to Host the Board of the Fund for Responding to Loss and Damage*. https://unfccc.int/news/philippines-selected-to-host-the-board-of-the-fund-for-responding-to-loss-and-damage
UNFCCC (2024f, September 21). *Ibrahima Cheikh Diong Selected as Inaugural Executive Director of the Fund for responding to Loss and Damage*. https://unfccc.int/news/ibrahima-cheikh-diong-selected-as-inaugural-executive-director-of-the-fund-for-responding-to-loss
UNFCCC. (2024g). *Introduction to Climate Finance*. Retrieved September 23, 2024, from https://unfccc.int/topics/introduction-to-climate-finance
UNFCCC COP. (2001a, July 23). *Decision 5/CP.6: Implementation of the Buenos Aires Plan of Action*. https://unfccc.int/cop6_2/documents/dec5cp6unedi tvers.pdf
UNFCCC COP. (2001b, November 10). *Decision 7/CP.7: Funding Under the Convention*. https://unfccc.int/files/cooperation_and_support/financial_mechanism/application/pdf/7_cp.7.pdf
UNFCCC COP. (2008, March 14). *Decision 1/CP.13: Bali Action Plan (FCCC/CP/2007/6/Add.1)*. https://unfccc.int/resource/docs/2007/cop13/eng/06a01.pdf
UNFCCC COP. (2011, March 15). *Decision 1/CP.16: The Cancun Agreements: Outcome of the Work of the Ad Hoc Working Group on Long-term Cooperative Action Under the Convention*. https://unfccc.int/resource/docs/2010/cop16/eng/07a01 pdf
UNFCCC COP. (2014, January 31). *Decision 2/CP.19: Warsaw International Mechanism for Loss and Damage Associated With Climate Change Impacts*. https://unfccc.int/documents/8106#beg
UNFCCC COP. (2016, January 29). *Report of the Conference of the Parties on its Twenty-First Session, Held in Paris from 30 November to 13 December 2015. Addendum. Part Two: Action Taken by the Conference of the Parties at Its*

Twenty-First Session (1/CP.21). Adoption of the Paris Agreement. https://unfccc.int/documents/9097

UNFCCC COP. (2022a, March 8). *Decision 1/CMA.3: Glasgow Climate Pact.* https://unfccc.int/sites/default/files/resource/cma2021_10_add1_adv.pdf

UNFCCC COP. (2022b, March 8). *Report: Addendum Part Two: Action Taken by the Conference of the Parties Serving as the Meeting of the Parties to the Paris Agreement at its Third Session* (FCCC/PA/CMA/2021/10/Add.3). https://unfccc.int/documents/460952

UNFCCC COP. (2022c, November 20). *Decision 2/CMA.4: Funding Arrangements for Responding to Loss and Damage Associated With the Adverse Effects of Climate Change, Including a Focus on Addressing Loss and Damage.* https://unfccc.int/documents/627487

UNFCCC COP. (2023, March 17). Decision 2/CP.27: Funding Arrangements for Responding to Loss and Damage Associated With the Adverse Effects of Climate Change, Including a Focus on Addressing Loss and Damage (FCCC/CP/2022/10/Add.1). *United Nations Framework Convention on Climate Change.* https://unfccc.int/sites/default/files/resource/cp2022_10a01_adv.pdf

UNFCCC COP. (2024a, March 15). *Decision 1/CMA.5: Outcome of the First Global Stocktake.* https://unfccc.int/documents/637073

UNFCCC COP. (2024b, March 15). Decision 1/CP.28: Operationalization of the New Funding Arrangements, Including a Fund, for Responding to Loss and Damage Referred to in Paragraphs 2–3 of Decisions 2/CP.27 and 2/CMA.4 (FCCC/CP/2023/11/Add.1). *United Nations Framework Convention on Climate Change.* https://unfccc.int/sites/default/files/resource/cp2023_09_cma2023_09.pdf

Uniper v. Netherlands, ICSID Case No. ARB/21/22 (2021, April 30).

United Kingdom (UK) Government. (2020, January 21). UK-Kenya Strategic Partnership 2020 to 2025: Joint Statement. *Press Release.* https://www.gov.uk/government/news/uk-kenya-strategic-partnership-2020-2025

United Nations (UN). (2023, December 2). *Opening Remarks by Ms. Rabab Fatima, Under-Secretary-General and High Representative for the Least Developed Countries, Landlocked Developing Countries and Small Island Developing States. COP28 Side Event: Resilience in the Face of Adversity: Addressing Loss and Damage in Small Island Developing States.* United Nations Office of the High Representative for the Least Developed Countries, Landlocked Developing Countries and Small Island Developing States. https://www.un.org/ohrlls/content/opening-remarks-ms-rabab-fatima-cop28-side-event-resilience-face-adversity-addressing-loss

United Nations (UN). (2024). *Pact for the Future.* https://www.un.org/sites/un2.un.org/files/sotf-pact_for_the_future_adopted.pdf

United Nations (UN) Climate Action Thought Leaders. (2022). *Loss and Damage: A Moral Imperative to Act.* https://www.un.org/en/climatechange/adelle-thomas-loss-and-damage

United Nations (UN) Climate Change Conference Baku. (2024, November 11–22). https://unfccc.int/cop29

United Nations (UN) Climate Press Release. (2023, December 13) *COP28 Agreement Signals "Beginning of the End" of the Fossil Fuel Era.* https://unfccc.int/news/cop28-agreement-signals-beginning-of-the-end-of-the-fossil-fuel-era.

United Nations (UN) Special Rapporteur on the Promotion and Protection of Human Rights in the Context of Climate Change. (2022, July 26). *Promotion and Protection of Human Rights in the Context of Climate Change Mitigation, Loss and Damage and Participation (A/77/226).* https://documents.un.org/doc/undoc/gen/n22/438/51/pdf/n2243851.pdf?OpenElement

United Nations (UN) Women. (2023, December 2). *New Report Shows How Feminism Can Be a Powerful Tool to Fight Climate Change.* https://www.unwomen.org/en/news-stories/feature-story/2023/12/new-report-shows-how-feminism-can-be-a-powerful-tool-to-fight-climate-change

United Nations (UN) Working Group on the Issue of Human Rights and Transnational Corporations and Other Business Enterprises. (2021, June 17). *A/HRC/47/39/Add.1: Report on Taking Stock of Investor Implementation of the Guiding Principles on Business and Human Rights.* https://www.ohchr.org/en/documents/thematic-reports/ahrc4739add1-report-taking-stock-investor-implementation-guiding

United Nations (UN) Working Group on the Issue of Human Rights and Transnational Corporations and Other Business Enterprises. (2024, May 2). *A/HRC/56/55: Investors, Environmental, Social and Governance Approaches and Human Rights.* https://www.ohchr.org/en/documents/thematic-reports/ahrc5655-investors-environmental-social-and-governance-approaches-and

United Nations (UN). (1999). *Methodology: Standard Country or Area Codes for Statistical Use (M49): Note on Developed and Developing Regions.* Retrieved October 17, 2024, from https://unstats.un.org/unsd/methodology/m49/

United Nations Commission on International Trade Law (UNCITRAL) (2024). *Working Group III: Investor-State Dispute Settlement Reform.* Retrieved September 10, 2024, from https://uncitral.un.org/en/working_groups/3/investor-state

United Nations Conference on Trade and Development (UNCTAD). (2014). *World Investment Report 2014: Investing in the SDGs: An Action Plan.* https://unctad.org/publication/world-investment-report-2014

United Nations Conference on Trade and Development (UNCTAD). (2022b). *International Investment in Climate Change Mitigation and Adaptation:*

Trends and Policy Developments. https://unctad.org/system/files/official-document/diaeinf2022d2_en.pdf

United Nations Environmental Programme (UNEP). (2022, November 1). *Adaptation Gap Report 2022*. https://www.unep.org/resources/adaptation-gap-report-2022.

United Nations Framework Convention on Climate Change (UNFCCC) Executive Committee of the Warsaw International Mechanism for Loss and Damage. (2024, March 22). *Loss and Damage Online Guide*. https://unfccc.int/documents/637576

United Nations Framework Convention on Climate Change (UNFCCC). (2013, October 9). *Non-Economic Losses in the Context of the Work Programme on Loss and Damage: Technical Paper*. https://unfccc.int/resource/docs/2013/tp/02.pdf

United Nations Office for Disaster Risk Reduction (UNDRR). (2024). *Small Island Developing States (SIDS)*. Retrieved October 16, 2024, from https://www.undrr.org/implementing-sendai-framework/sendai-framework-action/small-island-developing-states

United Nations Secretary-General (UNSG) Prime Minister of Antigua and Barbuda. (2024, May 28). *Call to Action on Mobilization of Resources for Small Island Developing States: Fourth International Conference on Small Island Developing States*. https://sdgs.un.org/sites/default/files/2024-05/Final%20-%20Call%20to%20Action%2C%20as%20at%2022%20May.pdf

United Nations. (2024). *Office of the High Representative for the Least Developed Countries, Landlocked Developing Countries and Small Island Developing States: Least Developed Countries Category*. Retrieved October 17, 2024, from https://www.un.org/ohrlls/content/ldc-category#:~:text=The%20United%20Nations%20defines%20LDCs,structural%20impediments%20to%20sustainable%20development

University of Zurich (UZH). (2024). *Climate Rights and Remedies (CRRP) Project: Climate Litigation Database*. Retrieved September 26, 2024, from https://climaterightsdatabase.com

Urgenda. (2024). *Global climate litigation*. Retrieved September 26, 2024, from https://www.urgenda.nl/en/themas/climate-case/global-climate-litigation

Vattenfall AB and others v. Federal Republic of Germany, ICSID Case No. ARB/12/12, Order (2021, November 9).

Warner, K., & Weisberg, M. (2023, January 19). A Funding Mosaic for Loss and Damage. *Science, 379*(6629), 219. https://doi.org/10.1126/science.adg5740

Warner, K., & Zakieldeen, S. A. (2012). *Loss and Damage due to Climate Change: An Overview of the UNFCCC Negotiations: European Capacity Building Initiative Background Papers*. https://oxfordclimatepolicy.org/publications/documents/LossandDamage.pdf

Warner, K., Van der Geest, K., Kreft, S., Huq, S., Harmeling, S., Kusters, K., & De Sherbinin, A. (2012). *Evidence from the Frontlines of Climate Change: Loss and Damage to Communities Despite Coping and Adaptation: Report No. 9*. United Nations University Institute for Environment and Human Security (UNU-EHS).

Wewerinke-Sing, M. (2023). The Rising Tide of Rights: Addressing Climate Loss and Damage Through Rights-Based Litigation. *Transnational Environmental Law, 12*(3), 537–566. https://doi.org/10.1017/S2047102523000183

Wijenayake, V. (2013). Addressing Loss and Damage with a Gender Lens. *CANSA*. https://cansouthasia.net/addressing-loss-and-damage-with-a-gender-lens/

World Bank Group. (2019, October). *Support to Small States*. https://pubdocs.worldbank.org/en/340031539197519098/SmallStates.pdf

World Bank Group (2024). *Gender-Inclusive Climate Investment Programme of the International Financial Corporation*. Retrieved September 25, 2024, from https://www.ifc.org/en/what-we-do/sector-expertise/gender/gender-Inclusive-climate-investment

World Bank. (2024a). *How Does the World Bank Classify Countries?* Retrieved October 17, 2024, from https://datahelpdesk.worldbank.org/knowledgebase/articles/378834-how-does-the-world-bank-classify-countries

World Bank. (2024b). *World Bank Country and Lending Groups*. Retrieved October 17, 2024, from https://datahelpdesk.worldbank.org/knowledgebase/articles/906519

World Economic Forum (WEF). (2024, January 18). *New Climate FDI Coalition Can Help Grow Climate Investment*. https://www.weforum.org/agenda/2024/01/the-new-ipa-climate-fdi-coalition-will-help-grow-climate-investment

Open Access This chapter is licensed under the terms of the Creative Commons Attribution 4.0 International License (http://creativecommons.org/licenses/by/4.0/), which permits use, sharing, adaptation, distribution and reproduction in any medium or format, as long as you give appropriate credit to the original author(s) and the source, provide a link to the Creative Commons license and indicate if changes were made.

The images or other third party material in this chapter are included in the chapter's Creative Commons license, unless indicated otherwise in a credit line to the material. If material is not included in the chapter's Creative Commons license and your intended use is not permitted by statutory regulation or exceeds the permitted use, you will need to obtain permission directly from the copyright holder.

INDEX

A
Actio popularis, 132, 133
AfCmHR. *See* African Commission on Human Rights
African Charter of Human and Peoples' Rights, 123, 130
African Commission on Human Rights, 130, 132
Anthropocene, 8, 171, 172, 175, 185, 190, 192

B
Bangladesh, 28, 32, 102, 173, 190

C
CEDAW. *See* Convention on the Elimination of All Forms of Discrimination against Women
CEDAW Committee. *See* Committee on the Elimination of Discrimination against Women
Chronic emergency. *See* Emergency
Climate action, 105, 207, 281, 307, 311
Climate change
  adaptation, 30, 176, 299
  claims, 124
  claims dataset, 218
  human-induced, 42
  induced migration. *See* Migration
  litigation, 3, 8, 202
  mitigation, 43, 205, 213, 303
  risks, 94, 96, 103
  Security nexus. *See* Security
Climate finance, 299, 302, 303, 313
Climate security. *See* Security
Climate violence. *See* Violence
Coastal, 5, 18, 19, 23–33, 35–40, 43, 44, 66, 69, 74, 76, 102, 106, 244, 278
  communities. *See* Communities
  regions, 23, 33, 42
Commission of Small Island States on Climate Change and International Law, 253, 254, 256, 280
Committee on international law and sea level rise. *See* International Law Association

Committee on the Elimination of Discrimination against Women, 151
Communities
  coastal, 6, 18, 23, 27, 31, 122
  coastal, relocation of, 32
  indigenous, 29, 42, 126
Conference of the Parties, 253, 292, 297, 300, 303
Convention on the Elimination of All Forms of Discrimination against Women, 151, 153
COP. *See* Conference of the Parties
COSIS. *See* Commission of Small Island States on Climate Change and International Law
Cultural heritage, 18, 29, 30, 42, 294

D
Disaster risk reduction, 105, 151, 153
Discrimination, 4, 146, 158, 236
  gender-based, 104
Displacement, 25, 28, 106, 153, 154, 176, 190, 294
Due diligence, 9, 279, 282, 283

E
Ecofeminist, 7, 146, 157, 160, 163
  method, 7, 146
ECtHR. *See* European Court of Human Rights
Emergency, 31, 128, 149, 154
  chronic, 148, 149, 157, 161
  environmental chronic, 146
Environmental sustainability, 173, 174, 176, 177, 179, 190, 191
Equality, 153, 313
  gender, 153, 239, 313
European Court of Human Rights, 8, 127, 130, 131, 133, 134, 137, 208, 210, 211, 262, 314, 316
Evolutionary interpretation of international law, 66

F
FDI. *See* Foreign investment
Fiji. *See* Pacific Island States
Foreign investment, 9, 10, 293, 303, 306, 309, 314, 315, 318
  climate-friendly, 10, 293, 302, 304

G
Gender-based discrimination. *See* Discrimination
GHG emissions. *See* Greenhouse gas emissions
Greenhouse gas emissions, 23, 207, 211, 265, 302

H
HRC. *See* Human Rights Committee
Human right
  committee, 125, 163
  council, 262
  justifications, 3, 8
  justifications, typologies of, 217
  of children, 127
  to a clean, healthy, and sustainable environment, 7, 123, 124, 130
  to life, 123, 125, 130, 140, 206, 224
  to private and family life, 123, 128, 131
  to private, family and home life, 126
  violations, victims of. *See* Victim status
Human Rights Committee, 146, 156, 313
Human security. *See* Security

I

IACtHR. *See* Inter-American Court of Human Rights
ICCPR. *See* International Covenant on Civil and Political Rights
ICJ. *See* International Court of Justice
ILA. *See* International Law Association
ILC. *See* International Law Commission
Indigenous, 30, 44, 126, 131, 137, 239, 300
  communities. *See* Communities
  knowledge, 29, 30, 44
Inequality, 102, 104, 152, 180
  gender, 102
Inter-American Court of Human Rights, 109, 124, 203, 254, 315
Intergovernmental Panel on Climate Change, 22, 23, 42, 77, 94, 106, 177, 234, 252, 253, 258, 281
International Court of Justice, 3, 64, 72, 79, 109, 124, 203, 214, 263, 270, 280, 315
International Covenant on Civil and Political Rights, 125, 126, 137
International Law Association, 4, 6, 69, 73, 74, 86, 122, 172, 204, 294
  committee on international law and sea level rise, 122, 204
International Law Commission, 2, 6, 9, 61, 62, 64–68, 70–73, 86, 121, 138, 155, 163, 242, 253, 270, 272
International Tribunal on the Law of the Sea, 109, 203, 214, 251, 261, 315
  advisory opinion of, 214
  jurisdiction of, 256
IPCC. *See* Intergovernmental Panel on Climate Change

ITLOS. *See* International Tribunal on the Law of the Sea

K

Kiribati. *See* Pacific Island States

L

Loss and damage
  fund, 298, 299
  mechanism, 9, 296
  Warsaw International Mechanism for, 292, 294, 296

M

Margin of appreciation, 137, 140, 210, 211, 316
Marine pollution, 149, 258, 265, 268, 273, 275
Migration, 4, 8, 10, 106, 139, 145, 146, 149, 150, 153, 155, 162, 172, 173, 175, 176, 178–183, 185–187, 192, 277
  as adaptation, 155, 178
  climate, 4
  forced, 28
  gendered, 146, 150
  socially sustainable, 174, 179

N

The Netherlands, 8, 37, 106, 123, 173, 184, 187, 189–191, 218
Non-refoulement, 124, 125, 136, 176
  principle of, 124

P

Pacific Islands. *See* Pacific Island States
Pacific Island States, 61, 62, 71, 74–76, 84, 86, 98

declaration of, 75
Paris Agreement, 149, 209, 222, 241, 242, 245, 267, 268
Progressive development of international law, 67
Protection of persons affected by sea level rise. *See* Sea level rise

S
Satellite, 20–22, 43
  altimetry, 20–22
  measurement, 20, 21
Sea level rise, 2, 6, 17, 25, 29, 30, 40, 62, 66, 67, 69, 70, 73, 75, 84, 85, 102, 242, 253, 292, 309
  adaptation strategies to, 40
  as a threat multiplier, 99
  causes of, 20
  direct costs of, 25
  indirect costs of, 26, 27
  protection of persons affected by, 3, 122
  scientific evidence of, 18, 94
  social implications of, 28, 30
  socio-economic realities of, 23
Security, 42, 80, 94, 96–102, 104, 107, 109, 110, 174, 190, 303
  climate, 110
  collective, 175
  council, role of, 6, 95, 98, 108
  human, 96, 108
  ontological, 174
Senegal, 33
SIDS. *See* Small island
Slow-onset events, 2, 4, 5, 309, 312
Slow violence. *See* Violence
SLR. *See* Sea level rise
Small island, 37, 84, 95, 102, 293, 295, 301, 309, 313
  developing states, 37, 256
Small island developing States. *See* Small island

Solomon Islands. *See* Pacific Island States
State, 3, 7, 65, 70, 73, 74, 76, 83, 84, 86, 87, 96, 106, 109, 126, 136, 139, 146, 213, 235, 256, 282, 318
  cooperation, 9, 234, 235
  sovereignty, 9, 172, 271
Statehood, 2, 67, 233, 244
Study Group on Sea-level Rise and International Law Pacific Island States. *See* International Law Commission
Sudden-onset events, 155, 159

T
TCN. *See* Third country national
Third country national, 173, 184, 187, 189–191

U
UNCLOS. *See* United Nations Convention on the Law of the Sea
UNFCCC. *See* United Nations Framework Convention on Climate Change
United Nations
  Convention on the Law of the Sea, 70, 75, 254
  Framework Convention on Climate Change, 155, 241, 291
  Human Rights Committee, 158, 279
United Nations Convention on the Law of the Sea, 70, 75, 254–258, 261, 267–269, 281, 283
United Nations Framework Convention on Climate Change, 99, 253, 256, 260, 292, 298, 299

Uruguay, 33, 316

V
Victim status, 134
Violence, 3, 98, 104, 129, 147, 148, 150, 154

climate, 7, 146–148
slow, 3, 146, 148, 150
Vulnerability
climate change, 103
coastal, 33
social-ecological, 103

MIX
Papier aus verantwortungsvollen Quellen
Paper from responsible sources
FSC® C105338

If you have any concerns about our products,
you can contact us on
ProductSafety@springernature.com

In case Publisher is established outside the EU,
the EU authorized representative is:
**Springer Nature Customer Service Center GmbH
Europaplatz 3, 69115 Heidelberg, Germany**

Printed by Libri Plureos GmbH
in Hamburg, Germany